Food Security: A Multidisciplinary Approach

Food Security: A Multidisciplinary Approach

Editor: Alexis Crum

MURPHY & MOORE
www.murphy-moorepublishing.com

Murphy & Moore Publishing,
1 Rockefeller Plaza,
New York City, NY 10020, USA

Visit us on the World Wide Web at:
www.murphy-moorepublishing.com

ISBN: 978-1-63987-234-3 (Hardback)

Cataloging-in-Publication Data

Food security : a multidisciplinary approach / edited by Alexis Crum.
p. cm.
Includes bibliographical references and index.
ISBN 978-1-63987-234-3
1. Food security. 2. Food supply. 3. Food. I. Crum, Alexis.
HD9000.5 .F66 2022
338.19--dc23

Table of Contents

Preface

The measure of the availability of food to individuals and their ability to access it is referred to as food security. It means that every human being has physical, economic and social access to food which is safe and nutritious. The dietary requirements of people should be met at all times to help them lead a healthy and active life. A major concern under this field is the availability of food to individuals of every class, region and gender without any bias. Food supply can be interrupted due to factors like shipping disruptions, economic instability, droughts, wars, fuel shortage, etc. The notable challenges to achieving food security are climate change, land degradation, population growth and global water crisis. This book is compiled in such a manner, that it will provide in-depth knowledge about the theory and practice of food security. Different approaches, evaluations, methodologies and advanced studies on this subject have been included herein. The book is appropriate for students seeking detailed information in this area as well as for experts.

The information contained in this book is the result of intensive hard work done by researchers in this field. All due efforts have been made to make this book serve as a complete guiding source for students and researchers. The topics in this book have been comprehensively explained to help readers understand the growing trends in the field.

I would like to thank the entire group of writers who made sincere efforts in this book and my family who supported me in my efforts of working on this book. I take this opportunity to thank all those who have been a guiding force throughout my life.

Editor

1

Climate change and eastern Africa: a review of impact on major crops

Umesh Adhikari[1], A. Pouyan Nejadhashemi[1, 2] & Sean A. Woznicki[1]

[1]Department of Biosystems and Agricultural Engineering, Michigan State University, East Lansing, Michigan
[2]Department of Plant, Soil, and Microbial Sciences, Michigan State University, East Lansing, Michigan

Keywords
Agriculture, East Africa, feed the future, food security

Correspondence
A. Pouyan Nejadhashemi, 524 S. Shaw Lane, Room 225, Department of Biosystems and Agricultural Engineering, Michigan State University, East Lansing, MI 48824.

E-mail: pouyan@msu.edu

Funding Information
The study was supported by Global Center for Food Systems Innovation at Michigan State University and funded through Higher Education Solutions Network of the United States Department for International Development (USAID).

Abstract

Global warming has become one of the major challenges in maintaining global food security. This paper reviews the impacts of climate change on fourteen strategic crops for eight sub-Saharan Africa countries. Climate change is projected to increase median temperature by 1.4–5.5°C and median precipitation by −2% to 20% by the end of the 21st century. However, large levels of uncertainty exist with temporal and spatial variability of rainfall events. The impact of climate change on crop yields in the region is largely negative. Among the grain crops, wheat is reported as the most vulnerable crop, for which up to 72% of the current yield is projected to decline. For other grain crops, such as maize, rice and soybean, up to 45% yield reductions are expected by the end of this century. Two grain crops, millet and sorghum, are more resilient to climate change for which projected impacts on crop yields are <20%. Root crops, such as sweet potato, potato and cassava are projected to be less affected than the grain crops with changes to crop yields ranging from about −15% to 10%. For the two major export crops, tea and coffee, up to 40% yield loss is expected due to the reduction in suitable areas caused by temperature increase. Similar loss of suitable areas is also expected for banana and sugarcane production, however, this reduction is due to rainfall variability in lowland areas. Other crops such as cotton and sugarcane are projected to be more susceptible to precipitation variation that will vary significantly in the region. In order to mitigate the long-term impacts of climate change on agricultural sectors, the development of small-scale irrigation systems and water harvesting structures seems promising, however, affordability of such measures remains a key issue.

Introduction

Evident from the increase in frequency of extreme climatological events, climate change has become unequivocal, impacting water resources, agricultural, and food systems (Bates et al. 2008; Brown and Funk 2008; Hartmann et al. 2013). Historical records show an increase in global temperature since the late 19th century (Hartmann et al. 2013). The last three decades have been reported to be successively warmer than all previous decades, and the first decade in 21st century has been the warmest decade on record. On an average, global temperature has increased 0.72°C since 1950 (Hartmann et al. 2013). With the increase in temperature, frequency of cold days, cold nights, and frost have decreased, while frequency of hot days, hot nights, and heat waves have increased. The Intergovernmental Panel on Climate Change (IPCC) Fifth Assessment Report (AR5) projects a continuous increase in global temperature if greenhouse gas (GHG) emissions persist unabated. IPCC further predicts an increase in temperature between 0.3°C and 0.7°C over the next two decades and an increase of 0.3–4.8°C by the end of the 21st century, depending upon emission scenarios (Collins et al. 2013; Kirtman et al. 2013). Along with increases in temperature, global warming is associated with changes in large-scale hydrological cycle elements, such as an increase in atmospheric water vapor, shifting precipitation patterns, changes in precipitation intensity and extreme events, reduced snow cover and extensive melting of ice, and variations in soil moisture content and runoff (Bates et al. 2008; Hartmann et al. 2013). Precipitation has generally increased from 1900 to the 1950s between 30°N and 85°N latitude and has significantly increased between 10°N to 30°N, but has declined after 1970 to present (Bates et al. 2008).

Sub-Saharan Africa (SSA) is one of the poorest regions in the world, where an estimated 386 million people (48% of the region's total population) live on less than $1.25 a day (Ravallion 2012). This group of people is considered the most vulnerable to climate change as they possess minimum financial and technical resources to cope with climate change (Eriksen et al. 2007; Wheeler and von Braun 2013). In SSA, hydro-meteorological disasters, especially droughts and floods, are the most common forms of natural disasters. As such, drought and floods represent 80% of the loss of life and 70% of economic losses related to natural hazards in SSA (Bhavnani et al. 2008). As a consequence of climate change, the frequency and intensity of floods and droughts are projected to increase in the future (Bernstein et al. 2008). Meanwhile, agriculture is a major economic sector in the region, employing 65% of the labor force and contributing 32% to the countries' national gross domestic products, and is characterized by low productivity and lack of modern farming technologies (Chauvin et al. 2012). Despite being a large sector of the national economies, agricultural production systems in SSA are largely rainfed and thus, their success is sensitive to climate variability. In addition, the rapidly growing population in the region has further increased pressure on the food production systems.

This paper reviews the impacts of climate change on fourteen staple and cash crops in SSA. Special focus is given to eight countries in East Africa: Ethiopia, Kenya, Malawi, Mozambique, Rwanda, Tanzania, Uganda, and Zambia, which were selected by the United States Agency for International Development (USAID) for the Feed the Future (FtF) program. These countries were selected by the USAID based on prevalence of poverty and hunger, potential for agricultural-led growth, opportunities for regional cooperation, commitment, will, leadership and governance of the host country, and availability of resources (Ho and Hanrahan 2011). This review includes projected climate change in East Africa, the current status of the staple and cash crops in the region, how climate change is projected to impact the yield of these crops, and possible climate change adaptation measures. By compiling numerous studies in a single review, this paper lays the foundation for planning adaptation measures in the region, especially concerning food security and climate change.

East African Agriculture

Agriculture is a major contributor to the national economies in the study area. Agriculture in East African countries is dominated by smallholders who contribute up to 90% of agricultural production (Salami et al. 2010; Wiggins and Keats 2013). Among the eight countries examined, agriculture accounts for 21–42% of the national gross domestic products (FAO 2013b). Currently, of all the crops grown in SSA, a cereal-legume mixed cropping pattern is the dominant system that includes maize, millet, sorghum, and wheat (Van Duivenbooden et al. 2000). As presented in Figure 1, the maize mixed cropping system covers over 40% of the area, followed by pastoral (14%), root crop (12%), and cereal-root crop mixed system (11%). Other major crops in the region include cassava, banana, and rice, while teff is a major crop in the Ethiopian highlands. In drier parts of East Africa, the mixed cropping system is based on millet; while in the humid regions mixed cropping systems are based on maize and cassava (Francis 1986). Coffee, tea, cotton, tobacco, and sugarcane are the major cash crops in most of the FtF countries in SSA.

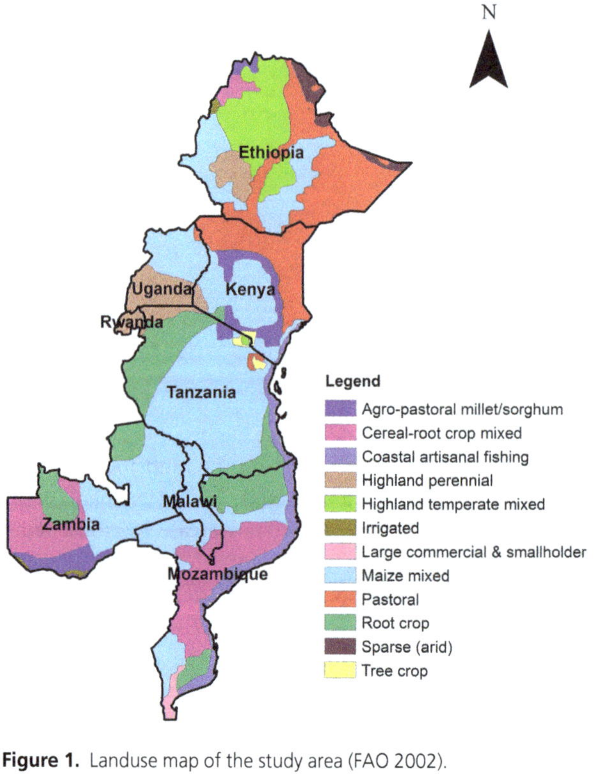

Figure 1. Landuse map of the study area (FAO 2002).

Despite the large contribution to national economies of SSA, yield of major crops remains below the global average (Fig. 2). In general, water availability is considered the primary limiting factor in semiarid sub-Saharan Africa (Barron et al. 2003). Only 18% of the potentially irrigable land or 3.5% of total cultivated area in SSA is irrigated (Foster and Briceno-Garmendia 2010). Lack of irrigation makes the agricultural system in the region vulnerable to rainfall variability and dry spells even during rainy seasons (Mupangwa et al. 2006). As such, low production of the rainfed agricultural systems in SSA are not due to low annual rainfall, but uneven distribution of the rainfall across the seasons (Barron et al. 2003). In addition to water availability, inadequate fertilizer application and continuous farming has resulted in removal of large quantity of nutrient from the soil, resulting in declined crop yield (Sanchez et al. 1997). Furthermore, current temperature in the region can be a limiting factor for some crops (such as wheat) if it is higher than optimum temperature (Liu et al. 2008). Lastly, majority loss of crop in East Africa is due to limited pest and weed control practices (Oerke 2006).

Climate Change in East Africa

Climate change is projected to increase temperature and precipitation variability in East Africa. Temperature in Africa is projected to rise faster than the rest of the world, which could exceed 2°C by mid-21st century and 4°C by the end of 21st century (Niang et al. 2014). Country-specific median projected change in temperature and precipitation for the 2090s (2080s for Rwanda) are presented in Figure 3. The projection for Rwanda was obtained from Cole (2011) and the rest were obtained from United Nations Development Programme (UNDP) Climate Change Country Profiles (McSweeney et al. 2010a ,b,c,d,e,f,g,h). Report for Rwanda represents an ensemble of 19 Global Circulation Models (GCMs) while UNDP studies comprise of an ensemble of 15 GCMs under three climate scenarios (B1: Rapid economic growth; focus on sustainability and environmental health; population growth peaks in 2050 and declines after; prevalent nonfossil fuel energy source use; relatively low increase in GHG emissions, A1B: Rapid economic growth and global economic development with balanced use of fossil fuels and nonfossil energy sources; population growth peaks in 2050 and declines after; moderate increases in GHG emissions, and A2: Economic development is regionally divided; global population continually grows; consistent fossil fuel use; relatively higher increase in GHG emissions). As shown in Figure 3, projected median increase in temperature by the end of this century is quite uniform across the region and ranges from 2°C to 3°C under the B1 scenario to above 4°C under the A2 scenario. However, large uncertainty exists in the temperature projection and the minimum and maximum projected rise in temperature range from 1.4°C to 5.5°C by 2090s.

IPCC AR5 reported that the future precipitation projections are more uncertain but likely to increase in the eastern Africa and decrease in the southern part (Niang et al. 2014). UNDP reports suggest general increase in annual precipitation in East Africa similar to IPCC AR5. Contrary to the general increase in median precipitation in the region, median precipitation is projected to decrease

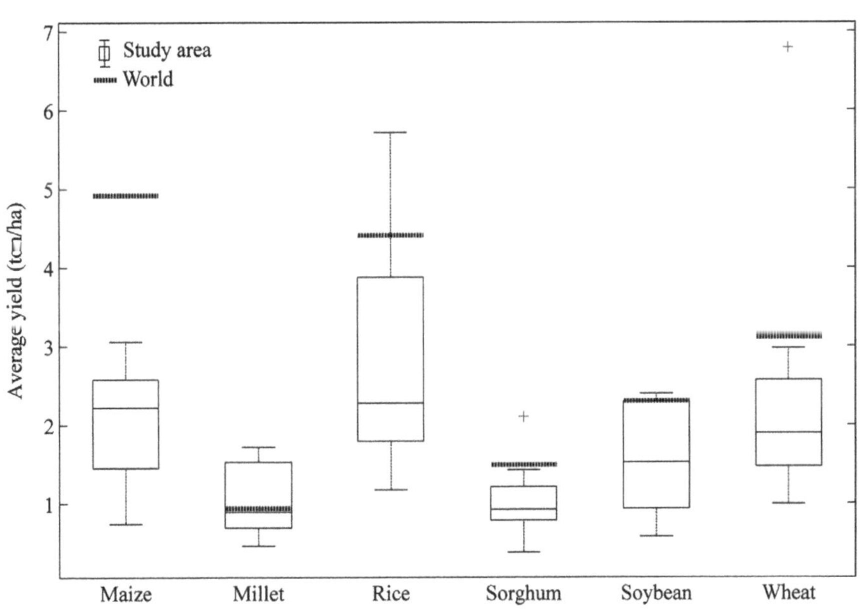

Figure 2. Average yield of major crops in study area and the world (2012) (FAO 2013a). (+ indicates outlier).

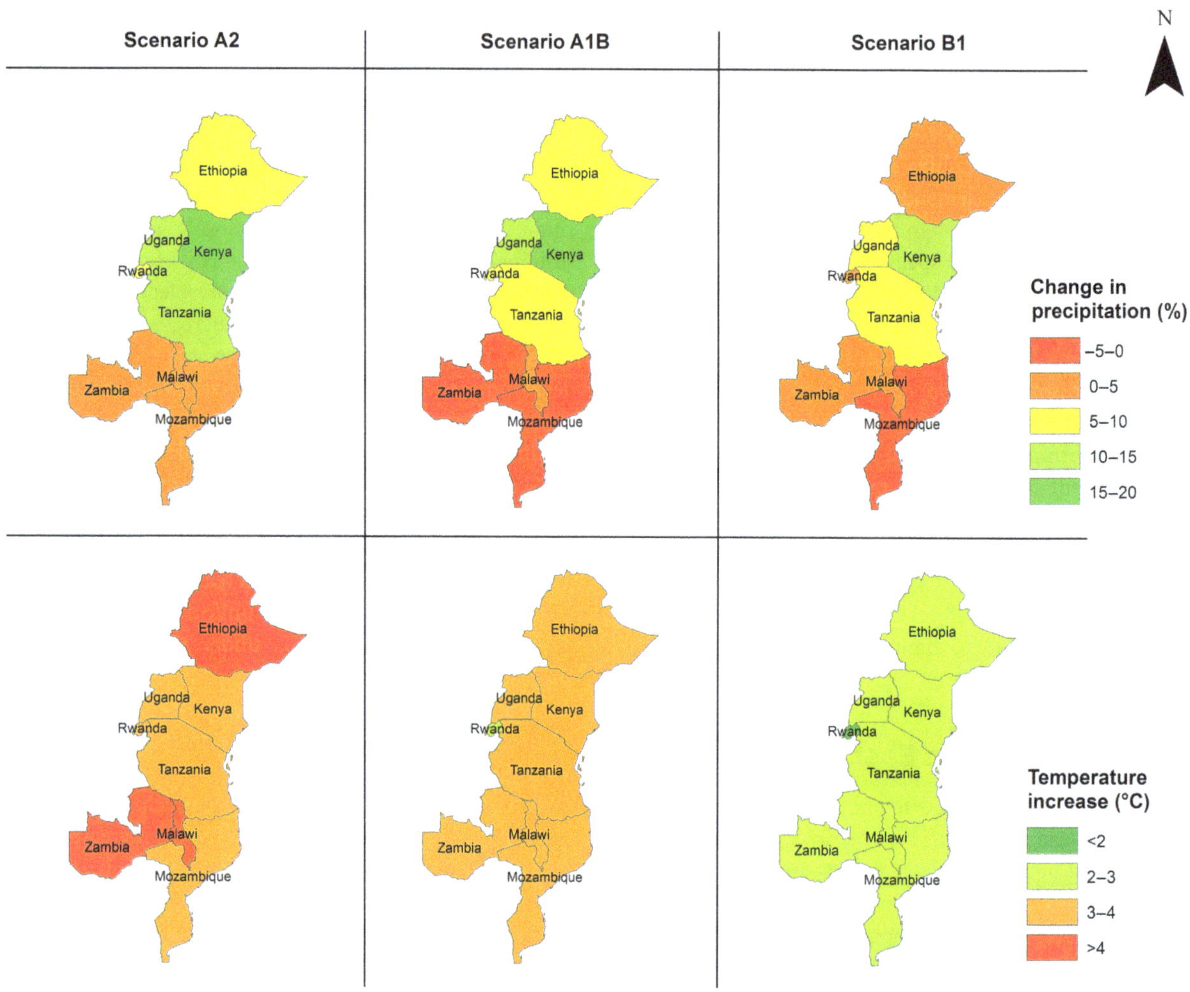

Figure 3. Median projected temperature and precipitation change in East Africa for 2090s (2080s for Rwanda). Temperature and precipitation projections are from an ensemble of 19 GCMs for Rwanda and 15 GCMs for the rest of the countries under B1, A1B and A2 emission scenarios (McSweeney et al. 2010a,b,c,d,e,f,g,h; Cole 2011).

in Mozambique (B1 and A1B scenarios) and Zambia (A1B scenario). In addition to median value, the range of precipitation changes are important to consider as large uncertainty exists in the precipitation projections. Among the countries studies by UNDP, projected change in precipitation range from −15% to +27% by 2060s and −16% to +49% by 2090s. Apart from uncertainty in annual precipitation, large variation in seasonal change has also been reported. For example in Ethiopia, projected annual changes in precipitation vary from −6% to +42%, while projected precipitation change in January-February-March months range from −42% to +78% under the three climate scenarios (B1, A1B, and A2). Likewise, in October-November-December, projected change in precipitation ranges from −10% to +70% in Ethiopia. These examples imply that even though projected annual change in precipitation is largely positive, seasonal variation should be closely monitored especially during the growing season. Since this has major impacts on water availability and crop production.

Climate Change Impact on Agriculture

Climate change is projected to overall decrease the yields of cereal crop in Africa through shortening growing season length, amplifying water stress and increasing incidence of diseases, pests and weeds outbreaks (Niang et al. 2014). Among the various environmental changes brought about by the climate change that limit crop yields, heat and water stresses are considered the most important (Prasad et al. 2008). Heat stress during development phases leads to fewer and smaller organs, reduced light interception due to shortened crop life, and altered carbon-assimilation processes including transpiration, photosynthesis, and respiration (Stone 2001). Heat stress during flowering and grain filling stages results in decreased grain count and weight, resulting in low crop yield and quality (Bita and Gerats 2013). Increase in temperature also increases saturation vapor pressure of air, thereby increasing evaporative demand. Plants close their stomata as a response to increased evaporative demand, thereby reducing photosynthesis rate and increasing vulnerability to heat injury (Lobell and Gourdji 2012). Even short duration heat shock can reduce crop yield substantially, especially if it coincides with the reproductive stage (Teixeira et al. 2013). Meanwhile, water stress leads to the shortening of the crop reproduction stage, reduction in leaf area and closure of stomata to minimize water loss, reducing crop yields (Barnabás et al. 2008). Water stress also increases pollen sterility that leads to reduced grain yield and quality (Alqudah et al. 2011). Water stress is frequently accompanied by heat stress, as dehydration of the plant tissue leads to overheating (Henckel 1964). The occurrence of water and heat stresses adversely impact plant growth and productivity, which is more pronounced than the individual impacts (Prasad et al. 2008). However, in areas with excess water and heat (due to climate change) it is projected that pathogen, weed, and insect infestation will further damage the agricultural systems (Ziska et al. 2011). In addition, rise in the CO_2 level is projected to benefit C3 crops, such as wheat, rice, and soybean, while no substantial effect is expected for C4 crops such as maize, sugar cane, millet, and sorghum (Conway 2009).

Concerning the impacts of climate change in the area of study, in high altitude regions such as mountainous lands in Ethiopia and Kenya, where temperature is the limiting factor for plant growth, a rise in temperature possibly will increase crop yield, but in lowland areas, will increase the risk of water stress (Thornton et al. 2009). Precipitation variability due to climate change results in increased irrigation water demand (Nelson et al. 2009). In addition, precipitation variability is expected to intensify the magnitude and frequency of flood and drought events that are both detrimental to agricultural industry. (Pachauri and Reisinger 2007). These events will likely further decrease crop water availability and threaten the productivity of the rainfed agriculture system in East Africa. Water and heat stress are projected to temporally decrease the length of the growing season while spatially shrinking the suitable areas for agricultural production, especially along the boundaries of arid and semiarid regions in Africa (Conway 2009; CIAT 2011b). Climate change is also projected to reduce the value of cropland by shifting high value agro-ecological zones to low value agro-ecological zones (Kurukulasuriya and Mendelsohn 2008). Increased rainfall intensity is projected to accelerate the rate of soil erosion in the future (Nearing et al. 2004), further threatening the crop productivity.

To better understand the spatial extent and magnitude climate change on agricultural production, we performed cluster analysis to assess the effect of temperature on five major crops. We used the temperature-crop yield relationships for maize (Muchow et al. 1990), wheat (Hatfield et al. 2008), rice (Baker and Allen 1993), sorghum (Hatfield et al. 2008), and soybean (Boote et al. 1998) to estimate the impacts of future climate scenarios (B1, A1B, A2) on the percent change (Table 1) and actual change in yields (Table S1). These values were applied to the spatial yield map obtained from HarvestChoice to compute the future yields (HarvestChoice 2015a,b,c,d,e). The multilevel thresholding method (Otsu 1975) was used to cluster the crop yields into three classes of high, medium, and low (Figs. 4 and S1–S4) and the results are presented in their respective sections.

Table 1. Projected percent change in yield under three climate change scenarios (B1, A1B and A2) by the 2090s (Muchow et al. 1990; Baker and Allen 1993; Boote et al. 1998; Hatfield et al. 2008; McSweeney et al. 2010a,b,c,d,e,f,g,h; Cole 2011).

Country	Maize			Rice			Wheat			Sorghum			Soybean		
	B1	A1B	A2	B1	A1B	A2	B1	A1B	A2	B1	A1B	A2	B1	A1B	A2
Ethiopia	−7	−10	−12	−6	−11	−14	−9	−14	−17	−18	−27	−33	−6	−9	−10
Kenya	−6	−9	−11	−7	−12	−15	−8	−12	−15	−16	−23	−29	−5	−8	−10
Malawi	−7	−10	−13	−5	−8	−11	−9	−14	−17	−17	−27	−34	−3	−6	−8
Mozambique	−6	−9	−12	−7	−11	−14	−8	−12	−16	−16	−24	−30	−5	−8	−10
Rwanda*	−6	−9	−10	−1	−2	−2	−13	−19	−20	−15	−23	−25	39	36	35
Tanzania	−6	−10	−12	−5	−9	−12	−8	−13	−16	−16	−25	−30	−5	−8	−10
Uganda	−7	−10	−11	−5	−9	−11	−9	−13	−15	−17	−26	−30	−5	−8	−9
Zambia	−7	−11	−13	−5	−9	−12	−10	−15	−18	−19	−29	−35	−3	−6	−8

*By the 2080s for Rwanda.

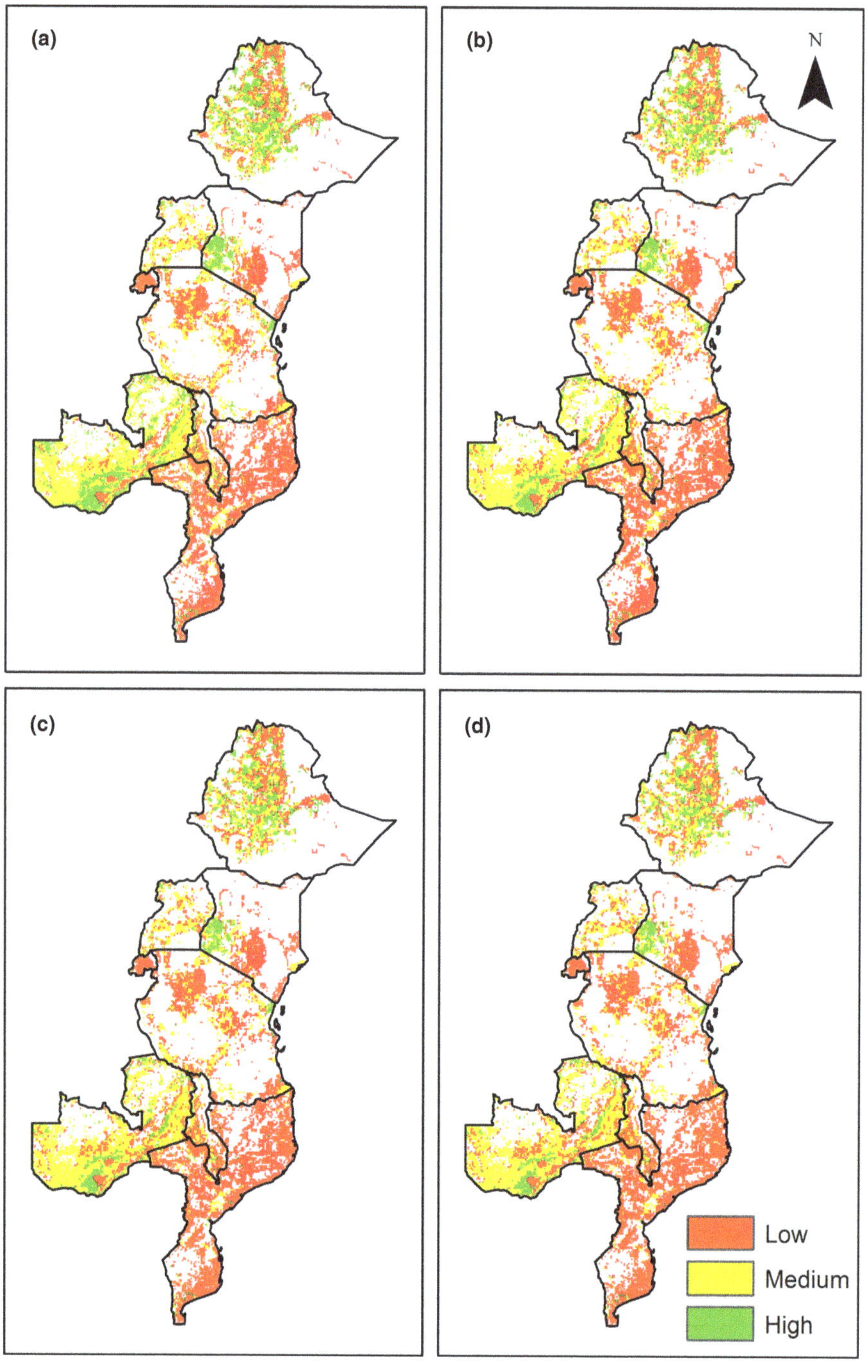

Figure 4. Current (a) and projected maize yield by the end of 21st century under B1 (b), A1B (c) and A2 (d) scenarios (Muchow et al. 1990; McSweeney et al. 2010a,b,c,d,e,f,g,h; Cole 2011; HarvestChoice 2015a).

Major Crops and the Impacts of Climate Change

Maize

Maize, the most widely cultivated staple crop in SSA, is primarily grown by smallholders and 77% of the total production in SSA (excluding South Africa) is consumed as food (Smale et al. 2011). Maize is cultivated on about 25 million hectares in SSA, which accounted for 27% of total cereal area and 34% of cereal production between 2005 and 2008 (Smale et al. 2011). Maize is the most important source of dietary protein and the second most important source of calories in eastern and southern Africa (Broughton et al. 2003).

Maize is one of the crops for which climate change impact has been widely studied. Schlenker and Roberts

(2009) reported increased maize yield with an increase in temperature up to 29°C followed by a sharp decline in yield with further temperature increases. However, Liu et al. (2008) reported the optimum maize-growing temperature to be 25°C. Each degree day spent above 30°C has been found to reduce final maize yield by 1% even under optimal rainfed conditions (Lobell et al. 2011). Other report suggests that a 1°C increase above norm reduces maize yield by 10% (Brown 2009). Using worldwide temperature and yield trends, Lobell and Field (2007) reported a decrease of 8.3% in maize yield per 1°C rise above normal. Easterling et al. (2007) plotted multiple crop yield projections against temperature rise, and reported that in low-latitude regions, where the majority of countries studied in this report lie, even a single degree increase in temperature results in about 3% loss in maize yield. Muchow et al. (1990) studied the effect of temperature in maize yield in the United States, where mean daily temperature ranged from 18°C to 29°C, and reported 4–8% yield reduction with a 2°C rise in temperature and 8–16% yield reduction with a 4°C rise in temperature. Runge (1968) found that daily maximum temperatures between 32.2°C and 37.8°C is beneficial to corn yield provided that adequate moisture is available. The authors reported that high temperature (when maximum temperature was 35°C) resulted in 9% reduction in maize yield per 1 inch reduction in rainfall. These findings suggest that at elevated temperature, maize not only suffers from temperature stress, but also becomes sensitive to moisture availability. Rainfed agriculture combined with potential variations in rainfall distribution under climate change, available soil moisture may not be able to meet increasing water demand.

Overall, maize production will decrease under future climate scenarios, though the degree of impact differs among simulations. Using the CERES-Maize model, Jones and Thornton (2003) predicted a 3–19% reduction in maize yield in the FtF countries by 2055 compared to 2000, where Ethiopia and Mozambique were projected to experience the least and greatest decrease in maize yield, respectively. Thornton et al. (2009) simulated maize and bean yield in East Africa, which extends to all FtF countries except Zambia, under A1FI (rapid economic growth and global economic development with intensive fossil fuel use; population growth peaks in 2050 and declines after; the highest increases in GHG emissions) and B1 storyline using the HadCM3 and ECHam4 models. A 3–15% reduction, depending on scenario and GCM, was projected on combined maize and bean production by 2050. Furthermore, the authors projected spatial heterogeneity for maize production with larger areas experiencing reduced yield as compared to the areas with yield gains in East Africa. By 2050, 9–33% of land was projected to observe reduced maize yield by more than 20%. Meanwhile, 39–68% of the simulated area was projected to increase yield by more than 5% depending on the scenario and GCM combination. Using the same models and scenarios, Thornton et al. (2010) predicted yield gains in Kenya and Rwanda and declines in Tanzania and Uganda in 2030 and 2050. Yield gains projected in Kenya and Rwanda were 15% and 11% by 2030 and 18% and 15% by 2050, respectively, while yield losses projected in Tanzania and Uganda were 3% and 2% by 2030 and 8% and 9% by 2050, respectively. Yield gains in Kenya and Rwanda were attributed to the beneficial temperature increases, which would bring growing season temperatures close to optimum temperature in temperate/tropical highlands. However, yield losses in Tanzania and Uganda occurred due to growing season temperatures rising past optimum temperatures. Without considering carbon fertilization, Nelson et al. (2009) projected 7–10% reduction in maize yield under the A2 storyline by 2050 in sub-Saharan Africa. Fischer (2009) included carbon fertilization and adapted crop varieties to estimate the yield of rainfed maize and projected a 1–9% increase in eastern Africa and 43–45% reduction in Southern Africa by 2050s under A2 climate change scenario. Even under optimum rainfed conditions, a 1°C increase in temperature is projected to reduce maize yield in 65% of the present maize cropping areas and under drought conditions, all areas are expected to experience yield loss (Lobell et al. 2011). Based on the relationship presented by Easterling et al. (2007) and considering the temperature projections as shown in Figure 3, yield reduction could be as much as 8–12% (B1), 16–30% (A1B), or 16–40% (A2) by the end of this century without adaptation. Based on the temperature-maize yield relationship presented by Lobell and Field (2007), maize yield in East Africa will be reduced by 16–20%, 24–31%, and 27–37% under B1, A1B and A2 storylines, respectively. Based on Muchow et al. (1990) and considering the area with similar mean daily temperature, maize yield in East Africa will be reduced by 6–7%, 9–11%, and 10–13% under B1, A1B and A2 storylines, respectively (Table 1). The maize yield percentage reduction translates into a loss of 38 kg/ha (Rwanda) to 225 kg/ha (Zambia) depending on field location (Table S1). In general, climate change has minimal impacts on shifting yield classes. The southern part of the study area has lower yield while the northern part has higher yield both under current and future climate scenarios (Fig. 4).

Despite large variations in projected impact on maize yield, there is a general consensus that climate change will adversely affect maize yield in East Africa. Multiple studies indicated that East Africa could lose as much as 40% of its maize production by the end of the 21st century.

Wheat

Wheat is generally cultivated as a winter rainfed crop in the highlands of Ethiopia, Kenya, Uganda, Rwanda, and Tanzania and as a winter irrigated crop in Zambia, Malawi, and Mozambique (Negassa et al. 2012). Wheat is a cool season crop and increasing temperature shortens its growth period by accelerating phonological development, resulting in reduced yield (You et al. 2005; Asseng et al. 2011). In SSA, average annual temperature in 1990 was 20.3°C in wheat harvest areas, which already exceeded the optimum wheat-growing temperature of 15–20°C (Liu et al. 2008). The exact level of the effects of climate change differ by location, but some studies suggest that a 1°C increase in temperature above norm reduces wheat yield by 10% (Brown 2009). Another study reported 3–4% reduction in wheat yield for every 1°C increase in temperature above 15°C (Wardlaw et al. 1989). Without considering the temperature effect on photosynthesis and grain-set, Hatfield et al. (2008) reported 7% yield reduction per 1°C increase in air temperature between 18 and 21°C and 4% reduction per 1°C increase above 21°C. Analyzing global wheat yield data, Lobell and Field (2007) reported a 5.4% wheat yield reduction per 1°C increase in temperature. By plotting the multiple simulation studies from low-latitude areas, Easterling et al. (2007) reported a linear relationship between an increase in temperature and the reduction in wheat yield with about 7% decrease for a 2°C rise in temperature and 34% yield decrease for a 4°C rise in temperature.

As wheat has a lower optimum temperature than rice, maize, millet, cassava, and sorghum (Liu et al. 2008), many simulation studies have projected a greater impact on wheat yield compared to other crops in East Africa (Liu et al. 2008; Fischer 2009; Nelson et al. 2009; Ringler et al. 2010). Fischer (2009), using HadCM3 under the A2 storyline, projected 63% and 44% yield reduction for rainfed wheat in eastern and southern Africa, respectively, by the 2050s even under carbon fertilization and adapted crop cultivars. Using the CSIRO climate model under the same storyline, the author projected 30%, 48%, and 72% yield reduction for rainfed wheat in eastern Africa by 2020s, 2050s, and 2080s, respectively, even considering carbon fertilization. Meanwhile, other crops such as rainfed maize and sorghum, were projected to experience 4% yield gains by the 2050s and a yield decline <2% by the 2080s. Liu et al. (2008) projected a 16–18% yield decrease in SSA by the 2030s, while increases in millet, rice, and maize yields were projected. Using the National Center for Atmospheric Research (NCAR) and CSIRO models and without considering carbon fertilization, Nelson et al. (2009) projected 34% to 36% reduction in wheat under the A2 scenario by 2050 in SSA. Tatsumi et al. (2011) modeled a number of crops using the Improved Global Agro-Ecological Zones method with the A1B storyline and projected a 15% yield loss in wheat in eastern Africa by 2090s compared to the 1990s. Using comprehensive climate change scenarios, which consisted of 17 GCMs, Ringler et al. (2010) simulated a number of crop yields and projected the greatest negative impact on wheat yield in SSA by 2050. Based on Easterling et al. (2007) and projected mean temperature increase (Fig. 3), wheat production in East Africa will decrease by 7–20%, 20–34%, and 20–40% under B1, A1B and A2 storylines, respectively, by the end of the 21st century. According to the report, even with adaptive measures, a 4°C rise in temperature will result in a 15% decrease in wheat production in low latitudes. Based on the global temperature-wheat yield relationship presented by Lobell and Field (2007) and temperature projections in Figure 3, East Africa could experience 10–13%, 16–20%, and 17–24% wheat yield decrease by the end of the 21st century under B1, A1B and A2 storylines, respectively. Considering the temperature projections (Fig. 3) and the relationship presented by Hatfield et al. (2008), wheat yield in East Africa will decrease by 8–13%, 12–19%, and 15–20% under B1, A1B, and A2 storylines, respectively (Table 1). The wheat yield percentage reduction translates into a loss of 69 kg/ha (Rwanda and Mozambique) to 1098 kg/ha (Zambia) depending on field location (Table S1). Spatial yield maps (Fig. S1) show lower yields in northern and southern regions of the study area with no major change in yield classes under future climate change scenarios considered.

The results presented above indicate that wheat is one of the most sensitive crops to climate change. Projected impacts in East Africa vary widely, but without climate change adaptation, eastern Africa could lose about two-thirds of the wheat productivity by the end of the 21st century.

Rice

Rice is a vital crop in East Africa where it is primarily grown by smallholder farmers as a rainfed crop (excluding Kenya, where the majority of rice is irrigated). It is the second most important crop in Tanzania and Malawi and the third most important crop in Kenya and Zambia (EUCORD 2002; Saka et al. 2006; FoDiS 2010). Rice is grown as upland rice, lowland rainfed rice, mangrove swamp rice, floating rice, and irrigated rice (EUCORD 2002). As most of the rice in East Africa is not irrigated, climate change-induced rises in temperature and increased variability in rainfall may impact production. The extent of this impact will vary by production system. Upland and lowland rice production, which constitute about 80% of rice production area in Africa, are projected to be the most vulnerable (Bimpong et al. 2011). Manneh et al. (2007) determined that climate change will affect rice production in Africa by increasing heat stress, drought, flooding and submergence, and salt stress. In rainfed production systems, drought is the most important production-limiting factor due to its sensitivity to moisture stress: yield is significantly reduced even under a mild drought (Guan et al. 2010). Similarly, heat stress results in high spikelet sterility, low tillering, stunting, and accelerated development, which eventually leads to reduced yield (Bimpong et al. 2011). A 1°C increase in temperature above norm reduces rice yield by 10% (Brown 2009). According to Liu et al. (2008), rice has an optimum growing temperature of 25°C, and temperature increases should result in higher rice yield in a majority of FtF countries. Correlating global rice yield with global temperature, Lobell and Field (2007) reported a 0.6% decrease in rice yield per 1°C increase in temperature. Easterling et al. (2007) plotted multiple simulation studies against temperature and projected a 2.6% yield increase with 1°C temperature rise to about 16% yield reduction for a 4°C rise in temperature. Baker and Allen (1993) found a quadratic relationship between temperature and rice yield at 330 ppm of CO_2 with zero yield at 37°C. Being a C3 crop, rice is expected to benefit from increased atmospheric CO_2 concentrations. However, increases in temperature have the potential to negate any benefit of elevated CO_2 in rice yield (Ainsworth 2008).

Projections of the impact of climate change on rice in the region vary among studies. Lobell et al. (2008) used 20 GCM models and projected a slight increase (<5%) in rice production in East Africa by 2030 as compared to 1998–2002. Using the Impact model, Ringler et al. (2010) projected a 0.24% increase in rice yield in eastern Africa and a 2.32% reduction in southern Africa by 2050. Another study by Tatsumi et al. (2011) using the Improved Global Agro-Ecological Zones method under A1B storyline also projected about 1% yield gain in rice in eastern Africa by 2090s compared to 1990s. Nelson et al. (2009) used the NCAR and CSIRO models and projected about a 15% reduction in yield under the A2 storyline by 2050 in SSA without considering carbon fertilization. Variations in rice yield projections among different studies are likely due to the difference in scenarios and models used in the prediction. Based on the temperature projection in Figure 3 and the relationship reported by Easterling et al. (2007), eastern Africa could lose 4–8%, 8–13%, and 8–16% of rice yield under B1, A1B and A2 scenarios, respectively by the end of the 21st century. However, based on Lobell and Field (2007) model, significantly lower yield loss is expected (between 1–3%) for B2, A1B, and A2. However, based on the Baker and Allen (1993) model and median temperature projections for the region, East African rice yield will decrease by 1–7%, 2–12%, and 2–15% under B2, A1B, and A2 storylines, respectively (Table 1). Compared with the current yield, the rice yield percentage reduction translates into a loss of 35 kg/ha (Malawi) to 397 kg/ha (Kenya) depending on field location (Table S1). Spatial yield maps show no major change in yield classes under future climate change. In general southern regions (Mozambique, Malawi, and Zambia) have lower yields while the rest of the region has higher yields (Fig. S2).

Overall, rice appears to be more resilient to climate change than maize or wheat. Nonetheless, climate change is projected to have negative impact on rice yield and eastern Africa could lose as much as 16% of its current rice production by the end of the 21st century.

Sorghum

In terms of quantity, sorghum is the second most important crop in Africa after maize and is the most important crop in the semiarid tropics (Obalum et al. 2011). Major sorghum-growing areas in FtF countries include much of north central, northwestern, western, and eastern mid-altitude areas of Ethiopia, Rwanda, northern and eastern Uganda, central Tanzania, and the areas in Kenya and Tanzania east of Lake Victoria (Wortmann et al. 2009). The importance of sorghum to Africa lies in its inherent ability to resist drought and withstand periods of high temperatures (Taylor 2003). Maiti (1996) reported the optimum vegetative growth temperature of sorghum is 26–34°C and an optimum reproductive growth temperature is 25–28°C. Currently, most of the sorghum in the region is grown under sub-optimum temperatures. About 54% of the sorghum is produced below 24°C (Wortmann et al. 2009). Based on Maiti (1996) and Wortmann et al. (2009), sorghum production should increase in the region with slight increases in temperature. However, some researchers have projected climate change to negatively impact sorghum yield. In a global simulation of sorghum yield, Lobell and Field (2007) reported an 8.4% decrease in sorghum yield for 1°C increase in temperature. Similarly, Hatfield et al. (2008) reported a 7.8% decrease in sorghum yield for 1°C rise in temperature from 18.5°C to 27.5°C. Water deficiency is another constraint in the region that has been cited as the most important sorghum production constraint (Wortmann et al. 2009). Water deficit may increase with increases in temperature and variability in rainfall and may offset yield gains associated with temperature increases. Being a C4 crop, sorghum may not benefit significantly from increased atmospheric CO_2 (Liu et al. 2008).

Combining the temperature projections in Figure 3 and the Lobell and Field (2007) projected yield model, sorghum production in East Africa will be reduced by as much as 16–20%, 24–31%, and 27–38% under B1, A1B, and A2 storylines, respectively, by the end of the 21st century. The Hatfield et al. (2008) crop yield model also produced similar results and showed decrease in sorghum production by 15–19%, 23–29%, and 25–35% under B1, A1B, and A2 storylines, respectively (Table 1). These yield losses are equivalent to 86 kg/ha (Mozambique) to 595 kg/ha (Uganda) when compared to the current crop yields (Table S1). Spatial yield maps show that high-yielding northern regions of the study area will be more impacted by the future climate change than the low-yielding southern regions (Fig. S3). However, comparing these findings to other studies conducted in the region, the yield losses are overpredicted. Knox et al. (2012) reported only a 15% sorghum yield reduction in the African continent by the 2050s. Most other simulation studies that focused on SSA or East Africa predict small changes in future sorghum production in East Africa. Liu et al. (2008) simulated sorghum yield in SSA and Lobell et al. (2008) simulated sorghum production in East Africa and both projected only small changes in sorghum production in the 2030s compared to the 1990s and 2000s. Fischer (2009) projected a 4% yield gain in rainfed sorghum in the 2020s and 2050s and a 2% reduction in 2080s in East Africa with CO_2 fertilization. Using NCAR and CSIRO models without carbon fertilization, Nelson et al. (2009) projected a 2–3% reduction in sorghum under the A2 storyline by 2050 in SSA.

These findings suggest that sorghum is more resilient to climate change than maize or wheat and will be minimally impacted (<5%) by the middle of this century.

Millet

Millet is cultivated mostly in the semiarid tropics and subtropics of Africa; however, it is also cultivated in other drought-prone sub-humid and medium-high altitude areas (Obilana 2003). Millet is a hardy crop that requires few inputs, is less susceptible to pests and diseases, and can be grown in the areas that are too hot and dry for sorghum (Cagley et al. 2009). Pearl millet and finger millet are the most commonly grown millet varieties, where pearl millet is grown in all sub-Saharan countries and finger millet is grown in eastern, southern and central Africa (Obilana 2003). Pearl millet is grown as a dry-land crop in semiarid regions, while finger millet is generally grown in uplands and sub-humid areas (Gari 2002). Millet is important for SSA's food security due to its adaptation to drought and heat, high nutritive, value and ability to be stored for long periods without quality degradation. In SSA, the average temperature in millet harvest areas in 1990 was 27.3°C, which was below the optimum millet-growing temperature of 30°C (Liu et al. 2008). Climate change is expected to raise the temperature in millet-growing areas closer to the optimum temperature, leading to a general increase in millet yield. However, similar to sorghum, millet is not expected to gain much from elevated atmospheric CO_2 levels.

Quantitative projections show both negative and positive impacts of climate change on millet yield. The discrepancy might be attributed to the difference in scenarios, models and time periods used in future projections and the extent of the area considered in the study. Ringler et al. (2010), using comprehensive climate change scenarios consisting of 17 GCMs, simulated crop yields and projected a slight increase (<5%) in millet yield in SSA in 2050. Compared to the 1990s, Liu et al. (2008) projected a 7–27% increase in millet yield in SSA by 2030s under the A1FI, A2, B1 and B2 storylines. Using NCAR and CSIRO models without considering carbon fertilization, Nelson et al. (2009) projected a 7–8% reduction in millet yield under A2 storyline by 2050 in SSA. Using 16 models under the A1B storyline, Schlenker and Lobell (2010) projected a median decrease of 17% in millet yield in SSA by 2050s. These findings are similar to those reported by Knox et al. (2012), who reported a 10% decline in millet yield in Africa by the 2050s.

Based on these studies, millet is more resilient to climate change than maize or wheat but less resilient than sorghum. It is expected that there will be about a 15% yield loss in East Africa by the middle of the century.

Dry beans and soybean

In eastern and southern Africa, beans are the second most important source of dietary protein following maize and the third most important source of calories after cassava and maize, with annual consumption exceeding 50 kg per person (Wortmann et al. 1998; Broughton et al. 2003). Beans are produced in over 20 countries in eastern and southern Africa covering over four million hectares, where Ethiopia, Kenya, Rwanda, Tanzania, and Uganda are among the major producers (Wortmann et al. 1998; Asfaw et al. 2009). Cultivation areas are concentrated in cooler highlands and warmer mid-elevation areas with altitudes greater than 1000 m above sea level. However, due to population pressure, the cropping area is being extended to lower elevations (Katungi et al. 2009). Schlenker and Roberts (2009) conducted a study on the effect of temperature on soybean production and reported a threshold temperature for soybean to be 30°C; increases in temperature up to 30°C gradually increased yield but temperatures beyond the threshold resulted in a sharp decline in soybean yield. In eastern and southern Africa, 80% of the bean-producing area has a mean annual temperature of 15–23°C, which is below the threshold temperature, favoring bean production (Wortmann et al. 1998). However, using global yield data and temperature, Lobell and Field (2007) reported 1.3% decline in soybean yield per 1°C increase in temperature. Boote et al. (1998) reported an optimum temperature of 22°C for soybean yield and a linear reduction thereafter until 30°C. Elevated CO_2 and ozone levels alter the type of soybean diseases prevalent and higher daily temperature has been found to increase disease incidence in soybean (Eastburn et al. 2010). As a C3 crop, beans are expected to benefit from elevated atmospheric CO_2 concentration. Ainsworth et al. (2002) reported a mean increase of 24% in soybean yield with elevated atmospheric CO_2, which was mainly due to pod number increases. Soybean yield is also affected by precipitation and subsequent moisture availability. In 65% of the bean-producing area in the region, the mean rainfall exceeds 400 mm during the three months after sowing, while in other areas yield is severely impacted by moisture deficit (Wortmann et al. 1998). When precipitation falls below 300 mm during the growing season, yield decline in beans is estimated to be 1000 kg/ha (Wortmann et al. 1998). Hence, in eastern and southern Africa, rainfall variability and soil moisture content, rather than rising temperature, are the crucial factors in determining the effect of climate change in soybean production.

Using 20 GCM models, Lobell et al. (2008) projected a slight decrease (<5%) in bean production in East Africa by 2030 as compared to 1998–2002. Thornton et al. (2009) used the HadCM3 and ECHam4 models and the A1FI and B1 storylines to model bean production in East Africa and projected that up to 13% bean cultivation area by 2030 and 56% of area by 2050 will experience greater than 20% yield reduction depending upon scenarios and models. Using the same models and scenarios, Thornton et al. (2010) projected yield gains of up to 36% in 2030 and up to 57% in 2050 in rainfed highlands. The authors projected mixed effects with minor variations in 2030 and up to 20% yield decline in 2050 in rainfed humid-sub-humid regions. Arid and semiarid regions of Tanzania and Uganda were predicted to gain yields in the 2030s and decline in the 2050s. Tatsumi et al. (2011) used the Improved Global Agro-Ecological Zones method under the A1B storyline and projected 12% decline in soybean production in East Africa by 2090 compared to 1990. However, only considering the impact of median temperature rise on soybean yield, the Lobell and Field (2007) model predicts a 2–6% reduction in East Africa by the end of this century across the B1, A1B, and A2 storylines. This is similar to the projections based on Boote et al. (1998) and temperature projection in East Africa (Fig. 3), which shows a yield reduction between 3% and 10% across all scenarios by the end of this century, except for Rwanda (Table 1). In Rwanda, the projections show a yield increase of 35–39% across all storylines as rise in temperature would bring the soybean-growing temperature close to the optimum temperature. These percentage changes would be equivalent to 226–252 kg/ha yield gain in Rwanda, and 11–134 kg/ha yield loss in rest of the countries (Table S1). Spatial cluster analysis shows shift from low yield to medium yield in Rwanda under climate change but no major change in yield class for other countries (Fig. S4).

These results suggest that bean yields in East Africa are more sensitive to rainfall variations than temperature rises. Projections based on climate change scenarios indicate that lowland areas could lose up to 20% of the current beans production but highland areas can gain up to 57% in bean productivity by the middle of the century. However, due to the ratio between lowland and highland areas, there will be a general decrease in dry beans and soybean productions.

Cassava

Cassava is the most important crop in SSA in terms of caloric intake (Rosenthal and Ort 2012). In East Africa, cassava is also the most important staple food crop in terms of total production, where production is concentrated in mid-altitude areas in the African Great Lakes region and the coastal zones of Tanzania and Kenya (Fermont 2009). Cassava is also traditionally cultivated in the northern part of Zambia, while about 70% of farmers in Mozambique cultivate cassava (Nielson 2009). The crop is more resilient to climate change due to its tolerance of high temperatures and intraseasonal drought (Jarvis et al. 2012). However, if a prolonged drought period (>2 months) falls during the root thickening initiation state, a root yield reduction of up to 60% may occur (Jarvis et al. 2012). Cassava shows better yield gain than grain crops at higher CO_2 concentrations, can recover from very long drought periods, and exhibits increases in optimum growth temperature under elevated CO_2 levels (Rosenthal and Ort 2012). These qualities make cassava a suitable crop in a future that is projected to experience elevated CO_2, increased temperature, and variable rainfall patterns. The findings are supported by other researchers who projected minimum impact, if not positive, or at least better performance of cassava than other crops.

Using a geographical information system-based model, Liu et al. (2008) projected small changes in sub-Saharan cassava production by 2030; a 2% reduction to 1% increase depending on scenario. Lobell et al. (2008) also projected a positive impact due to climate change with a slight increase on cassava production in in East Africa by 2030 compared to 2000. Using 16 models under the A1B storyline, Schlenker and Lobell (2010) projected 8% yield reduction for cassava compared to 17–22% reductions in maize, sorghum, millet, and groundnut by the mid-21st century in SSA. Tatsumi et al. (2011), using the Improved Global Agro-Ecological Zones method under the A1B storyline, projected a 10% increase in cassava yield in eastern Africa by 2090s compared to 1990s. Studies conducted by the International Food Policy Research Institute (IFPRI) using the IMPACT model, which takes global food supply, demand, price, etc. along with climate change into account, has also projected increase in cassava production between 40% and 100% in Malawi, Rwanda, and Uganda with no significant change in Mozambique and Tanzania in 2050 (Bashaasha et al. 2012; Kilembe et al. 2012; Maure et al. 2012; Saka et al. 2012; Tenge et al. 2012). Using the same model, Ringler et al. (2010) projected a 0.42% increase in cassava production in eastern Africa and a 0.75% reduction in cassava production in southern Africa in 2050. Furthermore, under different climate change scenarios, projected cassava yield performed better than or at least equal to (in the worst case scenario) maize, sorghum, millet, beans, potato, and banana (Jarvis et al. 2012) making it a potential candidate to ensure future food security in the region.

Based on different projection scenarios, cassava is projected to lose up to 8% or gain up to 10% of its yield in the future. Overall, cassava yield appears to be the least impacted by the climate change.

Sweet potato

One of the most widely grown crops in SSA, sweet potato is an important crop in the areas surrounding the Great Lakes in eastern and central Africa, such as Malawi and Mozambique (Shonga et al. 2013). Sweet potato is a major staple crop in Uganda, Rwanda, and parts of Tanzania, while it is a secondary food source in Kenya and most of Tanzania and Ethiopia (Smit 1997). Mainly cultivated by smallholders, sweet potato is the third most important source of carbohydrates in Uganda, while the country is the third largest producer of sweet potato in the world (Muyinza et al. 2012). Although the crop grows from semiarid lowlands to high-altitude zones, cultivation of sweet potato is most intense in altitudes of 800–1900 m (Smit 1997). Sweet potato is a tropical or a subtropical plant, which has an optimum-growing temperature of 20–25°C, but can be grown in temperatures ranging from 15°C to 33°C (Ramirez 1992). Lower nighttime temperature is required for tuber formation, while higher temperature in the day helps vegetative growth. Susceptibility of sweet potato to drought stress and lower nighttime temperature required for tuber formation makes the crop vulnerable to climate change (Ramirez 1992; Agili 2012). However, being a C3 crop, sweet potato benefits from elevated atmospheric CO_2 concentration levels (Mortley et al. 1996).

Information is sparse regarding the impact of climate change on sweet potato production. Using the IMPACT model, Ringler et al. (2010) projected 1.06% gain in eastern Africa and 1.14% gain in southern Africa in sweet potato yield by 2050. However, while considering the whole sub-Saharan region, the same crop was the second most impacted after wheat. In contrast to previous findings, using five GCM models under the A1B storyline and the Improved Global Agro-Ecological Zones model, Tatsumi et al. (2011) projected a 15% decline in sweet potato production in eastern Africa in 2090s compared to 1990s.

The studies on the impact of climate change on sweet potato in East Africa are not adequate to draw conclusions on the potential yield impact. However, susceptibility of sweet potato to high temperatures at night and climate-induced water stress suggest that the crop might be negatively impacted in the future.

Potato

Mostly grown by smallholders, Malawi was previously the biggest potato producer in SSA (Maganga 2012). Recent statistics shows that Kenya has surpassed Malawi with over 5 million tons of potato production in 2012 (FAO 2013a). In Kenya, potato is the second most important crop following maize. Here it is cultivated by about 500,000 smallholders, making it one of the most important sources of income and employment in rural areas (Obare et al. 2010). In Rwanda, potato is the second major food crop after banana, where it is grown mainly in the highlands in the southwest and north of the country (Muhinyuza et al. 2012). Potato is cultivated in all of the FtF nations, where cultivation is mainly concentrated in highland areas. In terms of total annual production, Rwanda (2.3 million tons) and Tanzania (1.8 million tons) rank third and fourth among the FtF nations, while Zambia is the lowest producer with only 30,000 tons per year (FAO 2013a). Optimum temperature for potato cultivation is 17°C; above which potato production is reduced through either lower plant development and productivity due to heat stress or decreased partitioning of assimilates to the tubers (Wolf et al. 1990). Likewise, moisture stress reduces potato yield by shortening the growing and dormancy period and by reducing the total number and size of potato tubers (Karafyllidis et al. 1996). Being a C3 crop, elevated atmospheric CO_2 levels increase potato yield by increasing the number of tubers.

However, under limited fertilization and a water stressed environment, actual yield gains may be insignificant (Miglietta et al. 1998; Ainsworth and McGrath 2010).

There are very few studies on the impacts of climate change on potato production in East Africa. Haverkort et al. (2013) used six coupled GCMs under the A2 storyline in four potato-growing agro-ecosystems in South Africa. The authors predicted that the positive effects of elevated CO_2 on water use efficiency and crop yield were more than adequate to compensate for the negative effects of increased temperatures and reduced water availability in 2050s. Similarly, IFPRI (Tenge et al. 2012) using the IMPACT model, predicted up to a 100% increase in potato yield and a 50% increase in cultivation area in Rwanda in 2050 compared to 2010. As a result, the authors predicted doubling or tripling the potato production by 2050. However, these predictions are based on IMPACT model, which considers global food supply, demand, price along with climate change. Contrary to previous findings, Jarvis et al. (2012), using 20 GCMs under the A1B storyline, projected about 15% reduction in potato yield in Africa by 2030. Similarly, Tatsumi et al. (2011), using five GCM models under the A1B storyline, projected a 17% decline in potato yield in eastern Africa in 2090s compared to 1990s. As a consequence of climate change, potato yield in most of East African countries (except Rwanda) will decrease due to heat and water stress.

Banana

The Great Lakes region in East Africa is the largest banana-producing and consuming region in Africa (AATF 2009). Banana is the most important food crop in Uganda and Rwanda, where annual per capita consumption exceeds 135 kg (FAO 2013a). Major banana-growing areas in East Africa include southwestern and central Uganda, most parts of Rwanda, the northern, southern, and eastern highlands in Tanzania, and the central and Kisii regions in Kenya (AATF 2009). Studies conducted in some of the FtF countries have revealed that drought stress is either the most important or the second most important constraint in banana production in the region (Van Asten et al. 2011). Optimal banana production is believed to require a constant and ample supply of water due to its permanent green vegetation and shallow root system (Robinson 1996). Though banana can survive water stress for long periods of time, low soil moisture and extended exposure to extreme temperatures (above 35°C) can reduce banana production (Thornton and Cramer 2012). As such, in the East African highlands, where annual rainfall is below 1100 mm, drought-induced yield reduction on rainfed bananas can reach up to 65% compared to wetter areas (Van Asten et al. 2011).

Eastern and Southern Africa, where banana production is limited by low temperature, are projected to experience increases in suitable area ranging from 1% to 11% by 2020 (Ramirez et al. 2011). However, variability in rainfall is expected to result in a loss of suitable areas in the lowlands (Ramirez et al. 2011). Though higher temperatures may increase suitable areas, rises in temperature also result in increased water demand, limiting banana cultivation to the areas projected to receive increased rainfall (Thornton 2012). Meanwhile, highland bananas are projected to observe significant yield loss due to increased risk of pest and diseases if the temperature increases by 2°C (Thornton and Cramer 2012).

Quantitative measures of banana yield under future climate change are limited. In general, moderate temperature rise in highlands would be beneficial for banana yield. However, rainfall variability in lowlands will limit the suitable areas for banana cultivation in the future.

Tea and coffee

Tea and coffee are grown throughout the FtF countries, mostly by smallholders as a cash crop. The production system employs millions of people in the region. Most of the countries export tea and coffee, which contributes significantly to national foreign exchange earnings. While Arabica and Robusta are the two dominant commercially produced coffee varieties grown around the world, Arabica is dominant in East Africa. Kenya, Ethiopia, Rwanda, Malawi, and Zambia primarily grow Arabica. Uganda predominantly grows Robusta, while Tanzania grows both types, mostly Arabica. Optimum-growing temperature for Arabica is 18–23°C while for Robusta is 22–26°C (ICO 2009). Hence, Arabica is cultivated in high altitudes while Robusta is cultivated in lowlands. Higher temperature hastens development and ripening of the cherry, impacting both productivity and quality of coffee.

Based on a study by the International Coffee Organization (ICO 2009), future coffee production is likely to decrease worldwide under the A1F1, A2, B1, and B2 storylines, although the magnitude varies between scenarios. Tanzania's Initial Communication to the UNFCCC (URT 2003) projected no impact with up to a 2°C rise in temperature in coffee production in the country. Variability in precipitation was identified as the decisive factor and the report concluded that an increase in precipitation could increase coffee yield by as much as 17%. In Kenya, the optimal coffee-producing zone is predicted to move upward from 1600 m to 1700 meter above sea level by 2050 (GIZ 2010). The report also projected a decrease in suitable coffee-producing area in the country from the current 50–70% to 30–50%. Jaramillo et al. (2011) predicted shrinkage in coffee-growing areas in most of Kenya, Uganda and Rwanda due to the prevalence of the coffee berry borer caused by increases in temperature, and expansion of suitable area in Tanzania and Ethiopia by 2050 under A2 and B2 storylines.

Similar to coffee, the optimum tea-growing area is projected to shift toward higher altitudes and total suitable area for tea cultivation is projected to shrink. In Kenya, as a result of rise in temperature and variation in precipitation, suitable areas for tea production are predicted to decrease from the current 60–80% to 35–55% and the current optimum tea production zone (1500–2100 m above sea level) is projected to shift to higher altitudes (2000–2300 m above sea level) by 2050 under the A2 storyline (CIAT 2011a). Another report projects that a rise in temperature by 2°C would render large areas in Kenya currently suitable for tea production unsuitable and that smallholders may not be able to cope with the effects of climate change (Simms 2005). Similar to Kenya, optimum tea-growing areas in Uganda are projected to shift from the current 1450–1650 meter above sea level to 1550–1650 meter above sea level under A2 storyline (CIAT 2011b). Due to the lack of high altitude areas, total areas available for tea cultivation are projected decrease by 2050 (CIAT 2011b). Total area suitable for tea cultivation is also projected to decrease from the current 60–80% to 20–40% by 2050. Similar shift in optimum tea and coffee-growing area and shrinkage in total suitable area can be expected in other coffee and tea-producing East African countries.

Overall, future climate change is projected to impact tea and coffee production negatively although quantitative results are limited. Optimum tea and coffee-growing area will shift to higher altitudes. This will limit the production of these crops due to several factors including land ownership, social preference, and land availability.

Cotton

Cotton is grown in most of the FtF countries and is an important cash crop for many smallholders. It is primarily grown in arid and semiarid regions in Kenya, northern and eastern regions in Uganda, Western and Eastern Cotton zones in Tanzania, the lowlands and mid-altitude regions in Ethiopia, southern parts of Malawi, central, eastern and southern provinces in Zambia, and many regions in Mozambique. Cotton is also among the major export commodities in Uganda, Tanzania and Mozambique. To some extent, cotton can tolerate high temperature and drought, although it is sensitive to water availability, especially during flowering and boll formation (Ton 2011). Higher temperatures are reported to decrease boll maturation period, reduced boll size, reduce the number of seeds per boll, reduce fiber length, and ultimately reduce cotton yield (Reddy et al. 1999; Meredith 2005; Pettigrew 2008). Optimum-growing temperature for cotton ranges from 23.5°C to 32°C (Burke et al. 1988). Hatfield et al. (2008) reported that cotton has an optimum-growing temperature of 25°C and the crop completely fails at a temperature of 35°C. Based on the findings and current mean annual temperature, a moderate increase in temperature is beneficial for cotton production in most of the FtF countries. Being a C3 crop, cotton is expected to experience yield increases with the increase in atmospheric CO_2 concentration as long as the temperature does not exceed optimum-growing temperature range (Yoon et al. 2009).

Data on the projected impacts of climate change on cotton production in the region is scarce. One study conducted in Zimbabwe (Gwimbi 2009) reported a decrease in cotton yield with decreasing precipitation and increasing temperature; an 80% decrease in precipitation resulted in a 38% decline in cotton yield. In another study conducted in Cameroon (Gerardeaux et al. 2013), cotton yields were projected to increase by 2050 under the A1B storyline as a result of a slight increase in temperature, and an increase in atmospheric CO_2 with no changes in rainfall. Studies conducted in Cameroon and Mali (Sultan et al. 2010) concluded that onset and duration of rainy season were the major drivers of temporal and spatial distributions of cotton productivity. The authors reported that the relationship between rainfall and cotton productivity depended on mean climatic conditions, as drier regions were most sensitive to climate variability. In Tanzania, a study projected no impacts on cotton production with an annual temperature rise of 2.7°C as the temperature still remained within the optimum-growing temperature (URT 2003). However, the areas projected to receive increased precipitation by 37% were predicted to experience up to a 17% increase in cotton yield, while the areas expected to receive decreased precipitation were projected to experience up to 17% yield loss.

Since it is expected that the rainfall variability will significantly increase by the end of the 21st century, there is a high probability that cotton production will be negatively impacted. However, there is an overall gap in quantitative analysis of climate change and cotton yield in the region.

Sugarcane

Sugarcane production is an important sector in all the FtF countries, where most of the sugarcane is grown either by large estates or by smallholders around the estates, also known as outgrowers. Exceptions are Ethiopia and

Kenya: in Ethiopia about 35% of the sugarcane is produced by smallholders, whereas in Kenya about 85% of sugarcane is produced by smallholders (KSB 2009; Assefa et al. 2010). Sugarcane is extensively grown in the Lake Victoria basin of Kenya and Uganda and is considered the most important cash crop in the region (Netondo et al. 2010). Sugarcane requires a temperature of 30–32°C during the main growing season, while temperature above 35°C is detrimental to plant growth (Blackburn 1984; Hunsgi 1993). Additionally, a 2–3°C increase in temperature was reported have a positive impact on sucrose yield, while decreases in rainfall had a negative impact on sucrose yield (Kiker 2002). Being a C4 crop, sugarcane will receive limited benefits from carbon fertilization, but De Souza et al. (2008) demonstrated increased sugarcane production under carbon enriched environments, indicating a positive impact of climate change.

Under the A1FI, A2, B1 and B2 storylines, Liu et al. (2008) projected increases in temperature ranging from 28.30°C to 28.54°C, which is below the optimum sugarcane-growing threshold. Hence, variation in precipitation and water availability seem to be the key factor in quantifying the effect of climate change on sugarcane in the region. In Ethiopia, sugarcane grown by smallholders is irrigated (Assefa et al. 2010), and hence climate change should have minimal effects on production. In Kenya, climate change is projected to increase frequency of drought (ROK 2002), which may result in moisture deficit affecting sugarcane production. Lukorito and Ouma (2010) also concluded that rainfall variability during the grand growth stage of sugarcane is a major factor affecting productivity in sugarcane-based areas. Studies conducted by IFPRI using the IMPACT model projected a 0.31% increase in yield for eastern Africa and a 1.09% increase in southern Africa by 2050 (Ringler et al. 2010). Using 20 GCM models, Lobell et al. (2008) projected a slight decrease (<5%) in sugarcane production in East Africa by 2030 as compared to 1998–2002. Reports on the effects of climate change on sugarcane production in other countries are lacking, although irrigated sugarcane production systems developed by estates should be more resilient in light of changing climate.

Sugarcane is resilient to temperature rise but is vulnerable to rainfall variability. Projected climate change impact on sugarcane is minimal and it is expected to be within 5% of the current yield.

Adaptation Options

The application of different adaptation techniques in East Africa should aim to improve several critical components including soil health, water conservation, livelihood diversification and the capacity of local institutions (FAO, 2014). In addition, effective management of land, water and soil resources should always be considered while planning adaptation measures, as these resources provide a safety net to the vulnerable groups (Paavola 2008).

It is evident from this review that some crops in eastern Africa will be primarily affected by heat stress in the future. For crops, such as wheat that are more impacted by heat stress switching to heat-resistant and drought tolerant crops, such as sorghum and millet, may mitigate temperature stress-related crop failure. As an adaptive measure to climate change, farmers in Africa have already begun selecting a combination of crops based on the prevailing climate (Kurukulasuriya and Mendelsohn 2006).

However, many crops will be more affected by water stress, caused by increased evapotranspiration and variability in rainfall, rather than heat stress. As such, rainfed farming in SSA is typically limited to 3–6 months during the rainy season and the crop yields are subjected to weather-driven fluctuations (Burney and Naylor 2012). Construction of new irrigation schemes and rehabilitation of existing schemes are possible options to reduce future water stress. Of the six million hectares equipped for irrigation in SSA, one million hectares need rehabilitation; more than 30%, 50%, and 60% of irrigation-equipped area in Uganda, Malawi, and Mozambique, respectively, are in need of irrigation rehabilitation (You 2008). In SSA, large-scale irrigation schemes have proven to be expensive and environmentally unsustainable (Ngigi 2003). Furthermore, the construction of large-scale irrigation projects is limited by both natural and social constraints such as water source availability, geography, and conflicting interests among stockholders (Fujiie et al. 2011). This has made the development of large-scale irrigation schemes implausible in SSA. However, the development of small-scale irrigation schemes is a more feasible option for maintaining future crop productivity. For smallholders, shallow and hand-dug wells could supplement water limitations, especially during the dry season (Ngigi 2009). Rainwater harvesting and storage from ground surfaces and rooftops for use during the dry season is another option to cope with rainfall variability (Elliot et al. 2011). The development and dissemination of drought tolerant crop cultivars could also help minimize the climate change impact in drought-prone areas.

Future climate change impacts on water resources in East Africa show that runoff will increase in the northern regions and decrease in the southern regions (Adhikari et al. 2015). Furthermore, groundwater sources will be less impacted by near-term climate change providing an alternative source of water for all regions (Bonsor et al. 2010). However, due to the lack of studies evaluating the impacts of climate change on groundwater by the end of the century, the feasibility of groundwater irrigation systems is uncertain. This means that for the near-future both northern and southern regions will have access to water for irrigation. However, various resources and social constraints may limit the capacity of the smallholders to adopt irrigation as an adaptation measure. With the rapid growth of population, there will be an increase in competition for water resources among various sectors such as municipal, industrial and agricultural. In addition, affordability remains a key issue in developing irrigation systems or water harvesting structures since more than 60% of the population in the region lives on less than $2 a day (World Bank 2015). In some places, geographic location necessitates the use of mechanical power to draw irrigation water and the lack of available power sources and associated costs may limit irrigation development (ADF 2006). Migration of the male population to urban areas in the region has increased the number of female-headed households, who lack financial resources and social support to adopt climate change adaptation measures (Olson et al. 2010; Hazell 2013). Furthermore, the migration of the population also disrupts the existing farming system and raises land ownership issues (Messina et al. 2014), thereby increasing the risk in investing in irrigation infrastructure.

Despite the fact that irrigation remains the key to cope with the climate change, several farm level adaptation measures can be implemented to minimize the risk of future conditions. These adaptation measures can be incremental, systems or transformational and are based on the severity of the climate impacts (Rickards and Howden 2012). Incremental adaptation is the fine tuning of the existing system to minimize impacts, which includes changing planting dates, crop varieties, plant density, and nutrient and water management practices (Rickards and Howden 2012). These changes can greatly reduce the adverse impact of climate change on crop yields, thereby minimizing the vulnerability (Easterling et al. 2007). In East Africa, farmers have already implemented marginal adaptation measures by changing their planting date and crop varieties, however, soil, water and land management practices have not been widely adopted (Kristjanson et al. 2012). Soil and water conservation are particularly important since these practices can enhance the effectiveness of irrigation, fertilizer and improved seeds (Kato et al. 2011). When the incremental adaptations become inadequate, farmers may have to implement systems adaptation (switching to resilient crops, crop diversification, or precision agriculture) or even to transformational adaptations (seek alternative income generation methods) (Rickards and Howden 2012).

Conclusions

This paper synthesizes more than 160 studies on the impacts of climate change on most of the economically viable crops in selected eastern African countries. A review of literature reveals that most of the impact studies have focused on cereal crops, while a number of noncereal crops (e.g., banana, sweet potato, cassava, potato), which serve as staple foods, are also important and should be considered in future research.

Current literature on the impacts of climate change on crops in East Africa suggested that wheat is likely to be the most negatively impacted crop. It is projected that wheat yield will decrease by up to 72% by the 2080s. Other grain crops such as maize, rice, millet and beans are projected to be moderately impacted, with yield changes ranging from −45% to +18%, −15% to +3%, −17% to +27% and −12% to +17%, respectively. Among the grain crops, sorghum is projected to be least impacted, where the projected impact is within 5% of the current yield. Root crops are projected to be less impacted by the climate change. For example, projected future change on the yields of cassava, sweet potato, and potato range from −8% to +10%, −15% to +1%, and −17% to −15%, respectively. Moderate temperature rise is beneficial for highland bananas, but higher increase in temperature would negatively impact its yield. Optimum production zones for tea and coffee are projected to decrease by as much as 40% due to rise in temperature. Cotton and sugarcane are less sensitive to temperature rise, but more prone to drought stress under climate change.

As most of the cropped area is rainfed, development of small-scale irrigation would be the most crucial step in ensuring future food security, but farmers' affordability remains a key problem. Soil and water conservation measure are important in maintaining crop productivity and building resilience against climate change. Development and adoption of drought tolerant varieties is one of the options to cope with the changing climate. Transitioning to heat tolerant and drought-resistant crops is would be necessary if the climate change renders current crops unsuitable in the future.

To better assist in formulating potential coping measures, further research is needed to quantify the impacts of climate change on regionally important staple food crops such as cassava, potatoes, sweet potatoes, and banana as well as cash crops such as tea, coffee, cotton, and sugarcane. As food production is also impacted by other factors such as landuse changes, disease incidence, and market demand, future research studies should take these factors into consideration when studying the effects of climate change on crop production. In addition, there is a need to incorporate uncertainty of climate models into consideration. This will provide better information in regard to the impacts of climate change in the region. In addition, the level of uncertainty can be reduced by repeating this type of analysis with the new set of climate change information, such as IPCC AR5 and future versions.

Acknowledgments

The study was supported by Global Center for Food Systems Innovation at Michigan State University and funded through Higher Education Solutions Network of the United States Department for International Development (USAID).

Conflict of interest

The authors declare that they have no conflict of interest.

References

AATF. 2009. Feasibility study on technologies for improving banana for resistance against bacterial wilt in sub-Saharan Africa. African Agricultural Technology Foundation, Nairobi, Kenya.

ADF. 2006. The Bugesera agricultural development support project (PADAP) appraisal report. African Development Fund, African Development Bank Group, Adidjan, Ivory Coast.

Adhikari, U., A. P. Nejadhashemi, and M. R. Herman. 2015. A review of climate change impacts on water resources in East Africa. Trans. ASABE In Review.

Agili, S. 2012. Selection, yield evaluation, drought tolerance indices of orange-flesh sweet potato (*Ipomoea batatas* Lam) hybrid clone. J. Nutr. Food Sci. 2:138.

Ainsworth, E. A. 2008. Rice production in a changing climate: a meta-analysis of responses to elevated carbon dioxide and elevated ozone concentration. Glob. Change Biol. 14:1642–1650.

Ainsworth, E. A., and J. M. McGrath. 2010. Direct effects of rising atmospheric carbon dioxide and ozone on crop yields. Pp. 109–130 *in* D. Lobell and M. Burke, eds. Climate change and food security: adapting agriculture to a warmer world. Springer, Dordrecht, the Netherlands.

Ainsworth, E. A., P. A. Davey, C. J. Bernacchi, O. C. Dermody, E. A. Heaton, D. J. Moore, et al. 2002. A meta-analysis of elevated CO2 effects on soybean (glycine max) physiology, growth and yield. Glob. Change Biol. 8:695–709.

Alqudah, A. M., N. H. Samarah, and R. E. Mullen. 2011. Drought stress effect on crop pollination, seed set, yield and quality. Pp. 193–213 *in* E. Lichtfouse, ed. Alternative farming systems, biotechnology, drought stress and ecological fertilisation. Springer, Dordrecht, the Netherlands.

Asfaw, A., M. W. Blair, and C. Almekinders. 2009. Genetic diversity and population structure of common bean (*Phaseolus vulgaris L.*) landraces from the East African highlands. Theor. Appl. Genet. 120:1–12.

Assefa, Y., D. E. Conlong, J. van den Berg, and A. Mitchell. 2010. Distribution of sugarcane stem borers and their natural enemies in small-scale farmers' fields, adjacent margins and wetlands of Ethiopia. Int. J. Pest Manag. 56:233–241.

Asseng, S., I. A. N. Foster, and N. C. Turner. 2011. The impact of temperature variability on wheat yields. Glob. Change Biol. 17:997–1012.

Baker, J. T., and L. H. Jr Allen. 1993. Contrasting crop species responses to CO_2 and temperature: rice, soybean and citrus. Pp. 239–260 *in* J. Rozema, H. Lambers and S. C. Geijn, eds. CO_2 and biosphere. Springer, Dordrecht, the Netherlands.

Barnabás, B., K. Jäger, and A. Fehér. 2008. The effect of drought and heat stress on reproductive processes in cereals. Plant, Cell Environ. 31:11–38.

Barron, J., J. Rockström, F. Gichuki, and N. Hatibu. 2003. Dry spell analysis and maize yields for two semi-arid locations in East Africa. Agric. For. Meteorol. 117:23–37.

Bashaasha, B., T. S. Thomas, M. Waithaka, and M. Kyotalimye. 2012. East African agriculture and climate change: a comprehensive analysis-Uganda. International Food Policy Research Institute, Washington, DC. Available at http://www.ifpri.org/sites/default/files/publications/aacccs_uganda_note.pdf. (assessed 22 December 2013).

Bates, B. C., Z. W. Kundzewicz, S. Wu, and J. P. Palutikof. 2008. Climate change and water. Technical Paper of the Intergovernmental Panel on Climate Change. Intergovernmental Panel on Climate Change, IPCC Secretariat, Geneva, Switzerland.

Bernstein, L., P. Bosch, O. Canziani, Z. Chen, R. Christ, O. Davidson, et al. 2008. Climate change 2007: synthesis report: an assessment of the intergovernmental panel on climate change. Intergovernmental Panel on Climate Change, Geneva, Switzerland.

Bhavnani, R., S. Vordzorgbe, M. Owor, and F. Bousquet. 2008. Status of disaster risk reduction in the sub-Saharan Africa region. The World Bank, Washington, DC.

Bimpong, K., B. Manneh, Z. Sander, K. Futakuchi, and T. Kumashiro. 2011. Climate change: Impacts and strategies on rice production in Africa. Paper presented at the Developing Climate-Smart Crops for a 2030 World Workshop, Addis Ababa, Ethiopia.

Bita, C. E., and T. Gerats. 2013. Plant tolerance to high temperature in a changing environment: scientific fundamentals and production of heat stress-tolerant crops. Front. Plant Sci. 4:273.

Blackburn, F. 1984. Sugar-cane. Longman Inc, New York.

Bonsor, H. C., A. M. MacDonald, and R. C. Calow. 2010. Potential impact of climate change on improved and unimproved water supplies in Africa. RSC Issues Environ. Sci. Technol. 31:25–50.

Boote, K. J., J. W. Jones, G. Hoogenboom, and N. B. Pickering. 1998. The CROPGRO model for grain legumes. Pp. 99–128 *in* G. Y. Tsuji, G. Hoogenboom and

P. K. Thornton, eds. Understanding options for agricultural production. Academic Publishers, Dordrecht.

Broughton, W. J., G. Hernandez, M. Blair, S. Beebe, P. Gepts, and J. Vanderleyden. 2003. Beans (*Phaseolus spp.*)–model food legumes. Plant Soil 252:55–128.

Brown, L.. 2009. Plan B 4.0: mobilizing to save civilization. W.W. Norton & Company, New York.

Brown, M. E., and C. C. Funk. 2008. Food security under climate change. Science 319:580–581.

Burke, J. J., J. R. Mahan, and J. L. Hatfield. 1988. Crop-specific thermal kinetic windows in relation to wheat and cotton biomass production. Agron. J. 80:553–556.

Burney, J. A., and R. L. Naylor. 2012. Smallholder irrigation as a poverty alleviation tool in sub-Saharan Africa. World Dev. 40:110–123.

Cagley, J. H., C. L. Anderson, and M. Klawitter. 2009. Gender and cropping: millet in sub-Saharan Africa. Evans School of Public Affairs, University of Washington, Seattle, WA.

Chauvin, N. D., F. Mulangu, and G. Porto. 2012. Food production and consumption trends in sub-Saharan Africa: Prospects for the transformation of the agricultural sector. UNDP Working Paper 2012–11. United Nations Development Programme, Regional Bureau for Africa. Addis Ababa, Ethiopia.

CIAT. 2011a. Future climate scenarios for Kenya's tea growing areas. International Center for Tropical Agriculture, Cali, Columbia.

CIAT. 2011b. Future climate scenarios for Uganda's tea growing areas. International Center for Tropical Agriculture, Cali, Columbia.

Cole, M.. 2011. Rwanda's climate: observations and projections. Smith School of Enterprise and the Environment, University of Oxford, Oxford, United Kingdom.

Collins, M., R. Knutti, J. Arblaster, J.-L. Dufresne, T. Fichefet, P. Friedlingstein, et al. 2013. Long-term climate change: projections, commitments and irreversibility. Pp. 1029–1136 *in* T. F. Stocker, D. Qin, G.-K. Plattner, M. Tignor, S. K. Allen, J. Boschung, A. Nauels, et al., eds. Climate change 2013: the physical science basis. Contribution of Working Group I to the fifth assessment report of the Intergovernmental Panel on Climate Change. Cambridge Univ. Press, Cambridge, United Kingdom and New York, NY, USA.

Conway, G. 2009. The science of climate change in Africa: impacts and adaptation. Discussion Paper, 1. Grantham Institute for Climate Change, Imperial College London, London, United Kingdom.

De Souza, A. P., M. Gaspar, E. A. Da Silva, E. C. Ulian, A. J. Waclawovsky, R. V. Dos Santos, et al. 2008. Elevated CO_2 increases photosynthesis, biomass and productivity, and modifies gene expression in sugarcane. Plant, Cell Environ. 31:1116–1127.

Eastburn, D. M., M. M. Degennaro, E. H. Delucia, O. Dermody, and A. J. Mcelrone. 2010. Elevated atmospheric carbon dioxide and ozone alter soybean diseases at SoyFACE. Glob. Change Biol. 16:320–330.

Easterling, W. E., P. K. Aggarwal, P. Batima, K. M. Brander, L. Erda, S. M. Howden, et al. 2007. Food, fibre and forest products. Pp. 273–313 *in* M. L. Parry, O. F. Canziani, J. P. Palutikof, P. J. van der Linden and C. E. Hanson, eds. Climate change 2007: impacts, adaptation and vulnerability. Contribution of Working Group II to the Fourth Assessment Report of the Intergovernmental Panel on Climate Change. Cambridge Univ. Press, Cambridge, U.K.

Elliot, M., A. Armstrong, J. Lobuglio, and J. Bartram. 2011. Technologies for climate change adaptation: the water sector. Roskilde: UNEP Risoe Centre, Roskilde, Denmark.

Eriksen, S., R. J. T. Klein, K. Ulsrud, L. O. Næss, and K. O'Brien. 2007. Climate change adaptation and poverty reduction: Key interactions and critical measures. GECHS Report 2007:1, Global Environmental Change and Human Security, University of Oslo, Norway.

EUCORD. 2002. Rice sector development in East Africa. The European Cooperative for Rural Development, Brussels, Belgium.

FAO. 2002. FAO-Geonetwork. Food and Agriculture Organization of the United Nationas, Rome, Italy.

FAO. 2013a. FAOSTAT online statistical service. United Nations Food and Agriculture Organization (FAO), Rome, Italy.

FAO. 2013b. FAO statistical yearbook 2013. United Nations Food and Agriculture Organization (FAO), Rome, Italy.

FAO. 2014. Adapting to climate change through land and water management in Eastern Africa: results of pilot projects in Ethiopia, Kenya and Tanzania. Food and Agricultural Organization of the United Nations, Rome, Italy.

Fermont, A. M. 2009. Cassava and soil fertility in intensifying smallholder farming systems of East Africa. [PhD dissertation], Wageningen University, the Netherlands.

Fischer, G. 2009.World Food and Agriculture to 2030/50: How do climate change and bioenergy alter the long-term outlook for food, agriculture and resource availability? Paper presented at FAO Expert meeting on how to feed the world in 2050, Rome, Italy.

FoDiS. 2010. Growing NERICA in Zambia-Upland rice. Ministry of Agriculture and Co-operatives/Japan International Cooperation Agency, Food Crop Diversification Support Project, Lusaka, Zambia.

Foster, V., and C. Briceno-Garmendia. 2010. Africa's infrastructure: a time for transformation. World Bank, Washington, DC.

Francis, C. A. 1986. Multiple cropping systems. Macmillan Publishing Company, New York.

Fujiie, H., A. Maruyama, M. Fujiie, M. Takagaki, D. J. Merrey, and M. Kikuchi. 2011. Why invest in minor projects in sub-Saharan Africa? An exploration of the scale economy and diseconomy of irrigation projects. Irrigat. Drain. Syst. 25:39–60.

Gari, J. A. 2002. Review of the African millet diversity. Pp. 19–22 *in* IPGRI's Programme for Neglected and Underutilised Species, ed. International workshop on fonio, food security and livelihood among the rural poor in West Africa. IPGRI/IFAD, Bamako, Mali.

Gerardeaux, E., B. Sultan, O. Palaï, C. Guiziou, P. Oettli, and K. Naudin. 2013. Positive effect of climate change on cotton in 2050 by CO_2 enrichment and conservation agriculture in Cameroon. Agron. Sustain. Dev. 33:485–495.

GIZ. 2010. Climate change adaptation and mitigation in the Kenyan coffee sector. Deutsche Gesellschaft für Internationale Zusammenarbeit (GIZ) GmbH, Eschborn, Germany.

Guan, Y. S., R. Serraj, S. H. Liu, J. L. Xu, J. Ali, W. S. Wang, et al. 2010. Simultaneously improving yield under drought stress and non-stress conditions: a case study of rice (*Oryza sativa L.*). J. Exp. Bot. 61:4145–4156.

Gwimbi, P.. 2009. Cotton farmers' vulnerability to climate change in Gokwe District (Zimbabwe): impact and influencing factors. J. Disast. Risk Stud., 2:81–92.

Hartmann, D. L., A. M. G. Klein Tank, M. Rusticucci, L. V. Alexander, S. Bronnimann, Y. Charabi, et al. 2013. Observations: atmosphere and surface. Pp. 159–254 *in* T. F. Stocker, D. Qin, G.-K. Plattner, M. Tignor, S. K. Allen, J. Boschung, et al., eds. Climate change 2013: the physical science basis. Contribution of Working Group I to the fifth assessment report of the Intergovernmental Panel on Climate Change. Cambridge Univ. Press, Cambridge, United Kingdom and New York, NY, USA.

HarvestChoice 2015a. Maize Yield (kg/ha, 2005). International Food Policy Research Institute, Washington, DC, and University of Minnesota, St. Paul, MN. Available at http://harvestchoice.org/data/maiz_y. (accessed 3 May 2015).

HarvestChoice 2015b. Rice Yield (kg/ha, 2005). International Food Policy Research Institute, Washington, DC, and University of Minnesota, St. Paul, MN. Available online at http://harvestchoice.org/data/rice_y. (accessed 3 May 2015).

HarvestChoice 2015c. Sorghum Yield (kg/ha, 2005). International Food Policy Research Institute, Washington, DC, and University of Minnesota, St. Paul, MN. Available at http://harvestchoice.org/data/sorg_y. (accessed 3 May 2015).

HarvestChoice 2015d. Soybean Yield (kg/ha, 2005). International Food Policy Research Institute, Washington, DC, and University of Minnesota, St. Paul, MN. Available at http://harvestchoice.org/data/soyb_y. (accessed 3 May 2015).

HarvestChoice 2015e. Wheat Yield (kg/ha, 2005). International Food Policy Research Institute, Washington, DC, and University of Minnesota, St. Paul, MN. Available at http://harvestchoice.org/data/whea_y. (accessed 3 May 2015).

Hatfield, J. L., K. J. Boote, B. A. Kimball, D. W. Wolfe, D. R. Ort, C. Izaurralde, et al. 2008. Agriculture. Pp. 21–74 *in* P. Backlund, A. Janetos, D. Schimel and M. Walsh, ed. The effects of climate change on agriculture, land resources, water resources, and biodiversity in the United States. U.S. Climate Change Science Program and the Subcommittee on Global Change Resources, Washington, D.C.

Haverkort, A. J., A. C. Franke, F. A. Engelbrecht, and J. M. Steyn. 2013. Climate change and potato production in contrasting South African Agroagro-ecosystems 1. Effects on land and water use efficiencies. Potato Res. 56:31–50.

Hazell, P. 2013. Comparative study of trends in urbanization and changes in farm size in Africa and Asia: implications for agricultural research. CGIAR Consortium, Montpellier Cedex, France.

Henckel, P. A.. 1964. Physiology of plants under drought. Annu. Rev. Plant Physiol., 15:363–386.

Ho, M. D., and C. E. Hanrahan. 2011. The Obama administration's feed the future initiative. Congressional Research Service, Washington, DC.

Hunsgi, G.. 1993. Production of sugarcane: theory and practice. McMillan India Ltd, Bangalore.

ICO. 2009. Climate change and coffee. International Coffee Organization, London, England.

Jaramillo, J., E. Muchugu, F. E. Vega, A. Davis, C. Borgemeister, and A. Chabi-Olaye. 2011. Some like it hot: the influence and implications of climate change on coffee berry borer (*Hypothenemus hampei*) and coffee production in East Africa. PLoS ONE 6:e24528.

Jarvis, A., J. Ramirez-Villegas, B. V. H. Campo, and C. Navarro-Racines. 2012. Is cassava the answer to African climate change adaptation? Trop. Plant Biol. 5:9–29.

Jones, P. G., and P. K. Thornton. 2003. The potential impacts of climate change on maize production in Africa and Latin America in 2055. Glob. Environ. Change 13:51–59.

Karafyllidis, D. I., N. Stavropoulos, and D. Georgakis. 1996. The effect of water stress on the yielding capacity of potato crops and subsequent performance of seed tubers. Potato Res. 39:153–163.

Kato, E., C. Ringler, M. Yesuf, and E. Bryan. 2011. Soil and water conservation technologies: a buffer against production risk in the face of climate change? Insights from the Nile basin in Ethiopia. Agri. Econ. 42:593–604.

Katungi, E., A. Farrow, J. Chianu, L. Sperling, and S. Beebe. 2009. Common bean in eastern and southern Africa: a situation and outlook analysis. International Centre for Tropical Agriculture, Cali, Columbia.

Kiker, G. A. 2002. CANEGRO-DSSAT linkages with geographic information systems: Applications in climate change research for South Africa. *Proceedings of*

International CANGRO Workshop, Mount Edgecombe, South Africa.

Kilembe, C., T. S. Thomas, M. Waithaka, M. Kyotalimye, and S. Tumbo. 2012. East African agriculture and climate change: a comprehensive analysis-Tanzania. International Food Policy Research Institute, Washington, DC.

Kirtman, B., S. B. Power, J. A. Adedoyin, G. J. Boer, R. Bojariu, I. Camilloni, et al. 2013. Near-term climate change: projections and predictability. Pp. 953–1028 *in* T. F. Stocker, D. Qin, G. K. Plattner, M. Tignor, S. K. Allen, J. Boschung, et al., eds. Climate change 2013: the physical science basis. Contribution of Working Group I to the fifth assessment report of the Intergovernmental Panel on Climate Change. Cambridge Univ. Press, Cambridge, United Kingdom and New York, NY.

Knox, J., T. Hess, A. Daccache, and T. Wheeler. 2012. Climate change impacts on crop productivity in Africa and South Asia. Environ. Res. Lett. 7:034032.

Kristjanson, P., H. Neufeldt, A. Gassner, J. Mango, F. B. Kyazze, S. Desta, et al. 2012. Are food insecure smallholder households making changes in their farming practices? Evidence from East Africa. Food Security 4:381–397.

KSB. 2009. Kenya sugar board strategic plan 2009. Kenya Sugar Board, Nairobi, Kenya.

Kurukulasuriya, P., and R. Mendelsohn. 2006. Crop selection: adapting to climate change in Africa. Centre for Environmental Economics and Policy in Africa (CEEPA), Discussion Paper No. 26. University of Pretoria, Pretoria, South Africa.

Kurukulasuriya, P., and R. Mendelsohn. 2008. How will climate change shift agro-ecological zones and impact African agriculture? Policy Research Working Paper 4717. The World Bank, Development Research Group, Washington, D.C.

Liu, J., S. Fritz, C. F. A. Van Wesenbeeck, M. Fuchs, L. You, M. Obersteiner, et al. 2008. A spatially explicit assessment of current and future hotspots of hunger in sub-Saharan Africa in the context of global change. Global Planet. Change 64:222–235.

Lobell, D. B., and C. B. Field. 2007. Global scale climate–crop yield relationships and the impacts of recent warming. Environ. Res. Lett. 2:014002.

Lobell, D. B., and S. M. Gourdji. 2012. The influence of climate change on global crop productivity. Plant Physiol. 160:1686–1697.

Lobell, D. B., M. B. Burke, C. Tebaldi, M. D. Mastrandrea, W. P. Falcon, and R. L. Naylor. 2008. Prioritizing climate change adaptation needs for food security in 2030. Science 319:607–610.

Lobell, D. B., M. Bänziger, C. Magorokosho, and B. Vivek. 2011. Nonlinear heat effects on African maize as evidenced by historical yield trials. Nat. Clim. Chang. 1:42–45.

Lukorito, C. B., and G. O. Ouma. 2010. Enhancing the capacity of Kenya on climate risk reduction and climate change adaptation for sustainable agricultural productivity and food security. Paper presented at Second RUFORUM Biennial Meeting, Entebbe, Uganda.

Maganga, A. M.. 2012. Technical efficiency and its determinants in irish potato production: evidence from Dedza District, Central Malawi. Afr. J. Agric. Res. 7:1794–1799.

Maiti, R. K.. 1996. Sorghum science. Science Publishers, Lebanon, NH.

Manneh, B., P. Kiepe, M. Sie, M. Ndjiondjop, N. K. Drame, K. Traore, et al. 2007. Exploiting partnerships in research and development to help African rice farmers cope with climate variability. J. SAT Agri. Res. 4:1–10.

Maure, G. A., S. Bila, S. Hachigonta, L. M. Sibanda, and T. S. Thomas. 2012. Southern African agriculture and climate change: a comprehensive analysis-Mozambique. International Food Policy Research Institute, Washington, DC.

McSweeney, C., M. New, and G. Lizcano. 2010a. UNDP Climate Change Country Profiles: Zambia. Available at http://country-profiles.geog.ox.ac.uk/. (accessed October 15, 2013).

McSweeney, C., M. New, G. Lizcano, and X. Lu. 2010b. The UNDP climate change country profiles improving the accessibility of observed and projected climate information for studies of climate change in developing countries. Bull. Am. Meteorol. Soc. 91:157–166.

McSweeney, C., M. New, and G. Lizcano. 2010c. UNDP Climate Change Country Profiles: Ethiopia. Available at: http://country-profiles.geog.ox.ac.uk/. (accessed October 15, 2013).

McSweeney, C., M. New, and G. Lizcano. 2010d. UNDP Climate Change Country Profiles: Kenya. Available at: http://country-profiles.geog.ox.ac.uk/. (accessed October 15, 2013).

McSweeney, C., M. New, and G. Lizcano. 2010e. UNDP Climate Change Country Profiles: Malawi. Available at: http://country-profiles.geog.ox.ac.uk/. (accessed October 15, 2013).

McSweeney, C., M. New, and G. Lizcano. 2010f. UNDP Climate Change Country Profiles: Mozambique. Available at: http://country-profiles.geog.ox.ac.uk/. (accessed October 15, 2013).

McSweeney, C., M. New, and G. Lizcano. 2010g. UNDP Climate Change Country Profiles: Tanzania. Available at http://country-profiles.geog.ox.ac.uk/ (accessed October 15, 2013).

McSweeney, C., M. New, and G. Lizcano. 2010h. UNDP Climate Change Country Profiles: Uganda. Available at http://country-profiles.geog.ox.ac.uk/. (accessed October 15, 2013).

Meredith, W. R. 2005. Influence of cotton breeding on yield and fiber quality problems *in Proceedings of Cotton Incorporated Annual Engineered Fiber Selection Conference*. Cotton Inc., Raleigh, NC.

Messina, J., U. Adhikari, J. Carroll, R. Chikowo, M. DeVisser, L. Dodge, et al. 2014. Population growth, climate change and pressure on the land – eastern and southern Africa. Global Center for Food Systems Innovation, Michigan State University, East Lansing, MI.

Miglietta, F., V. Magliulo, M. Bindi, L. Cerio, F. P. Vaccari, V. Loduca, et al. 1998. Free air CO_2 enrichment of potato (*Solanum tuberosum L.*): development, growth and yield. Glob. Change Biol. 4:163–172.

Mortley, D., J. Hill, P. Loretan, C. Bonsi, W. Hill, D. Hileman, et al. 1996. Elevated carbon dioxide influences yield and photosynthetic responses of hydroponically-grown sweetpotato. Acta Horticult. 440:31–36.

Muchow, R. C., T. R. Sinclair, and J. M. Bennett. 1990. Temperature and solar radiation effects on potential maize yield across locations. Agron. J. 82:338–343.

Muhinyuza, J. B., H. Shimelis, R. Melis, J. Sibiya, and M. N. Nzaramba. 2012. Participatory assessment of potato production constraints and trait preferences in potato cultivar development in Rwanda. Int. J. Dev. Sust. 1:358–380.

Mupangwa, W., D. Love, and S. Twomlow. 2006. Soil–water conservation and rainwater harvesting strategies in the semi-arid Mzingwane Catchment, Limpopo Basin, Zimbabwe. Phys. Chem. Earth, Parts A/B/C 31:893–900.

Muyinza, H., H. L. Talwana, R. O. Mwanga, and P. C. Stevenson. 2012. Sweetpotato weevil (*Cylas spp.*) resistance in African sweetpotato germplasm. Int. J. Pest Manag. 58:73–81.

Nearing, M. A., F. F. Pruski, and M. R. O'Neal. 2004. Expected climate change impacts on soil erosion rates: a review. J. Soil Water Conserv. 59:43–50.

Negassa, A., J. Koo, K. Sonder, B. Shiferaw, M. Smale, H. J. Braun, et al. 2012. The potential for wheat production in sub–Saharan Africa: analysis of biophysical suitability and economic profitability. International Maize and Wheat Improvement Center (CIMMYT), Addis Ababa, Ethiopia.

Nelson, G. C., M. W. Rosegrant, J. Koo, R. Robertson, T. Sulser, T. Zhu, et al. 2009. *Climate change: Impact on agriculture and costs of adaptation*. Food policy report. IFPRI International Food Policy Research Institute, Washington, DC.

Netondo, G. W., F. Waswa, L. Maina, T. Naisiko, N. Masayi, and J. K. Ngaira. 2010. Agrobiodiversity endangered by sugarcane farming in Mumias and Nzoia Sugarbelts of Western Kenya. Afr. J. Environ. Sci. Technol. 4:437–445.

Ngigi, S. N. 2003. What is the limit of up-scaling rainwater harvesting in a river basin? Phys. Chem. Earth, Parts A/B/C 28:943–956.

Ngigi, S. N.. 2009. Climate change adaptation strategies: water resources management options for smallholder farming systems in sub-Saharan Africa. The MDG Centre for East and Southern Africa, The Earth Institute at Columbia University, New York, NY.

Niang, I., O. C. Ruppel, M. A. Abdrabo, A. Essel, C. Lennard, J. Padgham, et al. 2014. Africa. Pp. 1199–1265 *in* V. R. Barros, C. B. Field, D. J. Dokken, M. D. Mastrandrea, K. J. Mach, T. E. Bilir, M. Chatterjee, K. L. Ebi, Y. O. Estrada, R. C. Genova, B. Girma, E. S. Kissel, A. N. Levy, S. MacCracken, P. R. Mastrandrea, L. L. White, eds. Climate change 2014: impacts, adaptation, and vulnerability. Part B: regional Aspects. Contribution of Working Group II to the Fifth Assessment Report of the Intergovernmental Panel on Climate Change. Cambridge Univ. Press, Cambridge, United Kingdom and New York, NY.

Nielson, H. 2009. The role of cassava in smallholder maize marketing in Zambia and Mozambique. [M. Sc. thesis], Michigan State University, East Lansing, MI.

Obalum, S. E., U. C. Amalu, M. E. Obi, and T. Wakatsuki. 2011. Soil water balance and grain yield of sorghum under no-till versus conventional tillage with surface mulch in the derived Savanna Zone of southeastern Nigeria. Exp. Agric. 47:89.

Obare, G. A., D. O. Nyagaka, W. Nguyo, and S. M. Mwakubo. 2010. Are Kenyan smallholders allocatively efficient? Evidence from Irish potato producers in Nyandarua North district. J. Develop. Agri. Econ. 2:79.

Obilana, A. B. 2003. Overview: importance of millets in Africa *in* P. S. Belton, J. R. N. Taylor, eds. Proceedings of the workshop on the proteins of sorghum and millets: enhancing nutritional and functional properties for Africa. Pretoria, South Africa. Available at http://www.afripro.org.uk/papers/paper02obilana.pdf. (assessed December 12, 2013).

Oerke, E. C. 2006. Crop losses to pests. J. Agri. Sci. 144:31–43.

Olson, J. M., D. Rubin, and E. Wangui. 2010. Gender, Agriculture, and Climate Change: A Regional Analysis for USAID/East Africa. Report prepared for USAID by DevTech Systems, Inc.

Otsu, N. 1975. A threshold selection method from gray-level histograms. Automatica 11:23–27.

Paavola, J. 2008. Livelihoods, vulnerability and adaptation to climate change in Morogoro, Tanzania. Environ. Sci. Policy 11:642–654.

Pachauri, R. K., and A. Reisinger. 2007. Climate change 2007: synthesis report. Contribution of Working Groups I, II and III to the Fourth Assessment Report of the Intergovernmental Panel on Climate Change. Intergovernmental Panel on Climate Change, Geneva, Switzerland.

Pettigrew, W. T. 2008. The effect of higher temperatures on cotton lint yield production and fiber quality. Crop Sci. 48:278–285.

Prasad, P. V. V., S. A. Staggenborg, and Z. Ristic. 2008. Impacts of drought and/or heat stress on physiological, developmental, growth and yield processes of crop plants *in* L. H. Ahuja, L. Ma and S. Saseendran, eds. Responses of crops to limited water: understanding and modeling water stress effects on plant growth processes. Advances in Agricultural Modeling Series 1, 301–355. ASA-CSSA, Madison, WI, USA.

Ramirez, G. P. 1992. Cultivation, harvesting and storage of sweet potato products. Pp. 203–215 *in* D. Machin and S. Nyvold, ed. Roots, tubers, plantains and bananas in animal feeding. FAO, Rome, Italy, 95.

Ramirez, J., A. Jarvis, I. Van den Bergh, C. Staver, and D. Turner. 2011. Chapter 20: changing climates: effects on growing conditions for banana and plantain (*Musa spp.*) and possible responses. Pp 426–438 *in* S. Yadav, B. Redden, J. L. Hattfield, H. Lotze-Campen, eds. Crop adaptation to climate change. Wiley-Blackwell, Chichester, U.K.

Ravallion, M. 2012. Benchmarking global poverty reduction. Policy Research Working Paper 6205. The World Bank, Washington, DC.

Reddy, K. R., G. H. Davidonis, A. S. Johnson, and B. T. Vinyard. 1999. Temperature regime and carbon dioxide enrichment alter cotton ball development and fiber properties. Agron. J. 91:851–858.

Rickards, L., and S. M. Howden. 2012. Transformational adaptation: agriculture and climate change. Crop Pasture Sci. 63:240–250.

Ringler, C., T. Zhu, X. Cai, J. Koo, and D. Wang. 2010. Climate change impacts on food security in sub-Saharan Africa: insights from comprehensive climate change scenarios (No. 1042). International Food Policy Research Institute (IFPRI), Washington, DC.

Robinson, J. C.. 1996. Banana and Plantains. CAB International, Wallingford, U.K.

ROK 2002. First national communication of Kenya to the conference of the parties to the United Nations Framework Convention on Climate Change, Republic of Kenya. Nairobi, Kenya. Available at http://unfccc.int/resource/docs/natc/kenncl.pdf. (assessed December 12, 2013).

Rosenthal, D. M., and D. R. Ort. 2012. Examining cassava's potential to enhance food security under climate change. Trop. Plant Biol. 5:30–38.

Runge, E. C. A.. 1968. Effect of rainfall and temperature interactions during the growing season on corn yield. Agron. J., 60:503–507.

Saka, A. R., A. P. Mtukuso, A. T. Daudi, M. H. P. Banda, and I. M. G. Phiri. 2006. Agricultural technologies released by the Ministry of Agriculture and Food Security: 2000–2005. Department of Agricultural Research Services, Ministry of Agriculture and Food Security, Lilongwe, Malawi.

Saka, J. D. K., P. Siable, S. Hachigonta, L. M. Sibanda, and T. S. Thomas. 2012. Southern African agriculture and climate change: a comprehensive analysis-Malawi. International Food Policy Research Institute, Washington, DC.

Salami, A., A. B. Kamara, and Z. Brixiova. 2010. Smallholder agriculture in east Africa: Trends, constraints and opportunities. Working Papers Series No. 105. African Development Bank, Tunis, Tunisia.

Sanchez, P. A., K. D. Shepherd, M. J. Soule, F. M. Place, and R. J. Buresh. 1997. Soil fertility replenishment in Africa. An investment in natural resource capital. Pp. 1–46 *in* R. J. Buresh, P. A. Sanchez and F. Calhoun, eds. Replenishing soil fertility in Africa. Soil Sci. Soc. Am, Madison, WI.

Schlenker, W., and D. B. Lobell. 2010. Robust negative impacts of climate change on African agriculture. Environ. Res. Lett. 5:014010.

Schlenker, W., and M. J. Roberts. 2009. Nonlinear temperature effects indicate severe damages to US crop yields under climate change. Proc. Natl Acad. Sci. USA 106:15594–15598.

Shonga, E., M. Gemu, T. Tadesse, and E. Urage. 2013. Review of entomological research on sweet potato in Ethiopia. Discour. J. Agri. Food Sci. 1:83–92.

Simms, A. 2005. Africa: up in smoke? The second report from the Working Group on Climate Change and Development, New Economics Foundation, London, United Kingdom.

Smale, M., D. Byerlee, and T. Jayne. 2011. Maize revolutions in sub-Saharan Africa. World Bank Policy Research Paper 5659. The World Bank, Washington, DC.

Smit, N. E. J. M. 1997. Integrated pest management for sweetpotato in eastern Africa. [PhD dissertation]. Agricultural University Wageningen, The Netherlands.

Stone, P. 2001. The effects of heat stress on cereal yield and quality. Pp. 243–291 *in* A. S. Basra, ed. Crop responses and adaptations to temperature stress. Food Products Press, Binghamton, NY.

Sultan, B., M. Bella-Medjo, A. Berg, P. Quirion, and S. Janicot. 2010. Multi-scales and multi-sites analyses of the role of rainfall in cotton yields in West Africa. Int. J. Climatol. 30:58–71.

Tatsumi, K., Y. Yamashiki, R. Valmir da Silva, K. Takara, Y. Matsuoka, K. Takahashi, et al. 2011. Estimation of potential changes in cereals production under climate change scenarios. Hydrol. Process. 25:2715–2725.

Taylor, J. R. N. 2003. Overview: importance of sorghum in Africa *in* P. S. Belton and J. R. N. Taylor, eds. Proceedings of the workshop on the proteins of sorghum and millets: enhancing nutritional and functional properties for Africa. Pretoria, South Africa. Available at http://www.afripro.org.uk/papers/paper01Taylor.pdf. (assessed December 12, 2013).

Teixeira, E. I., G. Fischer, H. van Velthuizen, C. Walter, and F. Ewert. 2013. Global hot-spots of heat stress on agricultural crops due to climate change. Agric. For. Meteorol. 170:206–215.

Tenge, N. G., M. Alphonse, and T. S. Thomas. 2012. East African agriculture and climate change: a comprehensive analysis-Rwanda. International Food Policy Research Institute, Washington, DC.

Thornton, P. 2012. Recalibrating food production in the developing world: Global warming will change more than just the climate. CCAFS Policy Brief no. 6. CGIAR Research Program on Climate Change, Agriculture and Food Security (CCAFS). Available at www.ccafs.cgiar.org. (assessed December 12, 2013).

Thornton, P., and L. Cramer. 2012. Impacts of climate change on the agricultural and aquatic systems and natural resources within the CGIAR's mandate. CCAFS Working Paper 23. CGIAR Research Program on Climate Change, Agriculture and Food Security (CCAFS), Copenhagen, Denmark.

Thornton, P. K., P. G. Jones, G. Alagarswamy, and J. Andresen. 2009. Spatial variation of crop yield response to climate change in East Africa. Glob. Environ. Change 19:54–65.

Thornton, P. K., P. G. Jones, G. Alagarswamy, J. Andresen, and M. Herrero. 2010. Adapting to climate change: agricultural system and household impacts in East Africa. Agric. Syst. 103:73–82.

Ton, P.. 2011. Cotton and climate change: impacts and options to mitigate and adapt. International Trade Center, Geneva, Switzerland.

URT. 2003. Initial national communication under the United Nations Framework Convention on Climate Change (UNFCCC). United Republic of Tanzania, Vice President's Office, Dar es Salaam, Tanzania.

Van Asten, P. J. A., A. M. Fermont, and G. Taulya. 2011. Drought is a major yield loss factor for rainfed East African highland banana. Agric. Water Manag. 98:541–552.

Van Duivenbooden, N., M. Pala, C. Studer, C. L. Bielders, and D. J. Beukes. 2000. Cropping systems and crop complementarity in dryland agriculture to increase soil water use efficiency: a review. NJAS-Wageningen J. Life Sci. 48:213–236.

Wardlaw, I. F., I. A. Dawson, and P. Munibi. 1989. The tolerance of wheat to high-temperatures during reproductive growth. 2. Grain development. Crop Pasture Sci. 40:15–24.

Wheeler, T., and J. von Braun. 2013. Climate change impacts on global food security. Science 341:508–513.

Wiggins, S., and S. Keats. 2013. Leaping and learning: linking smallholders to markets in Africa. Agriculture for Impact, Imperial College and Overseas Development Institute, London.

Wolf, S., A. Marani, and J. Rudich. 1990. Effects of temperature and photoperiod on assimilate partitioning in potato plants. Ann. Bot. 66:513–520.

World Bank. 2015. World development indicators. The World Bank, Washington, DC.

Wortmann, C. S., R. A. Kirkby, C. A. Eledu, and D. J. Allen. 1998. Atlas of common bean (*Phaseolus vulgalis* L.) production in Africa. CIAT-Pan-African Bean Research Alliance, Kampala, Uganda.

Wortmann, C. S., M. Mamo, C. Mburu, E. Letayo, G. Abebe, K. C. Kayuki, et al. 2009. Atlas of sorghum production in eastern and southern Africa. Univ. of Nebraska-Lincoln, Lincoln, NE.

Yoon, S. T., G. Hoogenboom, I. Flitcroft, and M. Bannayan. 2009. Growth and development of cotton (*Gossypium hirsutum* L.) in response to CO_2 enrichment under two different temperature regimes. Environ. Exp. Bot. 67:178–187.

You, L. Z. 2008. Africa infrastructure country diagnostic: Irrigation investment needs in sub-Saharan Africa. Summary of Background Paper 9. World Bank, Washington, DC.

You, L., M. W. Rosegrant, C. Fang, and S. Wood. 2005. Impact of global warming on Chinese wheat productivity. International Food Policy Research Institute, Washington, DC.

Ziska, L. H., D. M. Blumenthal, G. B. Runion, E. R. Jr Hunt, and H. Diaz-Soltero. 2011. Invasive species and climate change: an agronomic perspective. Clim. Change. 105:13–42.

2

Crop biotechnology: a pivotal moment for global acceptance

Eric Hallerman[1] & Elizabeth Grabau[2]

[1]Department of Fish and Wildlife Conservation, Virginia Polytechnic Institute and State University, Blacksburg, Virginia 24061
[2]Department of Plant Pathology, Physiology and Weed Science, Virginia Polytechnic Institute and State University, Blacksburg, Virginia 24061

Keywords
Bioregulation, genetically engineered crops, public perception of biotechnology.

Correspondence
Elizabeth Grabau, Department of Plant Pathology, Physiology and Weed Science, Virginia Polytechnic Institute and State University, Blacksburg, VA 24061.

E-mail: egrabau@vt.edu

Funding Information
USDA Foreign Agricultural Service. Virginia Agricultural Experiment Station

Abstract

The development and judicious use of agricultural biotechnology offers important contributions to food security and sustainability. Key contributions include improved yield, heightened pathogen and herbivore resistance, enhanced nutrient content, improved product quality, reduced spoilage, as well as entirely new traits. While a first wave of genetically engineered (GE) crop products has been commercialized and contributed to yield, other products – some posing significant benefits to target populations in the developing world – have become mired in controversy. Public misconception about nutritional and ecological risks, fears about multinational corporate dominance, as well as regulatory inaction have delayed the approval and use of GE crops. With new GE lines ready to pass through regulatory oversight, many of which originate from developing countries, we regard this as a pivotal moment for global acceptance of agricultural biotechnology. However, we note that some countries, international regulators, and even biotechnology companies seem willing to forego useful applications of GE crops. We conclude that educating and informing the public to combat misperception, and implementing review of regulatory guidelines based on decades of experience can help to realize the benefits of GE for food security, human well-being, and ecological sustainability.

Introduction

A major challenge facing mankind is how to increase world access to adequate food without depleting nonrenewable natural resources and causing environmental damage. The concept of food security has been defined (WHO 2015) as including both physical and economic access to food that meets people's dietary needs as well as their food preferences. Dramatic advances in food production, distribution, and access will be required if we are to meet future needs (National Academies 2000, National Research Council 2015). Since the first plant cell transformation successes over 30 years ago (Herrera-Estrella et al. 1983), transformation using *Agrobacterium tumefaciens* and other systems has been developed and extended to a widening variety of agricultural crops. This journal has previously reviewed successes, failures, and potential applications of plant biotechnology to agriculture (Halford 2012). In addition to support for basic science to advance plant biotechnology, there is also a critical need for translating plant science discoveries to the benefit of the poor in developing regions of the world (Delmer 2005).

Application of biotechnology approaches to crop improvement has resulted in increased yields, enhanced agronomic traits, better product quality and novel uses of crop products. After the first genetically engineered (GE) line – a Roundup-Ready soybean produced by Monsanto – was approved for commercial use in the United States in 1996, other lines followed and the percentage of GE crop production increased rapidly. However, commercial production is dominated by a few herbicide-resistant and insect-resistant crops, environmental benefits are disputed, and food safety is debated. Negative attitudes toward GE crops in some sectors of the developed world complicate their adoption in the developing world, where they are potentially most valuable for addressing the issue of food security (Whitty et al. 2013). Although these negative attitudes are generally at odds with scientific evidence, they persist in both the developed and developing worlds (Blancke et al. 2015).

In a landmark study, a working group representing several national and international academies of science (National Academies 2000) concluded that GE technology, coupled with important developments in other areas, should be used to increase the production of main food staples, improve the efficiency of production, reduce the environmental impact of production, and provide access to food for small-scale farmers. However, the potential utility of GE crop plants has been only partially realized in countries where food security is at issue. Success is partial because of different attitudes and because of the impacts of the respective regulatory oversight policies. In our review and synthesis, against the background of GE crops currently in production, we note examples of new products in the pipeline, some posing important implications for promoting food security and sustainability. Yet, we also note that regulatory and public acceptance issues complicate the adoption of GE plants for production agriculture. We regard this as a pivotal moment for regulatory action and practical adoption of agricultural biotechnology, a moment when decisions made within the next few years will have major bearing on the realization – or failure – of its potential for improving global food security. We focus on issues pertinent to GE crops; similar issues regarding genetically modified animals have been discussed elsewhere (Van Eenennaam et al. 2011, 2013). We call for thoughtful application of all agronomic and plant breeding technologies, including genetic engineering, to increase crop production and effectively address food security.

Adoption of Crop Biotechnology

Since they were first commercialized in 1996, GE crops have gained a rising share of agricultural production; 181.5 million ha of GM crops were planted in 2014 in 28 countries (ISAAA 2015), with most production in the United States (73.1 million ha), Brazil (42.2), Argentina (24.3), Canada (11.6), and India (11.6). A total of 18 million farmers planted GE crops, 16.9 million of whom were small farmers in developing countries. Most of the production was accounted for by four crops, soybean (82% of which was GE), cotton (68%), corn (30%), and canola (25%). Among 30 traits that have been engineered into crop plants (Nature 2013d), the most popular are herbicide tolerance, insect resistance, or both as stacked traits. Adoption of these GE crop lines has increased yields. Producing glyphosate-resistant corn and soybeans, farmers control most weeds with one herbicide rather than several and are better placed to adopt no-till soil conservation practices. However, many weed species recently have evolved resistance to glyphosate and other overused herbicides that are required by these GE crops, such that the long-term sustainability of facilitating low- and no-till agriculture is not certain (Mortensen et al. 2012). Production of *Bt* cotton and rice increases yield (9% in the case of cotton – Huang et al. 2003; 9–29% for rice – reviewed by Chen et al. 2011), whereas less pesticide is sprayed on the field, decreasing impacts on nontarget organisms (Huang et al. 2003; Li et al. 2015; Qiao 2015), including the farmers themselves (Huang et al. 2015a, 2015b). GE crops have been adopted faster than any other agricultural advance in the history of humanity (Alberts et al. 2013). The global value of GE seed in 2012 was US$15 billion (Nature 2013d).

Not all GE plants currently in production are row crops. One of the key early successes for GE agricultural plants – and one that is not widely recognized by the public – is disease-resistant papaya. Papaya is cultivated worldwide in the tropics and subtropics (GMO Compass 2013); the main producing countries are Brazil, Mexico, Nigeria, and India. In 2005, approximately 6.75 million tons were produced worldwide. Papaya ringspot virus (PRSV) is a major production problem, and can lead to dramatic losses of yield. GE papayas engineered for resistance to PRSV (Gonsalves 1998) have been produced commercially in Hawaii since 1998, and are credited with having saved the industry there. They have been approved by government regulators and sold commercially in the United States. These papayas are an important export product, mainly to Japan. After the success in Hawaii, other GE papaya lines are being developed in other regions for the viruses prevalent in those areas. In 2006, a GE virus-resistant papaya was approved and now is in commercial production in China. A cooperative project involving international companies and organizations from India, Indonesia, Thailand, Malaysia, the Philippines, and Vietnam is aimed at development of a GE virus-resistant papaya for Southeast Asia. However, despite the agricultural and commercial success of GE papaya, it remains controversial. Anti-GE activists destroyed papaya trees at least three times in Hawaii, and bills that would impose restrictions on agricultural biotechnology have been debated in Hawaii's legislature (Huffington Post 2013, Harmon 2014). Import of GE papayas into the European Union is not allowed (GMO Compass 2013). Despite virus-resistant papaya being a "pro-poor" GE crop, Greenpeace activists destroyed a test plot of transgenic papaya in Thailand in 2004, leading to a countrywide moratorium on all field testing of transgenic crops (Davidson 2008).

Adoption of GE crops varies widely among countries, with 32 now having approved lines for production or import for use as food and feed (Table 1). Not

Table 1. Numbers of GE crop events approved by given countries for commercialization, planting, or import for food and feed use.[1]

Country	Corn	Soybean	Canola	Rice	Potato	Other crops
Argentina	30	6	–			–
Australia	21	16	14	1	10	3 alfalfa, 2 sugar beet, 1 wheat
Bangladesh	–	–	–	–		1 eggplant[2]
Bolivia	–	1	–	–		–
Brazil	21	6	–	–		1 bean[3]
Canada	59	20	22	1	20	3 alfalfa, 1 papaya, 1 squash, 2 sugar beet, 4 tomato
Chile	1	1	1	–	–	1 tomato
China	16	8	12	2[4]	–	1 papaya, 1 sugar beet, 1 sweet pepper[5], 3 tomato[6]
Colombia	39	10	–	2	–	1 sugar beet, 1 wheat
Costa Rica	–	2	–	–	–	–
Egypt	1	–	–	–	–	–
European Union	39	8	11	–	1	1 sugar beet
Honduras	7	–	–	1	–	–
Indonesia	7	2	–	–	–	3 sugar cane[7]
Iran	–	–	–	1[8]	–	–
Japan	111	19	17	1[9]	8	3 alfalfa, 1 papaya, 3 sugar beet
Malaysia	6	6	–	–	–	–
Mexico	65	18	11	1	13	3 alfalfa, 1 sugar beet, 5 tomato
New Zealand	21	16	14	1	10	3 alfalfa, 2 sugar beet, 1 wheat
Panama	1	–	–	–	–	–
Paraguay	8	3	–	–	–	–
Philippines	46	9	1	1	8	2 alfalfa, 1 sugar beet
Russian Federation	12	7	–	1	2[10]	1 sugar beet
South Africa	33	9	4	1	–	–
South Korea	60	19	11	–	8	3 alfalfa, 1 sugar beet
Switzerland	3	1	–	–	–	–
Taiwan	63	18	–	–	–	–
Thailand	12	3	–	–	–	–
Turkey	16	3	–	–	–	–
United States of America	38	24	20	3	38	3 alfalfa, 3 chicory, 2 melon[11], 3 papaya[12,13], 1 plum[14], 2 squash, 3 sugar beet, 8 tomato, 1 wheat
Uruguay	11	7	–	–	–	
Vietnam	5					

[1]Data from ISAAA (2015) http://www.isaaa.org, accessed 7 January 2015.
[2]*Bt*-transgenic eggplant developed by Maharashtra Hybrid Seed Company.
[3]Antisense RNA viral disease resistance bean, Brazilian Ministry of Agriculture – Brazilian Agricultural Research Corporation (EMBRAPA).
[4]Two *Bt* rice lines, Huazhong Agricultural University.
[5]Pathogen-derived viral disease-resistant sweet pepper, Beijing University.
[6]Modified product quality tomato, Chinese Academy of Science; delayed-ripening tomato, Huazhong Agricultural University; pathogen-derived viral disease-resistant tomato, Beijing University.
[7]Drought-resistant sugar cane, PT Perkebunan Nusantara Xi (Persero).
[8]*Bt* rice, Agricultural Biotechnology Research Institute.
[9]Hypoallergenic rice, National Institute of Agrobiological Sciences.
[10]Two *Bt* potato lines protected from coleopteran insects, Centre Bioengineering, Russian Academy of Sciences.
[11]Delayed ripening melon, Agritope, Inc.
[12]Two pathogen-derived virus-resistant papaya lines, Cornell University and University of Hawaii.
[13]Pathogen-derived virus-resistant papaya line, University of Florida.
[14]Pathogen-derived virus-resistant plum line, U.S. Department of Agriculture – Agricultural Research Service.

surprisingly, adoption of GE crops is further advanced in developed countries than in developing ones. One crop in particular, insect-resistant *Bt* cotton, has been widely planted in many countries, developed and developing alike. For example, most cotton in India is GE and has provided economic benefit to many small producers. This is in stark contrast to sentiment against production of GE food crops in India (see below). While most GE crop lines currently in use were developed by multinational corporations, a growing number of GE lines developed by universities, government agencies, small companies, and nongovernmental organizations are now proceeding through regulatory approval for possible agricultural production (see footnotes for Table 1). These GE lines include *Bt* eggplant in India, rice in China and Iran, potato in Russia, disease-resistant pepper and tomato in China, and drought-resistant sugar cane in Indonesia.

Nonadoption of Crop Biotechnology

Despite their record of improved yield, improved profitability, heightened nutrition, and potential for addressing food security, GE plants remain controversial. Some countries or international unions, notably ones *not* facing food security issues, have chosen not to adopt GE crops. We explore selected cases.

European Union

Approvals of GE crops in Europe have proven highly controversial within countries and between member countries and the European Union itself. For GE crops to be approved in Europe, they must be deemed safe by the European Food Safety Authority (EFSA). The European Commission then must produce a draft decision within 3 months, to be voted upon by representatives of EU member states before approval can be finalized (Cressy 2013b). The EFSA has found eight crops safe, some as long ago as 2005. Only two GE crops have been approved by the European Union science advisory committees as safe for agricultural production, MON810 (a *Bt*-expressing, insect-resistant corn) and Amflora (a potato for use by the paper industry). However, eight EU nations have imposed a ban on cultivation of MON810 corn, and Poland has also banned Amflora (Nature 2013a, Science 2013a). These moves come despite EU legislation that requires all member states to permit cultivation of approved crops; European courts have already ruled against bans in two countries (Nature 2013c). France's highest court overturned the national ban on the GE corn, ruling that the government must prove that the crop causes health or environmental risks (Nature 2011c) and its State Council found the ban invalid for procedural reasons, but the agriculture minister said he would work to uphold the ban. The European Court of Justice ruled in 2011 that honey containing trace amounts of pollen from GM plants could no longer be sold in the European Union without a safety review (Nature 2011b), which could affect imports of honey from countries such as Argentina, where GE crops are produced at a large scale. Against this background, Germany-based BASF Corporation shifted its transgenic plant operations to the Americas in 2012 (Nature 2012a). Monsanto announced in mid-2013 that it would withdraw pending applications for GE crops including several lines of corn, a line of soybean, and a line of sugar beet (Science 2013a) and that it would focus on sales of conventionally bred seed in Europe. European GE plant researchers expressed regret, but not surprise (Cressy 2013b); many European GE plant researchers now do their work in more agbiotech-friendly countries outside the EU.

Developing Countries

Although food security is not perceived as a pressing issue within the European Union, it is important in many regions of the developing world. Yet, many developing countries have not adopted production of GE crops. Among sub-Saharan African countries, only South Africa has approved production of GE lines. For example, in Kenya, where more than one-quarter of the population is malnourished, the government banned the import of GE foods in 2012, although it did not ban research on GE crops (Owino 2012; Whitty et al. 2013). In Asia and Latin America, some countries have approved GE lines and others have not. Notable holdouts for production of GE food crops include India (Science 2013b), which is considering a 10-year moratorium on all field trials. We note that India also fought the adoption of Borlaug's selectively bred wheat lines that were the leading edge of the Green Revolution (Vietmeyer 2012). In South America, Ecuador banned GE crops in 2008 (Science 2011). Groups representing Peru's indigenous people campaigned for a ban on cultivation of GE crops, arguing that GE plants could damage native agriculture and threaten the country's flora, including over 2000 species of potato. Seeking more study, Peru's Congress passed a 10-year moratorium on the cultivation of GE crops (Science 2011); the legislation will not, however, stop research on GE crops. Meanwhile, Bolivian President Evo Morales opened the door to cultivation of GE lines there (Science 2011).

Case Studies of GE Plants Targeting Food Security and Sustainability

The first GE crops to be commercialized on a large scale offer herbicide tolerance and pest resistance traits, especially targeting developed-world farmers using high-input production methods and buying their seed each generation. While such production has indirect bearing on issues of food security, there also is a new wave of GE lines aimed at both staple and minor crops for production in the developing world. Another line of research aims to develop food products that are less prone to bruising or spoilage, thereby minimizing postharvest loss and promoting sustainability. Here, we present selected case studies of GE lines, noting their state of development and adoption. We note that many of these GE crops are at a pivotal stage of development or adoption, with some lines not being utilized because they are misunderstood by the public or delayed in regulatory review.

Golden Rice

The leading example of ongoing controversy and politicization of a GE plant intended for addressing food security and nutrition is the development of golden rice. Rice is the major staple grain for almost half of humanity. However, rice is usually milled to remove the oil-rich aleurone layer, and the remaining endosperm lacks several essential nutrients including provitamin A, or β-carotene. Heavy rice consumption promotes vitamin A deficiency, which is a serious health problem in at least 26 countries in Asia, Africa, and Latin America (WHO 2005) and is responsible for 1.9–2.8 million preventable deaths per year (Mayo-Wilson et al. 2011). In a technical tour de force, Ingo Potrykus and Peter Beyer and their research groups introduced multiple genes into rice to complete the β-carotene biosynthetic pathway. Heralding the original report in *Science*, Guerinot (2000) wrote of hope that this application of plant genetic engineering would ameliorate human misery and restore gene transfer technology to political acceptability.

Golden rice, however, proved the focus of intense controversy. Greenpeace (2002) claimed that an adult would have to consume at least 12 times the normal intake of 300 grams of rice in order to take in the daily recommended amount of provitamin A, calling claims that golden rice would provide nutritional benefits an "intentional deception". Since the first scientific publication reporting 1.6 μg of vitamin A precursor per gram of edible rice (Ye et al. 2000), collaborative work between the developers and the International Rice Research Institute has increased the level of β-carotene to 35 μg (Potrykus 2010), a sufficient quantity that a few ounces of cooked rice can provide enough β-carotene to eliminate the morbidity and mortality of vitamin A deficiency (Alberts et al. 2013). The inventors intend to make golden rice technology a public good without cost or license fees to public-sector rice-breeding institutions or smallholder farmers (Dubock 2013). No individual or organization involved in its development will benefit financially from its adoption. Using the empirical GTAP model of the world economy, Anderson et al. (2004) estimated that adoption of golden rice in Asia would lead to benefits of $15.2 billion per year through enhanced productivity of unskilled labor, exclusive of any gains due to welfare gains from reduced vitamin A deficiency.

The putative nutritional benefit arguably has been the focal controversy regarding golden rice. Golden rice was shown to be an effective source of vitamin A to adults in the United States (Tang et al. 2009). To prove most relevant to resolving the issue, however, a study would have to be conducted in a rice-consuming population and in children, who are most vulnerable to vitamin A deficiency. Against this background, Tang et al. (2012) fed 732 Chinese children golden rice, spinach, or capsules with β-carotene in oil and found that golden rice was as good a source of vitamin A as the capsules and better than spinach. However, upon publication of the study, Greenpeace China released a hostile message to the press, leading several of the Chinese collaborators on the study to distance themselves from the work; one was suspended from his position (Hvistendahl and Enserink 2012). While the study's results were not disputed, the investigator was cited for not documenting reviews and approvals in China and for having made changes to the study without approval of the Tufts University Institutional Review Board for research on human subjects (Science 2013b). Participants were found to have not been adequately informed about the GE nature of the rice, and Tang was banned from conducting human research for 2 years by the university (Nature 2013e). The Tang et al. (2012) article was retracted by the American Society for Nutrition (2015) because the authors were unable to demonstrate that the study had been approved by a local ethics committee in China or that parents of the children involved in the study had provided full consent.

Escalating requirements for testing have stalled the release of golden rice for more than a decade (Alberts et al. 2013). Golden rice is being evaluated in the Philippines; however, protestors from two anti-GE groups vandalized a field of golden rice there (Dubock 2013, Nature 2013e, Science 2013c). The destruction was condemned by over 6000 people in the scientific community (Chassy et al. 2013). Golden rice has not yet been approved for general production in any country. It could clear regulatory hurdles and reach farmers soon (Cressy 2013a), likely first in the Philippines and in Bangladesh (Nature 2011a, Dubock 2013), countries where food security is an issue. Despite media reports that golden rice had been approved for production in the Philippines, the International Rice Research Institute (2013) reported that data from 2 years of field trials had yet to be submitted to government regulators for evaluation as part of the biosafety approval process; golden rice will become broadly available to farmers and consumers if it is approved and shown to reduce vitamin A deficiency under community conditions, a process that was expected to take 2 years or more.

Golden rice is only one example of a crop that is GE to address a micronutrient deficiency (Mayer et al. 2008). The folate content of tomato (Diaz de la Garza et al. 2007) and rice (Storozhenko et al. 2007), as well as the iron content of rice (Goto et al. 1999; Lucca et al. 2001) have been increased dramatically. The Grand Challenges in Global Health initiative of the Bill and Melinda Gates Foundation has funded biofortification projects on banana, cassava, and sorghum in Africa (Mayer et al. 2008).

Bt Eggplant

Eggplant is an important vegetable in India, where it is valued for being a good source of fiber, calcium, phosphorus, folate, and vitamins B and C (ISAAA 2013b). Eggplant is grown on nearly 550,000 ha in India, an important cash crop for more than 1.4 million small farmers. Eggplant is prone to attack from insect pests and diseases, including the fruit and shoot borer (FSB) *Leucinodes orbonalis*; cultivation is often input intensive, especially regarding insecticide applications (George et al. 2002; ISAAA 2013b). However, since the larvae are concealed within shoots and fruits, they normally escape insecticide sprays; hence, farmers tend to overspray, contributing to cost, negative effects on the environment, high pesticide residues in vegetables, and risk to producers and consumers. Although several attempts have been made to develop resistant cultivars through traditional plant breeding, little or no success was achieved. Scientists at the Maharashtra Hybrid Seeds Company (Mahyco) have developed an eggplant line that can resist FSB attack and donated the technology to the Tamil Nadu Agricultural University and the University of Agricultural Sciences in Dharwad. Since its development in 2000, the crop has undergone eight assessments of food safety, environmental safety, human and animal health safety and biodiversity (Mahyco 2009). At a production level, the GE line yielded an average increase of 116% in marketable fruits over conventional hybrids, and a 166% increase over popular open-pollinated varieties. The significant decrease in insecticide usage reduced farmers' exposure to insecticides and resulted in a substantial decline in pesticide residues in the fruits. Scientists estimated that *Bt* eggplant would deliver farmers a net economic benefit ranging from US$330–397 per acre (Tamil Nadu Agricultural University 2008). *Bt* eggplant is the first food crop under evaluation for commercial release in India. The Genetic Engineering Approval Committee in 2009 recommended the commercial release of *Bt* eggplant event EE-1, the penultimate step before commercializing *Bt* eggplant hybrids and varieties in India. However, in 2010, the Ministry of Environment and Forest announced a moratorium on approval of *Bt* eggplant (USDA-FAS 2013), thereby preventing dissemination of this line to farmers and sales to consumers.

Mahyco donated the transgenic eggplant technology to public research institutions in the Philippines (Science 2012a) and Bangladesh (ISAAA 2014). However, Greenpeace vandalized field plots in the Philippines in 2011 and petitioned the Supreme Court, claiming that the crops are dangerous to humans and the environment and that the approval procedures were flawed. The Philippine National Academy of Science and Technology criticized Greenpeace actions (Science 2012a), with its president saying that scientists conducting the trials and government departments will file briefs explaining the biosecurity precautions.

Eggplant is also an important vegetable in Bangladesh, where it is grown by about 150,000 small farmers on about 50,000 ha (ISAAA 2014). In 2013, Bangladesh approved the release of four GE varieties of *Bt* eggplant for seed production and initial commercialization. In 2014, seedlings of four *Bt* eggplant varieties were distributed to 20 small eggplant farmers in four regions. The Bangladesh Agricultural Development Corporation undertook seed multiplication of four *Bt* eggplant varieties to be distributed to farmers in 2014. In 2015, *Bt* genes will be introduced in five other popular eggplant varieties. Over the next 5 years, the government of Bangladesh plans to introduce nine *Bt* eggplant varieties into cultivation on 20,000 ha across 20 districts (ISAAA 2014). Field-trial data indicate that *Bt* eggplant can improve yield by at least 30% and reduce the number of insecticide applications by 70–90%, resulting in a net economic benefit of US$1868/ha. At the national level, *Bt* eggplant is estimated to have the capacity to generate a net additional economic benefit of US$200 million per year for around 150,000 eggplant growers in Bangladesh (ISAAA 2014). Consumers will benefit from a blemish-free, more affordable food product.

GE Plums

Sharka or plum pox, caused by plum pox virus (PPV), is the most serious disease of *Prunus* sp. stone-fruit trees. Most cultivated *Prunus* species are highly susceptible, and conventional breeding has not produced highly resistant, commercially acceptable varieties. Transformation of plum (*Prunus domestica*) using DNA from the PPV coat protein gene led to production of transgenic plum lines resistant to the PPV (Ravelonandro et al. 2000). Field tests of GE transgenic lines in Poland, Romania, and Spain demonstrated that trees inoculated by bud-grafts allowed low-level PPV multiplication, from which they rapidly recovered. GE plants exposed to natural infection for 3 years did not become infected, whereas control trees were infected in the first year. Hybrid plums with the transgene inherited from the GE line were virus-resistant, demonstrating the usefulness of this line as a parent in developing new PPV-resistant plum varieties.

Outcrossing is expected in *P. domestica* because of high levels of self-incompatibility (CERA 2007). The risk of gene flow from transgenic to nontransgenic plums was evaluated in an 11-year study. Gene flow was detected in only four of the 11 years and then in only 0.31% of over 12,000 seeds tested. The results of the study suggested that a 400-m distance from non-GE plums would allow for coexistence (Scorza et al. 2013b). Bees were considered the most likely vector for pollen movement.

The GE plum line proceeded slowly through the regulatory process in the United States. Following application for nonregulated status (Scorza 2007), the U.S. Department of Agriculture reached a finding that its cultivation posed no significant impact (USDA-APHIS 2007a) and published notice in the Federal Register (USDA-APHIS 2007b). The U.S. Food and Drug Administration (2009a, 2009b) subsequently stated that it regarded the regulatory consultation process as complete. The Environmental Protection Agency registered the "HoneySweet" line in 2011; hence, cultivation of the product is approved within the United States (Scorza et al. 2013a). To our knowledge, the line is not yet in commercial production – this is a pivotal moment for its adoption. The Organic Consumers Association and other organic organizations opposed deregulation of the product because of concern for genetic "contamination" of conventional and organic stone-fruit varieties (Novak 2013). We note that outbreaks of plum pox pose an international problem to plum production, and that the GE plum project was the product of an international public institutional collaboration supported in part by the European Union. Yet, the EU, a major plum-producing region, seems unlikely to approve production of the product.

Simplot Potato

A new wave of GE food crops aims to provide benefits to consumers in terms of improved product quality and to sustainability through reduced loss to bruising or spoilage. The degree to which these benefits increase acceptability of GE food products is, however, yet to be seen. One example of a crop with improved product quality is Simplot's low acrylamide "Innate" potato, named to convey to the public that the modified product contains only potato DNA (Simplot Plant Sciences 2015).

Asparagine is the compound in starchy foods that is converted to acrylamide when products are baked, roasted or fried at high temperatures. The consumption of processed potato products contributes to approximately one-third of the average dietary intake of acrylamide (Boettcher et al. 2005). Concerns about potential health issues associated with dietary intake of acrylamide led Rommens et al. (2008) of Simplot Plant Sciences to reduce the accumulation of the asparagine in tubers of potato. A 20-fold reduction in asparagine yielded a similar decrease in the amount of acrylamide accumulated during heat processing and did not affect tuber yield or French fry quality. Rommens et al. (2008) estimated that replacement of existing potato lines by low-asparagine varieties would lower the ingestion of acrylamide by approximately 30%.

The U.S. Department of Agriculture approved commercial planting of Innate potatoes in late 2014 (Waltz 2015a). Simplot submitted the potato for a voluntary food safety review by the U.S. Food and Drug Administration, which it received in 2015 (Simplot Plant Sciences 2015). Growers will pay a premium for potato seed, but the crop will have fewer blemishes, and hence a higher percentage of the crops can be sold at a high price (Waltz 2015a). Three Innate varieties were expected to be available in limited quantities in 2015 in the fresh and fresh-cut markets. Fresh-cut potatoes are widely sold to institutional food service providers; in addition to decreased acrylamide production upon cooking, the Innate potato is advantageous because it does not require preservatives or additives to prevent browning.

Simplot is a major producer of potatoes for French fries, which raises questions as to whether fast-food chains will find the Innate potato an acceptable product. In the 1990s, previous lines of GE potatoes with virus and insect resistance traits were spurned by McDonalds and Frito-Lay, leading Monsanto to withdraw the line from the market (Waltz 2015a). Simplot is hoping that lack of foreign DNA in the Innate potato will address anti-GE sentiment. However, a company spokesman for the fast-food giant was widely quoted as saying that "McDonald's USA does not source GMO potatoes nor do we have current plans to change our sourcing practice." (Perkowski 2014). Simplot has secured commitments from farmers across multiple states for multiple years of production of Innate potatoes (Philpott 2014). The company's strategy is to market the potato directly to the consumer, and will introduce the product in test markets as early as summer 2015.

Concern for potato growers is not limited to consumer acceptance, however, and includes concerns about product segregation from export markets (Perkowski 2014). Monsanto previously received U.S. Department of Agriculture approval for five lines of GM potatoes offering resistance to disease and insects, but they did not succeed commercially. Monsanto's potatoes caused a trade disruption with Japan, which stopped importing U.S. potatoes when some were found in an imported shipment. A speaker for Simplot said that growers would have to keep GE potatoes separate from conventional potatoes and out of export channels (Pollack 2014). The University of Idaho recommended isolation distances for Innate fields and use of dedicated machinery for processing the potatoes

(Perkowski 2014). Simplot plans to apply for regulatory approval of the potatoes in major markets, including Canada, Mexico, Japan, and other parts of Asia (Pollack 2014). Lines of a second generation of Innate potatoes offering increased resistance to late blight disease and better storability are currently under review by the U.S. Department of Agriculture and the U.S. Environmental Protection Agency.

Arctic Apple

Okanagan Specialty Fruits of Canada has applied RNA interference technology in the Granny Smith and Golden Delicious varieties of apples to silence the expression of at least four genes involved in browning, yielding the Arctic apple (Waltz 2015b). Browning of cut or bruised fruits is caused by the enzyme polyphenol oxidase naturally present in the fruit. The enzyme catalyzes the oxidation of polyphenols to quinones, causing oxidative browning. While the damage is superficial, it affects appearance, taste, and texture of the fruit. Apples that do not brown could prove more appealing to consumers and reduce waste by minimizing disposal of bruised apples. Food service companies could cut and package apple products without using browning inhibitors such as calcium ascorbate.

The U.S. Department of Agriculture approved the Arctic apple in February 2015 (Waltz 2015b). About 22,000 trees will be planted in the United States during spring 2015, with the resulting fruit available in fall 2016. Okanagan has applied for regulatory approval of its apple in Canada. Whether the Arctic apple will become popular within the apple-production industry or will get caught up in anti-GE or labeling controversies is yet to be seen. Growers associations in the United States argued against regulatory approval for fear of disruption of export markets. Okanagan pledged to identify their product in marketing and packaging, to some degree addressing such concerns.

Table 2. Examples of GE crops in development whose production could enhance food security.

Crop	Trait	Comments
Pinto bean	Virus resistance	Bonfim et al. (2007), Tollefson (2011)
Corn	Drought resistance	Castiglioni et al. (2008)
Sugarcane	Drought tolerance	ISAAA (2013c)
Banana	Wilt resistance	Juma (2011a)
Cowpea	Insect resistance	ISAAA (2013a), Whitty et al. (2013)
Cassava	Virus resistance	Cressy (2013a), Whitty et al. (2013)
Banana	Increased β-carotene, iron, and other micronutrients, fungal wilt resistance	Cressy (2013a)
Sweet potato	Increased provitamin A	Hotz et al. (2012), Whitty et al. (2013)
Cassava	Increased vitamin A, iron, and protein	Nature (2011a)
Eggplant	Insect resistance	ISAAA (2013b)
Potato	Virus resistance	Mansoor et al. (2006), Qu et al. (2007)
Various crops	Insect and nematode resistance	Huang et al. (2006), Mao et al. (2007), Baum et al. (2007)
Tobacco, tomato	Male sterility	Rehman et al. (2007), Sandhu et al. (2007)
Tobacco, corn	Herbicide resistance	Townsend et al. (2009), Shukla et al. (2009)
Rice	Disease resistance	Li et al. (2012)

Other Examples in the Pipeline

While we have elaborated on a few selected case studies, in Table 2, we note other examples of GE plants whose approval and production would contribute to food security.

New GE Technologies – Highly Promising, but are Regulatory Oversight Authorities Ready?

We have cited a number of cases where GE plants produced through now-traditional transformation techniques have been approved – or not – by regulatory bodies. We have made the case that lack of timely passage through regulation has stymied the development and adoption of GE crops, especially those targeting food security in developing nations. Regulators will soon need to examine current guidelines and policies as new crops are developed using more recent approaches such CRISPR-Cas9 technology (Belhaj et al. 2013; Xie and Yang 2013; Doudna and Charpentier 2014; Hsu et al. 2014; Lozano-Juste and Cutler 2014; Bortesi and Fischer 2015). Using this RNA-guided genome editing approach, the resulting plants contain no genes from another species and leave no residual vector sequences that triggered regulatory oversight in first-generation GE crops. For example, regulatory oversight in the United States is triggered by the use of sequences from plant pathogens such as the *A. tumefaciens* Ti plasmid or the constitutive promoter from the cauliflower mosaic virus, both of which are widely used. These sequences subject transgenic crops to the authority of the U.S. Department of Agriculture's Animal and Plant Health Inspection Services (APHIS) Biotechnology Services (BRS) under the Plant Pest Act. Plants lacking pathogen or any vector-derived DNA sequences and showing additional no signs of genetic modification, may be treated differently in the regulatory process (Cressy 2013a). Since 2010, the

U.S. Department of Agriculture has told at least 10 groups that their products would not require regulation (Ledford 2013), removing a substantial financial barrier and speeding development; this has encouraged academic laboratories and small companies to pursue specialty crops that have been ignored by multinational seed companies, a development with implications for promoting food security in developing countries. Similar changes to regulatory philosophy in the developing world could improve prospects for GE crops to address food security.

A Pivotal Moment for Oversight and Adoption

Despite the clear benefits posed by cultivation of particular GE crops to increasing food security in the developing world, a number of issues have precluded widespread adoption. Key among these are: (1) regulatory issues including costs and delays and (2) public perception, corporate mistrust and misinformation. Against this background, we pose the question, "What must be realized for GE crops to contribute to food security?" We note that many of the complicating issues do not pertain to the products themselves.

Regulatory Issues and Cost

Current review of GE crops in the United States is governed by the Coordinated Framework for the Regulation of Biotechnology, which is based on extending the scope of existing legislation to the products of biotechnolgy. Tenets for review include an assessment of whether a crop poses a plant pest risk to agriculture (through the U.S. Department of Agriculture's Animal and Plant Health Inspection Service), whether a product is considered a pesticide with the potential to pose a risk to health and the environment (through the U.S. Environmental Protection Agency), and whether plant-derived food and feed products are safe for consumption (through the U.S. Food and Drug Administration). The original intent was that the safety of the product, rather than the method of generating the product, should be the guiding principle for regulatory oversight. It has been argued that the way forward in realizing benefits to global food security is to revise the regulatory framework based on biosafety record of GE crops for the past 25 years (Federoff and Beachy 2012). Rather than testing for the presence of DNA sequences that have been safely used for over 30 years as a trigger for regulatory oversight, issues that should guide review include health and environmental issues. Current regulatory protocols have delayed or prohibited potentially transformative crops from reaching human populations that would benefit from their cultivation. Revision of the regulatory protocols could guide wise adoption of public-domain varieties where they are most needed. Policy makers in developing countries should resist being affected by the politicized debate in Europe, instead starting with consideration of the food security problem before them and reaching their own judgment of the balance of pros and cons for their own context, guided by biosafety legislation (Whitty et al. 2013). We note that a recent action by the White House Office of Science and Technology Policy (OSTP) has called for a review of the Coordinated Framework for the Regulation of Biotechnology (OSTP 2015).

Another way in which current regulatory policy impacts global development and adoption of GE crops is the expense required to pass regulatory review. A recent review of the process of developing GE crops cited estimates of US $136 million and 13 years from concept to launch of a new product (Prado et al. 2014). This enormous investment excludes most public research institutions from the process and gives multinational companies a de facto monopoly on the technology. Governments should recognize the need for public-interest research in plant genomics, selective breeding, and biotechnology (National Academies 2000). Public-private collaboration is needed for the benefits of GE technology to be brought to all of the world's people (Nature 2013b). Incentives are needed to encourage the private sector to share with the public sector more of their capacity for innovation. Care should be taken so that research and development are not inhibited by overly protective intellectual property regimes. To enhance food security in the developing world, there will be instances where farmers are allowed to save seed for future use (National Academies 2000). Because broad intellectual property claims can stifle research, development and use of GE plants, agreements must be reached so that the benefits of GE crop research can reach field use in the developing world.

National governments must ensure that they have the capacity to implement biosafety guidelines and regulations (National Academies 2000). At present, only a few African countries have permitted cultivation of GE crops, only one for food purposes, partly because of restrictive national biosafety policies that impose regulatory barriers to the adoption of agricultural biotechnology (Juma 2011b).

Perception

Regulatory language also contributes to the perception that there is something unnatural or sinister about GE crops, and today we face a major public perception problem (Nature 2013b). A number of issues such as the scientific complexity, fast transition to utilization, and ethical, legal, and social issues all contribute to politicizing GE crops (NRC 2015). Blancke et al. (2015) applied ideas from the cognitive sciences, psychology, and culture to understand the appeal of opposition to GE crops, and proposed ways to address the situation. Education can, to some extent, abate the appeal of negative representations of GE crops. Although GE crops are at a disadvantage because they have been associated with unnaturalness by opponents, emphasis on benefits would trigger empathy. For example, it would be useful to inform the public that Bt corn contains less mycotoxin than conventional corn, that herbicide resistance crops improve soil quality by reducing tilling, that Bt crops enhance insect biodiversity, and that GE crops can reduce poverty. One recently published study should change the paradigm of what is considered "natural". The genomes of traditionally derived sweet potatoes have been found to contain the *A. tumefaciens* T-DNA as a result of the natural domestication process (Kyndt et al. 2015). Against this background, we recommend that scientists from both the public and private sectors reach out to the general public and to decision-makers to communicate the potential benefits and risks of GE crop production, and to advocate that we adopt well-targeted applications for GE crops. We support regulation based on the characteristics of the added trait rather than continuing our current protocols where common transformation vector sequences trigger a review even when decades of experience have yielded no concerns.

Well-conceived research and development, combined with well-chosen adoption of GE crop lines, will result in increased food security, human well-being, and ecological sustainability, whereas flawed research adds to the public concerns and misperceptions. For example, Séralini et al. (2012) investigated the health effects of Roundup herbicide and a Roundup-tolerant GE corn through the lifetime of rats, reporting altered sex hormone profiles and increased histological disruptions, tumor frequency, and mortality. The paper was promoted to the press, coinciding with release of a book and film (Nature 2012b). Predictably, the public response, especially in Europe, was hostile to GE food products. However, many scientists criticized the methodology and inferences drawn in the study, noting small sample sizes and use of a rat strain known to develop tumors spontaneously. After letters to the editor of the journal where the study was published, response by the authors, and critical reappraisal of the data, the journal retracted the publication as "inconclusive" (Food and Chemical Toxicology 2014). Similar findings had been reached by the EFSA and six European member states (Nature 2012c). While this case study shows the value of scientific debate and regulatory review, the public likely will remember only the breaking story (Nature 2012b), which serves only to extend the myth of the unnaturalness and potential safety issues of GE foods. A recent review of several key advances in GE crop improvement describes the fearmongering and fraud often perpetrated by GE opposition groups such as Greenpeace (Saletan 2015). The public – at least in Europe – perceives GE plant-derived foods differently from experts (Savadori et al. 2004), perceiving less benefit and greater harm to both humans and the environment. Such perceptions affect the adoption of GE crop technology not only in Europe, but also in the developing world where food security is at issue.

The potential environmental effects, both positive and negative, of GE plant technologies should be assessed within the context of specific applications, with effects assessed relative to those of conventional agricultural practices currently in use in places for which the GE crop was developed (National Academies 2000). Critics of GE crops often presume that most ecological consequences of their cultivation – for example, gene flow into wild relatives – are likely to prove negative. However, a review of 130 research projects by the European Commission (2010) showed the GE crops are not inherently more risky than conventionally bred plants. The application of crop biotechnology has had a number of ecological benefits (Juma 2011a). Production of *Bt* crops has reduced the manufacture of pesticides and farmers' exposure to them. Use of herbicide-tolerant plants has allowed farmers to reduce tilling and weeding, freeing their time for other activities, while reducing erosion and carbon release from the soil. For 2009, it was estimated that GE crops resulted in 7.6 billion kg of carbon dioxide sequestration, the equivalent of removing over 7 million cars from the road (Juma 2011a).

The first GE crops to be commercialized on a large scale have targeted herbicide tolerance and pest resistance traits as opposed to consumer benefits. Hence, the rewards of the first generation of GE crops mostly accrued to multinational corporations and large farmers, and were largely irrelevant to addressing world hunger (Nature 2010). This mismatch of interests has led to mistrust of multinational chemical and seed companies. Combined experience of over 50 years by the authors in discussing biotechnology in the classroom, with agricultural producers, regulators and the public has revealed a blending of issues in the minds of nonscientific audiences. Public mistrust of corporations that develop and market GE crops often is equated with the product itself. Traits with clear benefits to the end user, such as those developed in the pharmaceutical industry (e.g., recombinant insulin), might have been viewed more favorably as first-generation candidates in agriculture. Issues such as health, nutrition, environmental sustainability, and food security are much more compelling to the average consumer.

Conclusion

To ensure food security for all people, agriculture must evolve to meets the needs of a growing population while minimizing global environmental impacts. Above, we note many examples of how research and development on GE plants has resulted in lines that can improve the stability and yield of food production, provide nutritional benefits to the consumer, and reduce the environmental impacts of intensive and extensive agriculture. An agricultural model that combines the best features of traditionally bred and GE crops can make major contributions to global food security. We argue that revising regulations guided by experience of many thorough biosafety studies from the past 25 years will make it feasible to apply these technologies to crop improvement of minor crops and crops important to the nourishment and economic well-being of a greater segment of the world's population. Improving public communication and education are also critical to realizing the benefits of GE crops. A recent cover article in Newsweek magazine declared that "You are totally wrong about genetically altered food", which represents a change from past treatment of GE crops by the popular press. The article highlighted new technological breakthroughs that should help reduce unease the part of biotechnology skeptics. It revealed many examples of misinformation spread by polarizing opponents of new technology. In short, GE crops provide a valuable technical alternative among the variety of approaches that can and should be responsibly marshaled to feed a growing population. We close by repeating a quotation from John F. Kennedy, which although offered in another context, is fitting here: "The great enemy of truth is very often not the lie – deliberate, contrived and dishonest – but the myth – persistent, persuasive, and unrealistic. Too often we hold fast to the cliches of our forebears. We subject all facts to a prefabricated set of interpretations. We enjoy the comfort of opinion without the discomfort of thought".

Acknowledgments

The authors were supported in part by a grant from the U.S. Department of Agriculture-Foreign Agriculture Service. Funding for this work was provided in part by the Virginia Agricultural Experiment Station and the Hatch Program of the National Institute of Food and Agriculture, U.S. Department of Agriculture.

Conflicts of Interest

None declared.

References

Alberts, B., R. Beachy, D. Baulcombe, G. Blobel, S. Datta, N. Fedoroff, et al. 2013. Standing up for GMOs. Science 341:1320.

American Society for Nutrition. 2015. Retraction of Tang G, Hu Y, Yin S-A, Wang Y, Dallal GE, Grusak MA, and Russell RM. ß-carotene in Golden Rice is as good as ß-carotene in oil at providing vitamin A to children. Am J Clin Nutr 2012;96:658–64. Am. J. Clin. Nutr. 102:715.

Anderson, K., L. A. Jackson, and C. Pohl Nielsen. 2004. Genetically modified rice adoption: implications for welfare and poverty alleviation. Discussion Paper 0413, Center for International Economic Studies, Discussion Paper 0413. http://www.goldenrice.org/PDFs/Anderson_CIES_2004.pdf. Also available as World Bank Policy Research Working Paper 3380, August 2004 Available at http://elibrary.worldbank.org/doi/pdf/10.1596/1813-9450-3380 (accessed 16 June 2014).

Baum, J. A., T. Bogaert, W. Clinton, G. R. Heck, P. Feldmann, O. Ilagan, et al. 2007. Control of coleopteran insect pests through RNA interference. Nat. Biotechnol. 2511:1322–1326.

Belhaj, K., A. Chaparro-Garcia, S. Kamoun, and V. Nekrasov. 2013. Plant genome editing made easy: targeted mutagenesis in model and crop plants using the CRISPR/Cas system. Plant Methods 9:39.

Blancke, S., F. Van Breusegem, G. De Jaeger, J. Braeckman, and M. Van Montagu. 2015. Fatal attraction: the intuitive appeal of GMO opposition. Trends Plant Sci. 20:414–418.

Boettcher, M. I., T. Schettgen, B. Kütting, M. Pischetsrieder, and J. Angerer. 2005. Mercapturic acids of acrylamide and glycidamide as biomarkers of the internal exposure to acrylamide in the general population. Mutat. Res. 580:167–176.

Bonfim, K., J. C. Faria, E. O. P. L. Nogueira, É. A. Mendes, and F. J. L. Aragão. 2007. RNAi-mediated resistance to bean golden mosaic virus in genetically engineered common bean (*Phaseolus vulgaris*). Mol. Plant Microbe Interact. 20:717–726.

Bortesi, L., and R. Fischer. 2015. The CRISPR/Cas9 system for plant genomic editing and beyond. Biotechnol. Adv. 33:41–52.

Castiglioni, P., D. Warner, R. J. Bensen, D. C. Anstrom, J. Harrison, M. Stoecker, et al. 2008. Bacterial RNA chaperones confer abiotic stress tolerance in plants and improved grain yield in maize under water-limited conditions. Plant Physiol. 147:446–455.

CERA (Center for Environmental Risk Assessment). 2007. GM Crop Database: ARS-PLMC5-6 (C5). Available at http://cera-gmc.org/index.php?evidcode%5B%5D=C5&auDate1=&auDate2=&action=gm_crop_database&mode=Submit.

Chassy, B., N. Fedoroff, D. Kershen, M. McGloughlin, W. Parrott, and C. S. Prakash. 2013. Global scientific community condemns the recent destruction of field trials of Golden Rice in the Philippines. Change.org Available at http://chn.ge/143PyHo (accessed 12 January 2016).

Chen, M., A. Shelton, and G. Y. Ye. 2011. Insect-resistant genetically modified rice in China: from research to commercialization. Annu. Rev. Entomol. 56:81–101.

Cressy, D. 2013a. A new breed: the next wave of genetically modified crops is making its way to market – and might just ease concerns over 'Frankenfoods'. Nature 497:27–29.

Cressy, D. 2013b. Monsanto drops GM in Europe: region abandoned owing to stalled approval process. Nature 499:387.

Davidson, S. N. 2008. Forbidden fruit: transgenic papaya in Thailand. Plant Physiol. 147:487–493.

Delmer, D. P. 2005. Agriculture in the developing world: connecting innovations in plant research to downstream applications. Proc. Natl Acad. Sci. USA 102:15739–15746.

Diaz de la Garza, R. I., J. F. III Gregory, and A. D. Hanson. 2007. Folate biofortification of tomato fruit. Proc. Natl Acad. Sci. USA 104:4218–4222.

Doudna, J. A., and E. Charpentier. 2014. The new frontier of genome engineering with CRISPR-Cas9. Science 346:1077 DOI: 10.1126/science.1258096.

Dubock, A. 2013. Golden Rice: a long-running story at the watershed of the GM debate. Golden Rice Humanitarian Board. Available at http://b4fa.org/golden-rice-long-running-story-watershed-gm-debate/.

European Commission. 2010. A decade of EU-funded GMO research (2001–2010). Available at https://ec.europa.eu/research/biosociety/pdf/a_decade_of_eu-funded_gmo_research.pdf (accessed 15 January 2015).

Federoff, N. V., and R. Beachy. 2013. Facilitating market access for GE crops developed through public sector research. Pp. 377–390 *in* C. A. Wozniak, A. McHughen, eds. Regulation of agricultural biotechnology: the United State and Canada. Springer, Dordrecht, Netherlands.

Food and Chemical Toxicology. 2014. Retraction notice to "Long term toxicity of a Roundup herbicide and a Roundup-tolerant genetically modified maize" [Food Chem. Toxicol. 50:4221-4231]. Food Chem. Toxicol. 63:244.

George, S., H. S. Singh, and G. Naik. 2002. Brinjal shoot and fruit borer (*Leucinodes orbonalis*) status in coastal districts of Orissa. Pp. 121–123 *in* B. S. Babu, K. S. Varaprasad, K. Anitha, R. D. V. J. Prasada Rao, S. K. Chakrabarty, P. S. Chandurkar, eds. Resources management in plant protection during 21st century, volume II. Plant Protection Society of India, Hyderabad.

GMO Compass. 2013. Papaya. Available at http://www.gmo-compass.org/eng/database/plants/59.papaya.html (accessed 6 December 2013).

Gonsalves, D. 1998. Control of papaya ringspot virus in papaya: a case study. Annu. Rev. Phytopathol. 36:415–437.

Goto, F., T. Yoshihara, N. Shigemoto, S. Toki, and F. Takaiwa. 1999. Iron fortification of rice seed by the soybean ferritin gene. Nat. Biotechnol. 17:282–286.

Greenpeace. 2002. Genetically engineered "golden rice" is fool's gold. Pp. 52–54 *in* M. Ruse, D. Castle, eds. Genetically modified foods: debating biotechnology. Prometheus Books, Amherst, NY.

Guerinot, M. L. 2000. The green revolution strikes gold. Science 287:241–243.

Halford, N. G. 2012. Toward two decades of plant biotechnology: successes, failures and prospects. Food Energy Secur. 1:9–28.

Harmon, A. 2014. A lonely quest for facts on genetically modified crops. New York Times, January 4, 2014, A1.

Herrera-Estrella, L., A. Depicker, M. van Montagu, and J. Schell. 1983. Expression of chimaeric genes transferred into plant cells using Ti-plasmid-derived vector. Nature 303:209–213.

Hotz, C., C. Loechl, A. de Brauw, P. Eozenou, D. Gilligan, M. Moursi, et al. 2012. A large-scale intervention to introduce orange sweet potato in rural Mozambique increases vitamin A intakes among children and women. Br. J. Nutr. 108:163–176.

Hsu, P. D., E. S. Lander, and F. Zhang. 2014. Development and applications of CRISPR-Cas9 for genome engineering. Cell 157:1262–1278.

Huang, J., R. Hu, C. Pray, F. Qiao, and S. Rozelle. 2003. Biotechnology as an alternative to chemical pesticides: a case study of Bt cotton in China. Agric. Econ. 29:55–67.

Huang, G., R. Allen, E. L. Davis, T. J. Baum, and R. S. Hussey. 2006. Engineering broad root-knot resistance in transgenic plants by RNAi silencing of a conserved and essential root-knot nematode parasitism gene. Proc. Natl Acad. Sci. USA 103:14302–14306.

Huang, J., R. Chen, F. Qiao, and K. Wu. 2015a. Biosafety management and pesticide use in China's Bt cotton production. China Econ. Rev. 33:67–75.

Huang, J. K., R. F. Hu, F. B. Qiao, Y. H. Yin, H. J. Liu, and Z. R. Huang. 2015b. Impact of insect-resistant GM rice on pesticide use and farmers' health in China. Sci. China Life Sci. 58:466–471.

Huffington Post. 2013. Possible eco-terrorists attack genetically modified papayas on the Big Island. Available at http://www.huffingtonpost.com/2013/09/29/ecoterrorism-papayas-hawaii_n_4013292.html?view=print&comm_ref=false (accessed 6 December 2013).

Hvistendahl, M., and M. Enserink. 2012. Charges fly, confusion remains of golden rice study in Chinese children. Science 337:1281.

International Rice Research Institute. 2013. Clarifying recent news about Golden Rice. Available at http://irri.org/blogs/golden-rice-blog/clarifying-recent-news-about-golden-rice.

ISAAA (International Service for the Acquisition of Agri-biotech Applications). 2013a. Nigeria develops *Maruca*-resistant cowpea to reduce farm loss. Crop Biotech Update, April 3, 2013. Available at http://www.isaaa.org/kc/cropbiotechupdate/article/default.asp?ID=10799 (accessed 24 December 2013).

ISAAA (International Service for the Acquisition of Agri-biotech Applications). 2013b. Pocket K No. 35: *Bt* brinjal in India. Available at http://www.isaaa.org/resources/publications/pocketk/35/(accessed 24 December 2013).

ISAAA (International Service for the Acquisition of Agri-biotech Applications). 2013c. Sugarcane - *Saccharum* sp: 3 events. Available at: http://www.isaaa.org/gmapprovaldatabase/advsearch/default.asp?CropID=27&TraitTypeID=6&DeveloperID=Any&CountryID=Any&ApprovalTypeID=Any (accessed 24 December 2013).

ISAAA (International Service for the Acquisition of Agri-biotech Applications). 2014. Top ten facts about Bt brinjal in Bangladesh. ISAAA Brief 47-2014: Executive summary. Available at https://www.isaaa.org/resources/publications/briefs/47/executivesummary/default.asp. (accessed 26 May 2015).

ISAAA (International Service for the Acquisition of Agri-biotech Applications). 2015. Pocket K No. 16: global status of commercialized biotech/GM crops in 2014. Available at http://www.isaaa.org/resources/publications/pocketk/16/(accessed 29 May 2015).

Juma, C. 2011a. Preventing hunger: biotechnology is key. Nature 479:471–472.

Juma, C. 2011b. Science meets farming in Africa. Science 334:1323.

Kyndt, T., D. Quispe, H. Zhai, R. Jarret, M. Ghislain, Q. Liu, et al. 2015. The genome of cultivated sweet potato contains *Agrobacterium* T-DNAs with expressed genes: an example of a naturally transgenic food crop. Proc. Natl Acad. Sci. USA, 112:5844–5849.

Ledford, H.. 2013. US regulation misses some GM crops. Nature 500:389–390.

Li, T., B. Liu, M. H. Spalding, D. P. Weeks, and B. Yang. 2012. High-efficiency TALEN-based gene editing produces disease-resistant rice. Nat. Biotechnol. 30:390–392.

Li, Y., E. Hallerman, Q.S. Liu, Y. Peng, and K. Wu. 2015. Risk analysis and governance issues regarding development of Bt rice lines in China. Plant Biotechnol. J., in press, doi: 10.1111/pbi.12464.

Lozano-Juste, J., and S.R. Cutler. 2014. Plant genome engineering in full bloom. Trends Plant Sci. 19:284–287.

Lucca, P., R. F. Hurrell, and I. Potrykus. 2001. Genetic engineering approaches to improve the bioavailability and the level of iron in rice grains. Theor. Appl. Genet. 102:392–397.

Mahyco. 2009. Biosafety data of Bt brinjal containing cry1Ac (EE1) event developed by M/s Maharastra Hybrid Seeds Company Limited. Available through Government of India Ministry of Environment, Forest, and Climate Change at http://www.envfor.nic.in/divisions/csurv/geac/bt_brinjal.html (accessed 7 October 2015).

Mansoor, S., I. Amin, M. Hussain, Y. Zafar, and R. W. Briddon. 2006. Engineering novel traits in plants through RNA interference. Trends Plant Sci. 11:559–565.

Mao, Y.-B., W.-J. Cai, J.-W. Wang, G.-J. Hong, X.-Y. Tao, L.-J. Wang, et al. 2007. Silencing a cotton bollworm P450 monooxygenase gene by plant-mediated RNAi impairs larval tolerance of gossypol. Nat. Biotechnol. 25:1307–1313.

Mayer, J. E., W. H. Pfeiffer, and P. Beyer. 2008. Biofortified crops to alleviate micronutrient malnutrition. Curr. Opin. Plant Biol. 11:166–170.

Mayo-Wilson, E., A. Imdad, K. Herzer, M. Y. Yakoob, and Z. A. Bhutta. 2011. Vitamin A supplements for preventing mortality, illness, and blindness in children aged under 5: systematic review and meta-analysis. Br. Med. J. 343:d5094.

Mortensen, D. A., J. F. Egan, B. D. Maxwell, M. R. Ryan, and R. G. Smith. 2012. Navigating a critical juncture for sustainable weed management. Bioscience 62:75–84.

National Academies (Royal Society of London, U.S. National Academy of Sciences, Brazilian Academy of Sciences, Chinese Academy of Sciences, Indian National Science Academy, Mexican Academy of Sciences, and Third World Academy of Sciences). 2000. Transgenic plants and world agriculture. National Academy Press, Washington, DC. Available at www.nap.edu.

National Research Council. 2015. Public engagement on genetically modified organisms: when science and citizens connect. Workshop Summary. National Academy Press, Washington, DC. Available at www.nap.edu.

Nature. 2010. How to feed a hungry world. Nature 466:531–532.

Nature. 2011a. Golden rice funds. Nature 472:265.

Nature. 2011b. GM honey concern. Nature 477:254.

Nature. 2011c. No to French GM ban. Nature 480:10.

Nature. 2012a. Transgenic flight. Nature 481:243.

Nature. 2012b. Poison postures: researchers working on controversial topics must take care how they promote their results. Nature 489:474.

Nature. 2012c. GM study rebutted. Nature 492:12.

Nature. 2013a. Poland GM ban. Nature 493:138.

Nature. 2013b. Fields of gold: research on transgenic crops must be done outside industry if it is to fulfil its early promise. Nature 497:5–6.

Nature. 2013c. Tarnished promise: genetically modified crops generate hype and hatred. Nature 497:21.

Nature. 2013d. GM crops: a story in numbers. Nature 497:22–23.

Nature. 2013e. GM-rice fallout. Nature 501:467.

Novak, S. 2013. USDA moving to deregulate genetically modified plums. Available at http://recipes.howstuffworks.com/usda-moving-to-deregulate-genetically-modified-plums.htm (accessed 9 December 2013).

Office of Science and Technology Policy. 2015. Improving transparency and ensuring continued safety in biotechnology. Available at https://www.whitehouse.gov/blog/2015/07/02/improving-transparency-and-ensuring-continued-safety-biotechnology (accessed 30 July 2015).

Owino, O. 2012. Scientists torn over Kenya's recent GM food ban. Nature. Available at http://dx.doi.org/10.1038/nature.2012.11929.

Perkowski, M. 2014. McDonald's not interested in GMO potatoes. Capital Press, November 19, 2014.

Philpott, T. 2014. Will this new GMO potato take off: McDonald's has spoken. Mother Jones, Friday November 14, 2014.

Pollack, A. 2014. U.S.D.A. approves modified potato. Next up: French fry fans. New York Times, November 7, 2014:B1.

Potrykus, I. 2010. Regulation must be revolutionized. Nature 466:561.

Prado, J. R., G. Segers, T. Voelker, D. Carson, R. Dobert, J. Phillips, et al. 2014. Genetically engineered crops: from idea to product. Annu. Rev. Plant Biol. 65:769–790.

Qiao, F. 2015. Fifteen years of Bt cotton in China: the economic impact and its dynamics. World Dev. 70:177–185.

Qu, J., J. Ye, and R. Fang. 2007. Artificial microRNA-mediated virus resistance in plants. J. Virol. 81:6690–6699.

Ravelonandro, M., R. Scorza, A. Callahan, L. Levy, C. Jacquet, M. Monsion, et al. 2000. The use of transgenic fruit trees as a resistance strategy for virus epidemics: the plum pox (sharka) model. Virus Res. 71:63–69.

Rehman, N., M. Shah, S. Mansoor, A. A. Khan, Y. Zafar, and R. W. Briddon. 2007. RNAi-mediated male sterility of tobacco by silencing *TA29*. Mol. Biotechnol. 36:159–165.

Rommens, C. M., H. Yan, K. Swords, C. Richael, and J. Ye. 2008. Low-acrylamide French fries and potato chips. Plant Biotechnol. J. 6:843–853.

Saletan, W. 2015. Unhealthy fixation: the war against genetically modified organisms is full of fearmongering, errors and fraud. Labeling them will not make you safer. Slate, July 15, 2015. Available at http://www.slate.com/articles/health_and_science/science/2015/07/are_gmos_safe_yes_the_case_against_them_is_full_of_fraud_lies_and_errors.html. (accessed 30 July 2015).

Sandhu, A. P. S., R. V. Abdelnoor, and S. A. Mackenzie. 2007. Transgenic induction of mitochondrial rearrangements for cytoplasmic male sterility in crop plants. Proc. Natl Acad. Sci. USA 104:1766–1770.

Savadori, L., S. Savio, E. Nicotra, R. Rumiati, M. Finucane, and P. Slovic. 2004. Expert and public opinion of risk from biotechnology. Risk Anal. 24:1289–1299.

Science. 2011. Peru set to forbid GM crops. Science 332:1365.

Science. 2012a. Philippine scientists defend transgenic crops. Science 336:654.

Science. 2013a. Monsanto withdraws E.U. patent applications. Science 341:324.

Science. 2013b. Scientists: freeze GM trials. Science 341:324–325.

Science. 2013c. Activists destroy 'golden rice' field trial. Science 341:699.

Scorza, R. 2007. Application for determination of non-regulatory status of C5 ('HoneySweet') plum (*Prunus domestica* L.) resistant to plum pox virus. Identifier ARS-PLMC5-6, USDA-ARS Appalachian Fruit Tree Research Station, Kearneysville, WV. 93 pp. Available at http://cera-gmc.org/docs/decdocs/07-313-003.pdf (accessed 18 December 2013).

Scorza, R., A. Callahan, C. Dardick, M. Ravelonandro, J. Polak, T. Malinowski, et al. 2013a. Genetic engineering of plum pox virus resistance: 'HoneySweet' plum – from concept to product. Plant Cell Tissue Organ Cult. 115:1–12.

Scorza, R., A. B. Kriss, A. M. Callahan, K. Webb, M. Demuth, and T. Gottwald. 2013b. Spatial and temporal assessment of pollen- and seed-mediated gene flow from genetically engineered plum *Prunus domestica*. PLoS One 8:e75291.

Séralini, G.-E., E. Clair, R. Mesnage, S. Gress, N. Defarge, M. Malatesta, et al. 2012. Long term toxicity of a Roundup herbicide and a Roundup-tolerant genetically modified maize. Food Chem. Toxicol. 50:4221–4231.

Shukla, V. K., Y. Doyon, J. C. Miller, R. C. DeKelver, E. A. Moehle, S. E. Worden, et al. 2009. Precise genome modification in the crop species *Zea mays* using zinc-finger nucleases. Nature 459:437–441.

Simplot Plant Sciences. 2015. Innate™ potato receives FDA safety clearance. Available at http://www.simplot.com/news/innate_potato_receives_fda_safety_clearance (accessed 24 April 2015).

Storozhenko, S., V. De Brouwer, M. Volckaert, O. Navarrete, D. Blancquaert, G. F. Zhang, et al. 2007. Folate fortification of rice by metabolic engineering. Nat. Biotechnol. 25:1277–1279

Tamil Nadu Agricultural University – Department of Plant Molecular Biology and Biotechnology. 2008. Report of the multilocation research trial on Bt brinjal open-pollinated varieties conducted at Coimbatore and

Madurai 2007–2008. 57 pp. Available at http://www.moef.gov.in/divisions/csurv/geac/Bt_Brinjal/TNAU-Bt%20 brinajl_MLR_report.pdf (accessed 7 October 2015).

Tang, G., J. Qin, G. G. Dolnikovski, R. M. Russell, and M. A. Grusak. 2009. Golden rice is an effective source of vitamin A. Am. J. Clin. Nutr. 89:1776–1783.

Tang, G., Y. Hu, S. Yin, Y. Wang, G. E. Dallal, M. A. Grusak, et al. 2012. β-carotene in golden rice is as good as β-carotene in oil at providing vitamin A to children. Am. J. Clin. Nutr. 96:658–664.

Tollefson, J. 2011. Brazil cooks up transgenic bean. Nature 478:168.

Townsend, J. A., D. A. Wright, R. J. Winfrey, F. Fu, M. L. Maeder, J. K. Joung, et al. 2009. High-frequency modification of plant genes using engineered zinc-finger nucleases. Nature 459:442–445.

USDA-FAS (U.S. Department of Agriculture – Foreign Agriculture Service). 2013. India agricultural biotechnology annual 2013. GAIN Report #IN3083. Available at http://gain.fas.usda.gov/Recent%20GAIN%20Publications/Agricultural%20Biotechnology%20Annual_New%20Delhi_India_7-15-2013.pdf (accessed 28 May 2015).

U.S. Food and Drug Administration – Center for Food Safety and Nutrition, Office of Food Additive Safety. 2009a. Biotechnology consultation: note to the file BNF No. 000101. Available at http://cera-gmc.org/docs/decdocs/09-055-002.pdf (accessed 18 December 2013).

U.S. Food and Drug Administration – Center for Food Safety and Nutrition, Office of Food Additive Safety. 2009b. Biotechnology consultation: agency response letter BNF No. 000101. Available at http://cera-gmc.org/docs/decdocs/09-055-001.pdf (accessed 18 December 2013).

USDA-APHIS (U.S. Department of Agriculture - Animal and Plant Health Inspection Service). 2007a. Approval of USDA-ARS request (04-264-01P) seeking a determination of nonregulated status for C5 plum resistant to plum pox virus.

USDA-APHIS (U.S. Department of Agriculture - Animal, Plant Health Inspection Service). 2007b. Determination of nonregulated status for plum genetically engineered for resistance to plum pox virus. Fed. Reg. 72:38556–38557, Friday, July 13, 2007.

Van Eenennaam, A. L., E. M. Hallerman, and W. M. Muir 2011. The science and regulation of food from genetically engineered animals. CAST Commentary QTA2011-2, Council on Agricultural Science and Technology, Ames, IA.

Van Eenennaam, A. L., W. M. Muir, and E. M. Hallerman. 2013. Are unaccountable regulatory delay and political interference undermining the FDA and hurting American competitiveness? Food Drug Policy Forum 3:1–20.

Vietmeyer, N. 2012. Our daily bread: the essential Norman Borlaug. 284 pp. Bracing Books, Lorton, VA.

Waltz, E. 2015a. USDA approves next-generation GM potato. Nat. Biotechnol. 33:12–13.

Waltz, E. 2015b. Nonbrowning GM apple cleared for market. Nat. Biotechnol. 33:326–327.

Whitty, C. J. M., M. Jones, A. Tollervey, and T. Wheeler. 2013. Africa and Asia need a rational debate on GM crops. Nature 497:31–33.

WHO (World Health Organization). 2005. Global prevalence of vitamin A deficiency in populations at risk 1995–2005. WHO Global Database on Vitamin A deficiency. Available at http://www.who.int/vmnis/vitamina/prevalence/report/en/ (accessed 7 January 2015).

WHO (World Health Organization). 2015. Trade, foreign policy, diplomacy and health: food security. Available at http://www.who.int/trade/glossary/story028/en (accessed 6 October 2015).

Xie, K., and Y. Yang. 2013. RNA-guided genome editing in plants using a CRISPR-Cas system. Mol. Plant 6:1975–1983.

Ye, X., S. Al-Babili, A. Kloti, J. Zhang, P. Lucca, P. Beyer, et al. 2000. Engineering the provitamin A (β-carotene) biosynthetic pathway into (carotenoid-free) rice endosperm. Science 287:303–305.

3

Alternate wetting and drying irrigation maintained rice yields despite half the irrigation volume, but is currently unlikely to be adopted by smallholder lowland rice farmers in Nepal

Katharine R. Howell[1], Pitambar Shrestha[2] & Ian C. Dodd[1]

[1]Lancaster Environment Centre, Lancaster University, Lancaster LA1 4YQ, UK
[2]Local Initiatives for Biodiversity, Research and Development (LIBIRD), PO Box 324, Pokhara, Nepal

Keywords
Alternate wetting and drying, early vigor, Nepal, rice, smallholders, technology adoptiontechnology adoption

Correspondence
Katharine R. Howell, Lancaster Environment Centre, Lancaster University, Lancaster LA1 4YQ,
E-mail: k.howell@lancaster.ac.uk

Funding Information
ICD thanks BBSRC (BB/J003336/1) for continued support of research into AWD.

Abstract

Alternate wetting and drying (AWD) irrigation can save water while maintaining rice yields, but in some countries its adoption by farmers remains limited. Key knowledge gaps include the effect of AWD on early vegetative vigor and its relationship with yield; the effects of AWD on yield and water use efficiency of local cultivars used by smallholder farmers; and the socio-economic factors influencing current irrigation scheduling. To address these questions, an on-farm field trial of dry-season (chaite) rice, comparing two locally important cultivars (Hardinath-1 and CH-45) under AWD imposed from 1 week after transplanting to flowering and continuous flooding (CF), was carried out in Agyauli in the central Terai region of Nepal, and triangulated with social research methods exploring the rationale for current irrigation scheduling and perceptions of AWD. Although AWD plots received on average 57% less irrigation water than CF plots, yields did not significantly differ between irrigation treatments, indicating that AWD could considerably enhance crop water use efficiency in this region. In the earlier flowering, more vigorous CH-45, there were no treatment differences in any yield component while in the later flowering Hardinath-1, an 11% decrease in filled grain number was compensated by a 14% increase in the percentage of effective tillers per hill. Although leaf elongation rate on the main tiller did not differ between treatments, tillering and green fraction (a measure of canopy closure) were significantly higher under AWD. Surveys established that most local farmers are already using a local adaptation of AWD to modify irrigation volumes, in some cases in response to a limited and unreliable water supply. However, farmers have few direct incentives to reduce overall water use under current water governance, and formal AWD practices are therefore unlikely to be adopted despite their viability as a water-saving irrigation technique.

Introduction

Concerns are mounting about the production of sufficient food to feed a growing global population, due to increasingly unreliable and contested water supplies and lower overall water availability in many of the world's regions. Irrigation played a key role in the yield gains of the Green Revolution, but this was often reliant on unsustainable water withdrawals, and further yield gains on this scale therefore need to be accompanied by increases in water productivity (Rockström et al. 2007). Of particular concern is rice (*Oryza sativa* L.) production, a staple crop for over 3 billion people worldwide, which is usually either produced in water-intensive irrigated paddy systems or is dependent on high levels of rainfall (Van Nguyen and Ferrero 2006).

To allow farmers to actively reduce their water use and associated costs, or save water for other purposes or times (Bouman et al. 2007), various water-saving irrigation techniques have been developed. One such technology, developed in the 1990s by the International Rice Research Institute (IRRI), is alternate wetting and drying (AWD) irrigation (Price et al. 2013). As most lowland rice varieties can tolerate a 30% reduction in total irrigation volume without significantly decreasing yield (Richards and Sander 2014), this practice allows the soil to dry out partially before re-irrigating. IRRI's "safe AWD" recommendations, intended to minimize yield reductions, are that the soil is dried until soil water depth reaches 15 cm below the surface, and that the field is re-irrigated to a standing water depth of around 5 cm (Bouman and Lampayan 2009). Crops may be continuously flooded during and after flowering to prevent yield reductions, or AWD irrigation may be continued through this period (Bouman et al. 2007).

In field experiments and on-farm trials, AWD significantly reduced water use whilst maintaining or even increasing yields, compared to continuous flooding (CF) (Mishra et al. 1990; Bouman and Tuong 2001; Tabbal et al. 2002; Belder et al. 2004; Mandal et al. 2009; Mishra and Salokhe 2010). The mechanisms underpinning these responses are not fully understood, but AWD has been linked to improved root growth, enabling greater access to water and nutrients at depth in the soil profile (Yang et al. 2009). Rhizosphere drying caused by AWD can also alter plant hormone signaling (Davies et al. 2011) and enhance grain filling rate, particularly in inferior spikelets (Zhang et al. 2010, 2012), thereby increasing plant water use efficiency (Tabbal et al. 2002).

However, despite these advantages, formal adoption of AWD by farmers in many rice-producing regions has been limited (Blanke et al. 2007; Senthilkumar et al. 2009). Exploring the reasons for this is crucial to understand how AWD can be more successfully developed. A major barrier to the widespread adoption of AWD is that its effects on yield (positive or negative) vary considerably between different soils, climates, cultivars and management practices (Bouman and Tuong 2001; Price et al. 2013), as well as over time, between seasons and years (Mandal et al. 2009).

AWD may be unsuitable in sandy soils as water drains quickly and water savings are limited, whilst in heavy clays with shallow water tables AWD may be unnecessary since soil water table depth never drops below the deepest roots (Belder et al. 2004).

There may also be socio-economic, cultural, and political reasons why farmers are reluctant to adopt new irrigation technologies (Burnham et al. 2014), especially without the scientific and visual evidence of successful local field trials (Alcon et al. 2014). Considerable research into some of the social factors determining the uptake of AWD has taken place at a regional scale and focused on economic modeling, speculative cost-benefit analyses (Mushtaq et al. 2006; Blanke et al. 2007; Li and Li 2010; Cai et al. 2012) or analyses of the impacts of AWD adoption on farmers' income, labor costs and profitability (Moya et al. 2004). What has often been missing is a consideration of farmers' rationales for current irrigation systems, which is particularly relevant in the case of smallholders in less developed countries where factors such as land and labor may be more limiting than water or yield (Kürschner et al. 2010), farmers may be more risk averse (Koundouri et al. 2006) or water is more likely to be managed at a community level (Li and Li 2010).

Consequently, a triangulation (Nuijten 2011) of scientific and social scientific methods was identified as a suitable means to determine reasons for possibly limited adoption of AWD in Nepal. An on-farm field trial within lowland, smallholder rice production in Agyauli Village Development Committee (VDC), Nepal, investigated local perceptions of AWD (including people's responses to the field trial) using social science research methods to determine current irrigation scheduling techniques and whether farmers need or want to change water management. Furthermore, this trial investigated the physiological effects of AWD (as opposed to CF) on early plant growth and yields of two locally grown rice cultivars, which differed dramatically in flowering date, and which were managed (with the exception of irrigation scheduling) according to local practices. In agreement with other studies, it was hypothesized that AWD would allow significant water savings without decreasing crop yields.

Materials and Methods

Four replications of two irrigation treatments (CF and AWD) and two locally commonly grown improved cultivars (CH-45 and Hardinath-1) were laid out in a randomized arrangement of 4 m x 4 m square plots, with 2 m wide irrigation channels and 1 m spacings, and earth bunds built around each plot according to local practices. The field, which was fallow over the winter, was prepared using a buffalo-pulled plow and puddler and farmyard manure and 46 kg ha^{-1} each of diammonium phosphate and urea were applied prior to transplanting. Transplanting was carried out using local planting densities and spacings (mean number of seedlings per hill = 5, mean spacing between hills = 160 mm). The crop was top dressed with a further 60 kg ha^{-1} urea and a stemborer infestation treated with chemical insecticide at 36 days after transplanting. Heading occurred in CH-45 and Hardinath-1 from 32 and 51 days after transplanting, respectively, and both cultivars were harvested on Day 88. During the

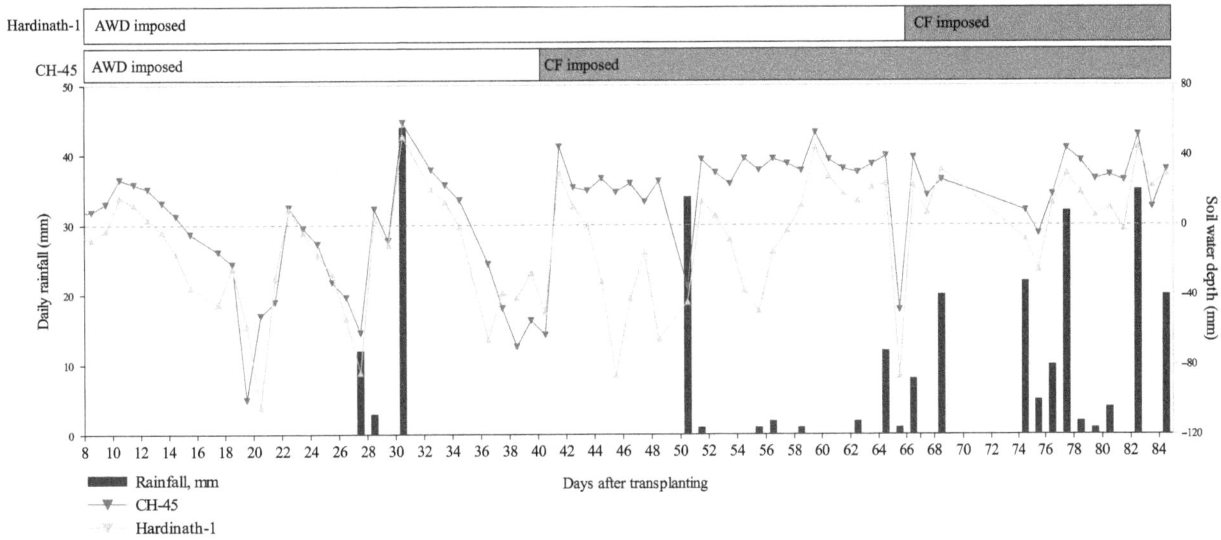

Figure 1. Daily rainfall in mm for Agyauli (bars) and treatment imposition of alternate wetting and drying (AWD) in two cultivars (Hardinath-1 and CH-45) from 8 to 84 days after transplanting. Lines show mean daily soil water depth in mm in AWD plots in the two cultivars, with dashed grey line indicating soil surface. Horizontal bars at top show periods for which AWD (white) was imposed for the two cultivars, and when they were continuously flooded (CF) during flowering (grey), in line with IRRI recommendations.

production cycle, mean maximum and minimum temperatures were 37°C and 26°C respectively; and mean relative humidity was 72% (meteorological data courtesy of Nepal Department of Meteorology and Hydrology). Rainfall was infrequent during the first 60 days after transplanting and more frequent thereafter as the monsoon arrived (Fig. 1).

Irrigation scheduling

CF plots were irrigated sufficiently to maintain a mean ponded water depth of 20–30 mm, an average of twice every three days, from the local irrigation canal (*kulo*) with supplementation via buckets as necessary. AWD was imposed from 10 to 66 days after transplanting in Hardinath-1, and from 10 to 40 days in CH-45 (Fig. 1). The maximum water depth following rewatering of AWD plots was 78 mm above the soil surface at 30 days after transplanting, while the minimum water depth before rewatering of AWD plots was 150 mm below the soil surface at 20 days after transplanting. In line with IRRI recommendations, all plots were continuously flooded from approximately one week before, to one week after, flowering (when 50% heading was visually estimated to have been reached). IRRI recommendations for "safe AWD" (Bouman and Lampayan 2009) were adapted according to the local context. As the plow pan was approximately 15 cm below the soil surface, plots were flooded when water reached, or was just above, the plow pan. Soil water tubes (radius 52 mm) were made using cheap and readily available materials, namely plastic bottles perforated with a safety pin. Daily measurements of ponded water depth in CF plots, and soil water depth in AWD plots, were recorded to ensure adherence to these protocols. Following lodging, plots were drained, and soil moisture (but no standing water) maintained, to prevent grain germination. Irrigation volumes were calculated as a function of water depth in the inlet tube, water flow rate and duration of flooding. AWD plots received on average less than half the irrigation volume of CF plots (Table 1).

Quantifying early vigor and late lodging

Weekly manual tiller counts and daily manual measurements of the growing leaf of the main tiller of sample plants were carried out for six randomly selected border plants (to allow nondestructive access) per plot. For each leaf, its elongation rate was calculated as the difference between its lengths on subsequent days. Leaf elongation

Table 1. Irrigation volume per unit area ($m^3 m^{-2}$) applied to 20 rice plots from transplanting to harvest, under two irrigation treatments (AWD and CF), and cultivars (CH-45 and Hardinath 1), for five replicates for each cultivar/treatment combination. Data are means ± SE, with different letters showing significant difference to $P < 0.05$ (ANOVA).

Treatment	Mean ± SE
AWD/CH-45	0.86 ± 0.13 a
AWD/Hardinath-1	0.54 ± 0.08 a
CF/CH-45	1.6 ± 0.14 b
CF/Hardinath-1	1.7 ± 0.33 b

was determined to have finished when its daily elongation rate fell below 10% of its maximum, whereupon a new leaf was selected. Green fraction (GF) per plot was calculated on a weekly basis. Eight digital photographs were taken of each plot, according to the recommendations (Casadesús and Villegas 2014), and analyzed using BreedPix 2.0, which calculates the proportion of green pixels (hue 60–180°) in each image. This nondestructive technique (Lee and Lee 2013) visualized weed as well as crop growth, so fresh weed weight per plot (following manual weeding) was calculated 31 days after transplanting.

Strong winds and heavy rain associated with intense thunderstorms caused lodging in CH-45 plots at 66, 68, and 82 days after transplanting. Following each lodging event, each plot was given a visual damage score from 0 to 5, whereby 0 represents no detectable damage, and subsequent numbers were ranked in 20% increments.

Measuring yield components

Four days before harvest, the proportion of effective tillers was calculated from the number of tillers and panicles per sample hill. Grain size and weight were measured according to an IRRI protocol (Gomez 1972). Three random samples of four hills per plot were selected and cut three days before harvest. From each of these samples, four panicle samples were selected, and the grain from each of these stripped and stored in an envelope. The remaining grain was stripped and placed in a large envelope, and these grain samples and the straw were air-dried for four days, after which all samples were weighed. Number of grains and number of filled grains were manually counted for each panicle sample. The remaining crop was cut at 88 days after transplanting, threshed and air dried. The grain for each plot was weighed, and moisture content estimated using a Wile 55 moisture meter, taking five moisture readings per plot. These were used to calculate adjusted grain weight at 14% moisture.

Statistical analysis

Data were analyzed in R using quasipoisson models for time series data, and in SPSS (IBM, Portsmouth, UK) with t-tests (tillering) and nonparametric tests (Kruskal–Wallis and Mann–Whitney U-tests, for LER and GF, respectively) for individual days. Weed and yield data were analyzed in R with two-way ANOVAs and general linear models, respectively.

Social research methods

A household survey was devised and piloted with four local households. Hundred and one households were then selected at random from a list of all 321 households growing *chaite* rice in Agyauli VDC, and surveys were carried out during four weeks by the secretary of the village farmers' committee, who was identified as a suitable researcher due to his familiarity with Agyauli, fluency in Nepali and Tharu, and the respect and trust in which he is held locally (Mullings 1999; Budruk 2010). He was accompanied for the pilot surveys (by the first 2 authors of this manuscript) to provide feedback on his questioning style and ensure that he satisfied the ethical requirements of the research. The majority of respondents have limited literacy skills, so the surveys were carried out in a structured interview format. Respondents were usually interviewed at a time and location suitable to them, and were fully appraised of the purpose of the research and required to give verbal consent before the interview began.

Surveys gathered basic social information (gender, age, and farm size) followed by more specific questions about irrigation sources and management, current irrigation scheduling techniques, and irrigation-related challenges. Respondents were also asked about their perceptions and experiences of AWD and asked to select an irrigation scheduling technique from a choice of two in a series of three choice experiments, adapting a previous model (Brouwer et al. 2009). Survey data were anonymized and translated from Nepali into English by local NGO staff (Budruk 2010). Qualitative answers were given a numerical code for analysis, and basic descriptive statistics generated using SPSS.

Follow-up interviews were arranged to explore surprising responses about uses and perceptions of AWD. Four respondents who had said in the survey that they had used AWD were selected to give a cross-section of ages and gender, and were interviewed when convenient to them, by fitting around their farm work. Caste and gender inequalities, although important mediators of farming practice and access to water (Panta and Resurrección 2014) were not considered here due to the focus on household-scale experiences within a relatively homogenous ethnic group. Interviews were semistructured, with NGO staff acting as translators. Responses were recorded in writing, qualitatively analyzed and used to inform answers provided in the survey. Additionally, participant observation (Willis and Trondman 2002) of irrigation management and maintenance was carried out throughout the field trial.

Results

Early vigor and yield

Tiller number was significantly higher under AWD than CF from 21 days after transplanting onwards (Table 2, Fig. 2A). Tiller number was significantly higher in CH-45

Table 2. *P*-values for the effect of irrigation and cultivar, and interaction of irrigation and cultivar (irrigation*cultivar) where including this significantly improved the fit of the statistical model, on components of early vigor and crop development. Asterisks denote statistical significance to $P < 0.05$ (*), <0.01 (**) and 0.001 (***).

Growth characteristics	Period (days after transplanting)	*P*-values Irrigation	Cultivar	Irrigation*Cultivar
Leaf elongation rate (mm/day)	7–66	0.028*	<0.001***	Not in model
Number of tillers per plant	7–56	<0.001***	<0.001***	Not in model
Green fraction	9–53	0.020*	<0.001***	Not in model
Weed fresh weight per square meter (kg m^{-2})	31	0.002**	0.18	0.24
Lodging score	66–82	<0.001***	Not in model	Not in model

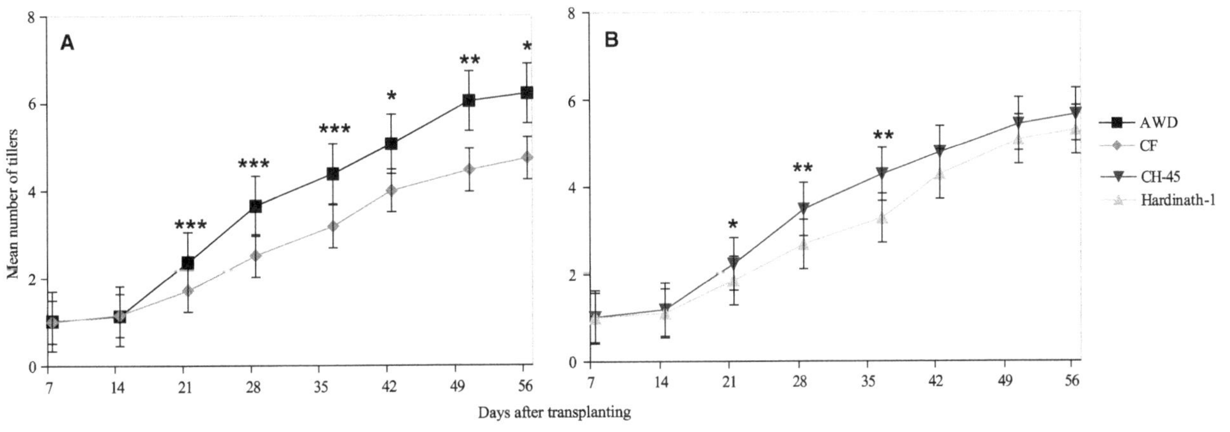

Figure 2. Mean tiller number, (A) Under two irrigation treatments (alternate wetting and drying, AWD and continuous flooding, CF) averaged across genotypes, and (B) In two different cultivars (CH-45 and Hardinath-1), averaged across treatments. Bars show ±1 standard error of 30 replicates per genotype/treatment combination. Asterisks denote statistical significance to $P < 0.05$ (*), <0.01 (**) and 0.001 (***) (*t*-test).

from 21 to 35 days after transplanting, and remained higher thereafter.

Irrigation and cultivar significantly affected leaf elongation rate (LER) over the whole vegetative period (Table 2) and on some individual days (Fig. 3). Plants under both AWD and CF had higher mean LER at different times with no systematic pattern. CH-45 had consistently faster rates of leaf growth than Hardinath-1, particularly towards the end of the vegetative phase.

Green fraction (GF) was significantly affected by irrigation regime, and especially cultivar (Table 2). Under AWD, CH-45 had significantly higher GF throughout the vegetative period (Fig. 4A). Under CF, CH-45 had significantly higher GF, but less so in the first three weeks after transplanting (Fig. 4B). Although weeding decreased GF in all treatments and cultivars, GF was less reduced by weeding in CH-45 under both irrigation treatments. Weed weight was negatively correlated with difference in GF before and after weeding (Kendall's tau = −0.16, P = 0.004). Mean GF values of nearly 1, indicating complete canopy closure, were achieved following weeding in CH-45 under both treatments, but not in Hardinath-1.

Weed fresh weight per plot was significantly higher, but much more variable, under CF than AWD in CH-45, but there was no significant difference between treatments in Hardinath-1 (Table 3). Weed fresh weight and irrigation volume were strongly positively correlated (Kendall's tau = 0.504, P = 0.003). Those weeding the plots commented that weeding was slightly more difficult in AWD plots because the soil was harder. Visual lodging damage scores in CH-45 plots were significantly higher under AWD following the second lodging event (Table 3).

Grain weight per plot was not significantly different between treatments or cultivars (Table 4, Fig. 5A). The number of filled grains per panicle was significantly higher (by 89%) in Hardinath-1, and under CF (by 12%; Fig. 5B). The percentage of filled grains per panicle was significantly higher in CH-45 (by 18%) and again significantly higher under CF (by 6.7%; Fig. 5C). Individual grain weight was also significantly higher in CH-45 (by 21%), but did not differ between treatments (Fig. 5D). Dry straw weight was significantly greater in CH-45 (by 39%), although in Hardinath-1 it was slightly higher (by 13%) under AWD

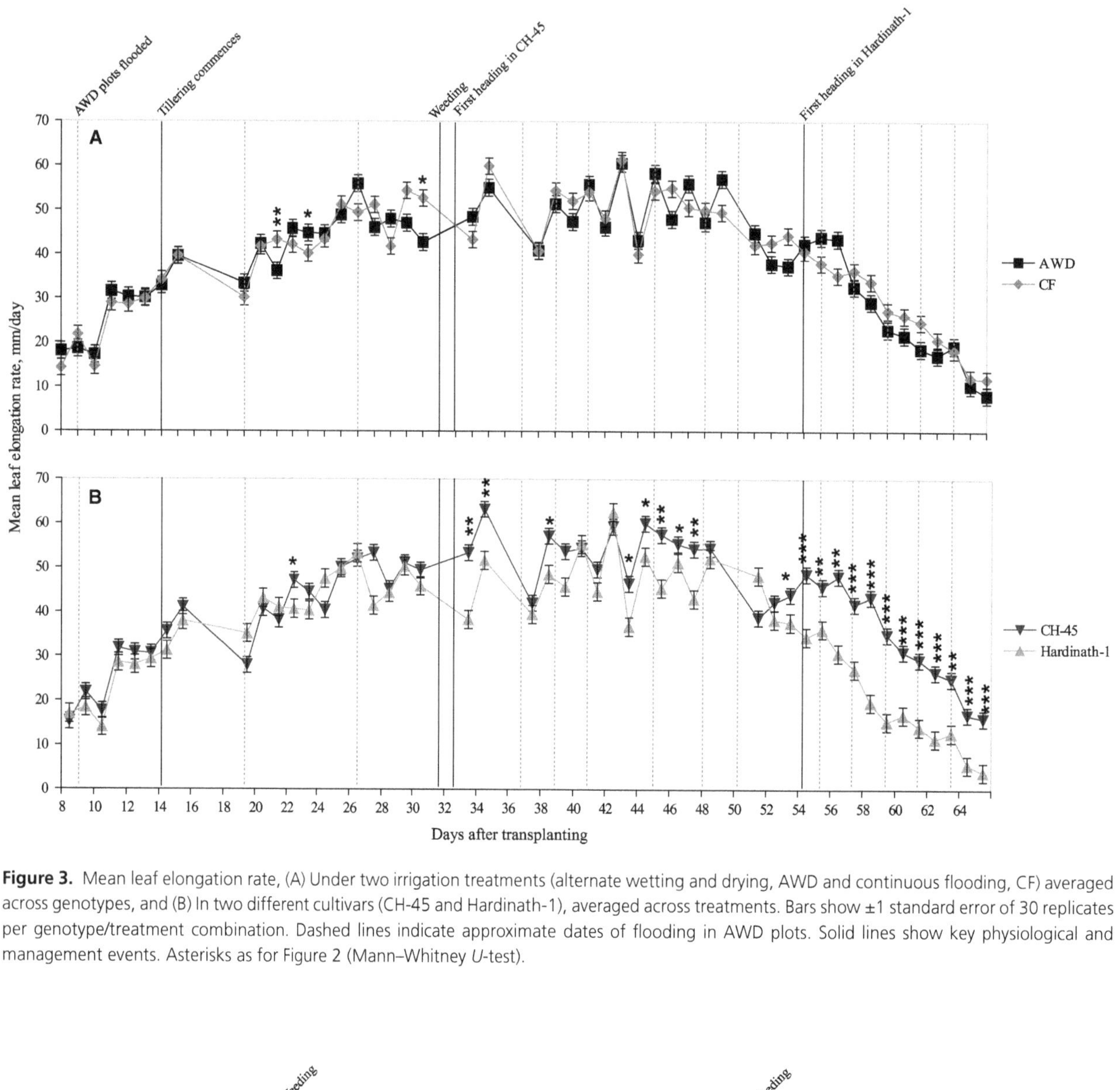

Figure 3. Mean leaf elongation rate, (A) Under two irrigation treatments (alternate wetting and drying, AWD and continuous flooding, CF) averaged across genotypes, and (B) In two different cultivars (CH-45 and Hardinath-1), averaged across treatments. Bars show ±1 standard error of 30 replicates per genotype/treatment combination. Dashed lines indicate approximate dates of flooding in AWD plots. Solid lines show key physiological and management events. Asterisks as for Figure 2 (Mann–Whitney *U*-test).

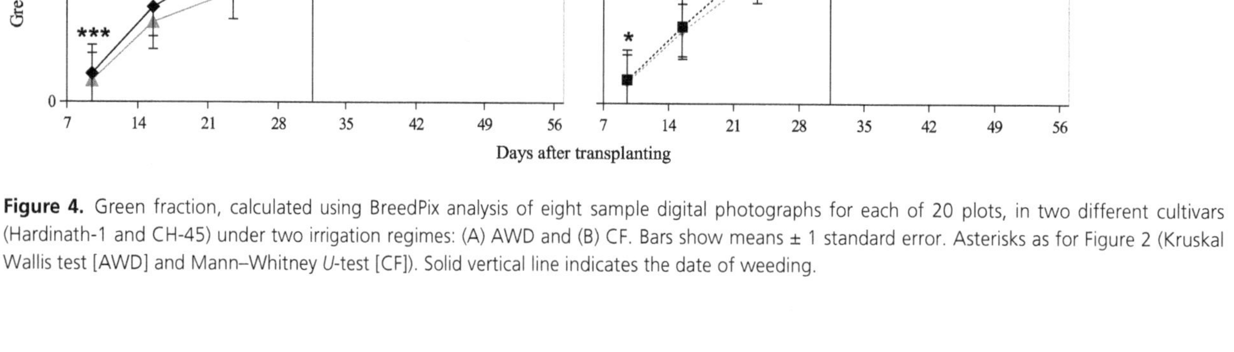

Figure 4. Green fraction, calculated using BreedPix analysis of eight sample digital photographs for each of 20 plots, in two different cultivars (Hardinath-1 and CH-45) under two irrigation regimes: (A) AWD and (B) CF. Bars show means ± 1 standard error. Asterisks as for Figure 2 (Kruskal Wallis test [AWD] and Mann–Whitney *U*-test [CF]). Solid vertical line indicates the date of weeding.

than CF (Fig. 5E). Harvest index was therefore significantly higher in Hardinath-1 (by 26%), and slightly but non-significantly higher under CF in both cultivars (Fig. 5F). Water use efficiency (WUE), calculated as moisture-corrected grain weight per plot divided by the total irrigation water received per plot during the whole crop cycle, was significantly higher under AWD (by 133%), particularly in Hardinath-1 under AWD (Fig. 5G).

There was a significantly higher percentage of effective tillers (by 2%) under AWD overall (Fig. 5H). There was a strong, significant interaction with cultivar ($P = 0.0056$; Table 4): CH-45 plots had a higher percentage of effective tillers than Hardinath-1 regardless of irrigation regime. In Hardinath-1 plots, percentage of effective tillers under AWD was significantly lower than CH-45/CF but significantly higher than Hardinath-1/CF.

Table 3. Weed fresh weight per square meter under two irrigation treatments (AWD and CF), and cultivars (CH-45 and Hardinath-1), weeded 31 days after transplanting, and visual damage scores for CH-45 plots (Hardinath-1 did not lodge), under the above irrigation treatments, following lodging events 66, 68 and 82 days after transplanting (dat). Data are means ± SE of five replicates, with different letters showing significant difference to $P < 0.05$ (ANOVA).

Treatment	Weed weight (kg m^{-2})	Lodging score 66 dat	Lodging score 68 dat	Lodging score 82 dat
AWD/CH-45	0.35 ± 0.10 a	3.6 ± 0.5 a	5.0 ± 0 b	5.0 ± 0 b
AWD/Hardinath-1	0.45 ± 0.14 ab			
CF/CH-45	0.69 ± 0.28 ab	2.2 ± 0.5 a	3.6 ± 0.5 a	5.0 ± 0 b
CF/Hardinath-1	0.62 ± 0.07 b			

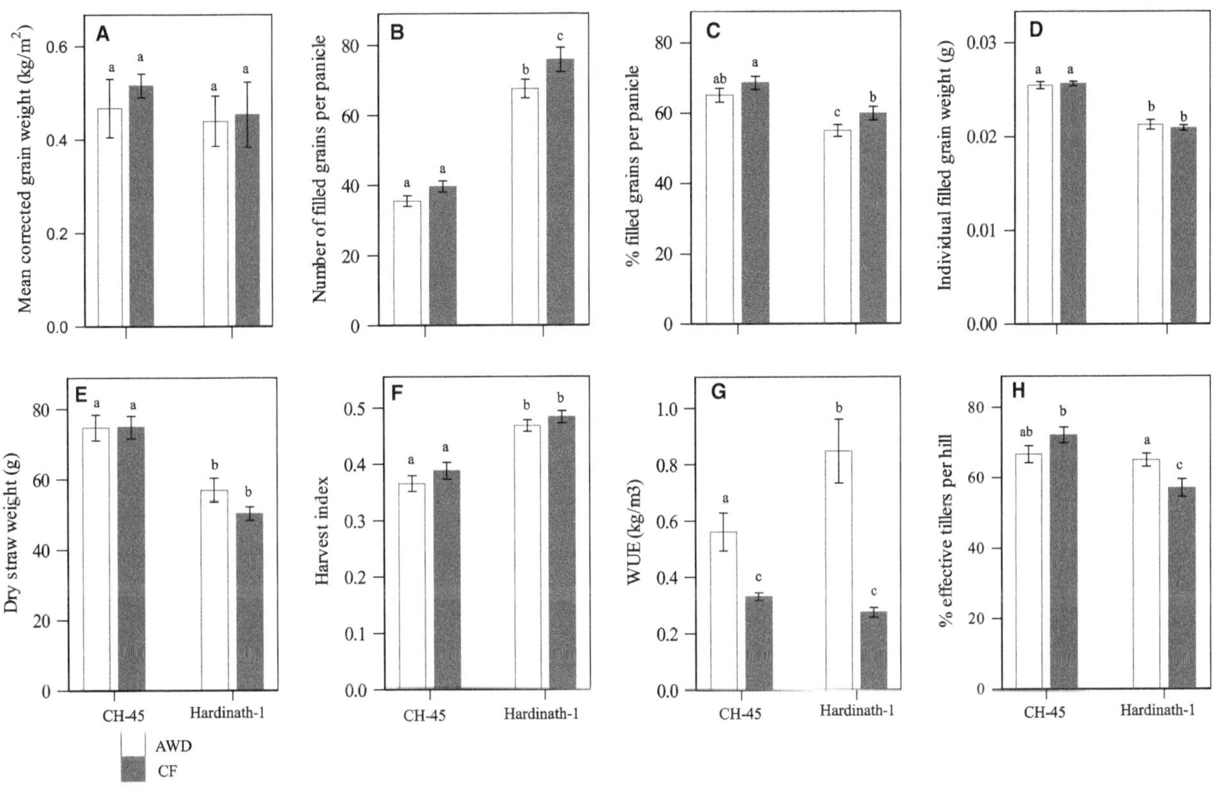

Figure 5. Yield components under alternate wetting and drying (AWD) and continuous flooding (CF) in two rice cultivars, CH-45 and Hardinath-1. Different letters show significant ($P < 0.05$) differences between means. Bars show means ± standard error. Yield components are as follows: (A) Grain weight in kg per square metre corrected to 14% moisture ($n = 20$); (B) Number of filled grains per panicle ($n = 240$); (C) Percentage of filled grains per panicle ($n = 240$); (D) Individual filled grain weight in g, calculated using per panicle filled grain weight and number ($n = 240$); (E) Dry straw weight in g per 4-hill sample ($n = 60$); (F) Harvest index, calculated as ratio of grain weight and total biomass weight per 4-hill sample ($n = 60$); (G) Water use efficiency per plot, expressed in kg/m^3 and calculated from corrected grain weight per plot and total volume of irrigation water applied from transplanting to harvest excluding rainfall ($n = 20$); and (H) Percentage effective tillers per hill ($n = 120$).

Social dimensions of irrigation scheduling

Survey respondents derive their irrigation either from a small shared gravity- fed irrigation canal, a *kulo* (55% of respondents, with *kulo*s shared between a mean number of 99 households); a shared or private shallow tubewell (36%); or a combination of *kulo* and tubewell (8%, although more farmers can access a community tubewell in times of severe water scarcity). *Kulo*s and tubewells are central to agriculture and daily life in Agyauli, as a source of irrigation, drinking and washing water and of protein such as crabs, fish and snails. Water use is charged proportionally to the area of land under irrigation, at a fixed price per season for tubewell users, and via labor or financial contributions to *kulo* maintenance for *kulo* users. Water use for both sources is regulated by community-appointed committees.

Most respondents flood their rice fields every ~2–3 days (55%). Some 15% irrigate daily, and 21% weekly (Fig. 6A), while others irrigate depending on soil type and elevation

Table 4. *P*-values for effect of irrigation, cultivar, and interaction of irrigation and cultivar (irrigation*cultivar) on yield components. Asterisks as described in Table 2.

	P-values		
Yield components	Irrigation	Cultivar	Irrigation*Cultivar
Corrected grain weight per square meter (kg m^{-2})	0.22	0.49	0.51
No. filled grains per panicle	0.030*	<0.001***	0.95
% filled grains per panicle	0.024*	<0.001***	0.72
Individual grain weight (g)	0.12	<0.001***	0.15
Dry straw weight per 4-hill sample (g)	0.31	<0.001***	0.27
Harvest index per 4-hill sample	0.21	<0.001***	0.75
% effective tillers per 4-hill sample	0.028*	0.57	0.0056**
Water use efficiency per square meter (kg m^{-3} m^{-2})	0.0023**	0.0086**	0.0069**

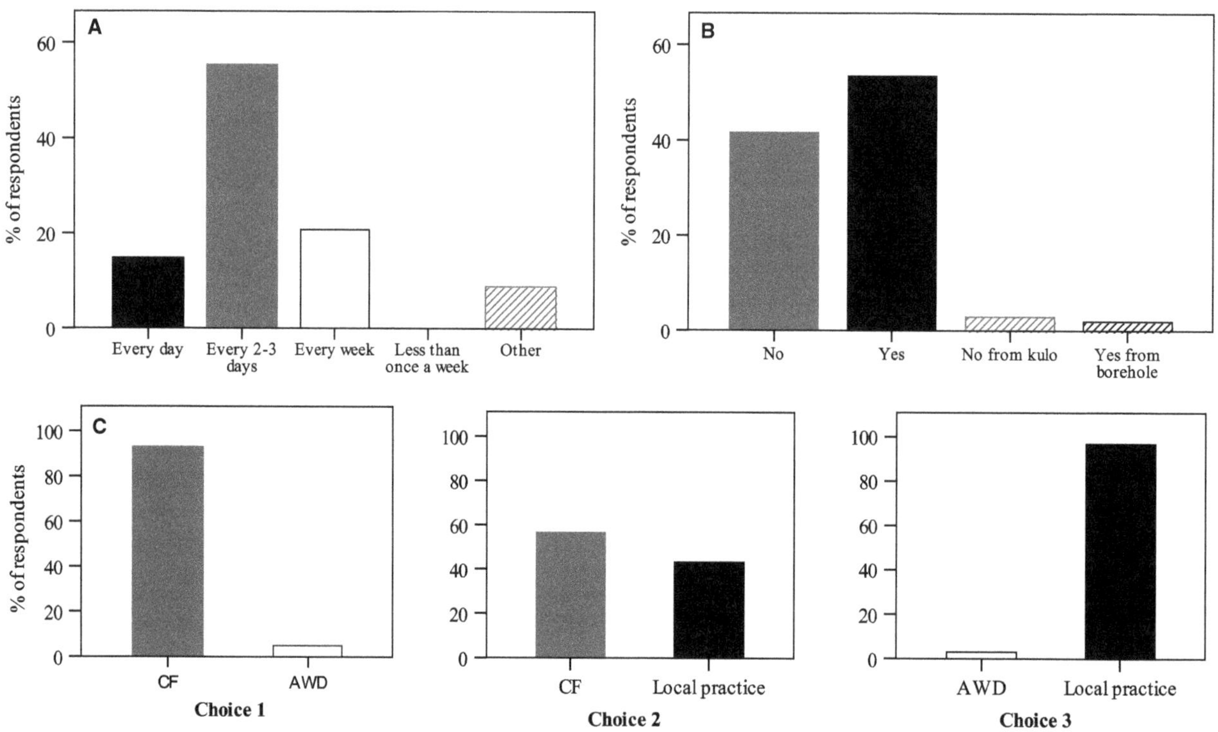

Figure 6. Graphs showing responses as the percentage of 101 respondents giving that response to the following questions: (A) How often do you flood the field? (B) Can you always access the irrigation water you need? (C) Choices 1, 2 and 3 were choice experiments in which respondents were asked to choose between continuous flooding (CF), alternate wetting and drying (AWD) and local practices; two respondents gave no replies, and one respondent answered "don't know" to Choice 1.

relative to water sources. Most respondents flood their field based on changes in soil water status (62%) or the availability of water (35%, particularly high amongst kulo users). An overwhelming 93% of respondents use the same irrigation scheduling every year. Those who do change their scheduling are influenced by interannual variability in the timings of rain and panicle development, and noted that their scheduling depends on timetabling of load-shedding (nonavailability of electricity) and turn-taking between farmers. Crucial times for the crop to have access to water were stated as during tillering (88% of respondents), a few days after transplanting (25%) and after weeding (14%). Respondents were aware of the need to increase the efficient use of available water by building bunds properly (74%), mending holes in them (65%) and irrigating the field carefully (12%). No respondents store water for irrigation.

More than half (56%) of respondents said they can always get access to enough irrigation water for *chaite* rice, with 5% qualifying that they can always access water from their tubewell but not the *kulo* (Fig. 6B). The weather was considered to affect water availability by 70% of respondents. This was linked to hotter and drier weather in the summer causing a decline in the water table (44%) and in natural sources of the *kulo* such as springs and rivers (49%), which in turn can affect electricity production and availability and hence the use of tubewells (3%).

When asked whether they felt their ability to irrigate *chaite* rice was restricted, the majority of respondents (63%) said that insufficient water was a problem, although 35% said it was not restricted. Other restrictions (19%) include having to wait a long time to irrigate if there is no water available during their turn, and lack of fuel or electricity for operating the tubewell, as fuel is expensive and there is extensive load-shedding, particularly at the height of the dry season.

For respondents who cannot always access enough irrigation water, ways of preventing yield reductions comprise technical strategies, such as using an electric or diesel motor pump or earth dams in the *kulo* to ensure water reaches the field; using a tubewell to supplement their water supply (mentioned by 24% of respondents answering this question); and social strategies, such as co-operating with and borrowing turns from other farmers (13%) and speeding up the rotation system (12%). If there is insufficient water at crucial points in crop physiological development, respondents tend to feel powerless. Some 70% said that if this occurred the crop would die or yield would be reduced. However, respondents 2 and 3, relating the experiences of a local farmer whose *chaite* rice crop is frequently destroyed by inadequate water, explained that the benefits of getting a harvest in a good year outweighs the labor and input costs of growing an ineffective crop in a water-scarce year, particularly as straw can be used to feed livestock.

Surprisingly, given the limited formal promotion of AWD in Nepal (Uprety 2011), 91% of respondents had heard of AWD, and 90% said they had tried it on their farm. The benefits of AWD mentioned by these respondents included more effective management of pests such as stem-borer and leaf-roller (64%), increased tillering (27%); improved general crop health (27%); and preventing crop decay (22%). However, interview data and participant observation strongly suggested that the concept of AWD is interpreted differently by some local farmers. Interview respondents 2, 3 and 4 drain the field in response to pest infestation (stemborer and leaf-roller), re-irrigating the field after no more than one or two days. Respondent 1 reduces ponded water depth just before, and during, the tillering stage, which was believed to increase tillering, while increasing ponded water depth again at flowering. All four respondents said that these were well-established practices which they had learned from older relatives.

The popularity of these long-standing practices, and perceived need to maintain soil moisture were reflected in responses to the choice experiment questions in the survey (Fig. 6C). The overwhelming majority of respondents to these questions preferred the CF option, or flooding every 2–3 days as in local practice, over AWD with flooding every 7–10 days (95:5 and 97:3 percent respectively), whilst opinion was more divided between the former two (57:43).

Discussion

Early vigor is not related to final grain yield

Despite applying over 50% less water, the AWD treatment maintained a similar crop yield to CF plots (Table 4, Fig. 5A), in spite of numerous physiological changes that occurred during the growing season. Accelerated canopy closure (Fig. 4) and less frequent standing water on the soil surface under AWD (Fig. 1), coupled with partial stomatal closure during soil drying cycles (Zhang et al. 2010), likely limited soil evaporation and transpiration respectively in our experiment. Conversely, the increased GF under AWD (Fig. 4) may enhance canopy water loss early in the season, and further studies of leaf and canopy gas exchange seem necessary to reconcile these apparently conflicting findings, in explaining the lower water requirements of rice crops grown with AWD. Since there was little systematic difference in LER between irrigation treatments throughout the vegetative stage (Fig. 3) and GF following weeding (Fig. 4), AWD had a limited impact on vegetative growth except for tillering.

Contrary to expectations that soil drying (Bouman and Tuong 2001) would decrease tiller initiation and cause more frequent tiller death under AWD (Yang and Zhang 2010), tiller number was significantly higher under AWD than CF throughout development (Fig. 2A, Tables 2 and 4). Since higher tillering is often associated with higher tiller abortion later in vegetative growth (Peng et al. 1994), the higher percentage of effective tillers in Hardinath-1 grown under AWD (Fig. 5H) is unexpected. Planting density may have a critical effect on tiller dynamics (Borzogi et al., 2011) 'Effect of plant density on yield and yield components of rice.' *World Applied Sciences Journal.* 12: 2053-2057.) and may interact with irrigation treatment. Further work on the regulation of tiller production and survival seems necessary (Kariali and Mohapatra 2007), but the concentrations, dynamics and root to shoot ratios of different plant hormones (such as auxins, abscisic acid, and cytokinins) in response to drying and re- wetting cycles appear to play a role (Zhang et al. 2010; Liu et al. 2011; Dodd et al. 2015).

Although AWD delayed heading by 8 to 17 days in previous field trials (Gill et al. 2011), there were no differences in heading date between treatments, and plots under AWD had slightly, but not significantly, lower yields than CF plots. This clearly indicates the potential for AWD irrigation to be used in this and similar contexts to reduce water use without affecting yield or requiring changes in other aspects of management, but further trials are needed to account for interannual variability in effects on yield and WUE. Elsewhere in Nepal, AWD significantly decreased yields in some years (possibly due to erratic rainfall) but not others (Mandal et al. 2009). Each cultivar maintained yield under AWD in a slightly different way: with no significant variation in yield components in CH-45 which was exposed to a shorter duration of AWD (Fig. 1), while more effective tillers in Hardinath-1 compensated for a decrease in grain number (Fig. 5). Significantly higher numbers of tillers and effective tillers under AWD may have caused more competition between tillers and panicles for plant resources, resulting in significantly lower grain weight, number and filling (Peng et al. 1994). This contrasts with previous findings that lower tiller number under AWD was compensated for by higher grain weight and a greater proportion of grain filling per panicle (Bouman and Tuong 2001; Zhang et al. 2010).

There were also significant interactions between irrigation treatment and cultivar (Table 4). Despite much earlier heading of CH-45 (Yoshida 1981), higher grain weight and filling occurred, perhaps related to the slower decline in LER in this cultivar (Fig. 3). GF was significantly higher in CH-45 throughout the vegetative period under AWD, probably related to its enhanced shoot vigor as a pre-IR8 (a widely released semidwarf) variety, which may contribute to its greater weed competitiveness. There was also an apparent trade-off between cultivars between grain weight and filling, and number of grains. The resource trade-off between grain weight and number in cereals is well-established (Gambín and Borrás 2010), but a study in longer duration rice varieties found no significant relationships with grain filling rate, concluding that this is genetically determined (Venkateswarlu et al. 1981).

The relationship between early vigor and yield components is complex. There appears to have been a trade-off between tillering, and panicle size and grain development, as found previously (Peng et al. 1994). This greater early vigor in CH-45 significantly increased dry straw weight and significantly lowered harvest index, but had no effect on overall grain yield. Interestingly, both genotypes and irrigation treatments produced similar yields regardless of differences in early vigor.

AWD has been proposed to increase resilience to lodging, since the crop may develop deeper roots to access soil water (Yang et al. 2009). However, in heavy soils such as the clays in Agyauli, drying can lead to particle cementation (Rao and Revanasiddappa 2006) and soil compaction (Sanchez 1973). This may restrict root growth and render a crop under AWD more vulnerable to lodging, as observed here (Table 3). Whether limited root development enhances tillering but slows grain development under AWD (Figs. 2 and 5) merits further studies.

Although more weeds have been reported (but not quantified) in on-farm trials of AWD (Kürschner et al. 2010), in this experiment the opposite occurred. This could be related to the relatively shallow ponded water depths in CF plots (20–30 mm) and to soil contraction under AWD restricting weed growth as well as crop root development. Enhanced crop tillering can decrease weed competitiveness (Zhao et al. 2006) but the choice experiment and conversations with local farmers suggest that weed competitiveness is not considered important in selecting cultivars and irrigation scheduling. Weed weight was negatively correlated with the difference in GF calculated immediately before and after weeding (Kendall's tau = −0.16, $P = 0.004$), indicating that the accuracy of BreedPix in measuring crop development would benefit from detecting the color and texture of weeds in digital images (Gée et al. 2008; Meyer and Neto 2008).

From promising science to a socially viable agronomic technique?

The agronomic results, when combined with the findings that many local farmers experience problems accessing sufficient water to irrigate *chaite* rice (Fig. 6B), suggest that AWD shows considerable promise. However, although respondents are familiar with the concept of AWD and may even have tried it, they overwhelmingly said that they would be unwilling to adopt AWD as their principal irrigation scheduling technique. Determining the reasons for this provides insight into the wider challenges of promoting the adoption of particular agricultural technologies, and of pursuing science and technology developments which will most benefit farmers.

AWD, as recommended by IRRI, relies on two key assumptions about the socio-economic context in which it is to be adopted. Firstly, that adopting AWD is a straightforward and rational choice for farmers, as long as water supply is reliable and can be controlled separately for each field or farm. Secondly, that there are strong, immediate incentives for reducing on-farm water consumption, for example if water is paid for per volume per farmer (Moya et al. 2004; Blanke et al. 2007). However, in Agyauli and many other regions in South and East Asia (Mushtaq et al. 2006; Kürschner et al. 2010), one of the main problems facing irrigated agriculture, and for the adoption of AWD, is unreliability, which could result in soil drying beyond the "safe AWD" limits at key points in crop development, or require more frequent, pre-emptive flooding, thereby reducing possible water savings. Likewise, many farmers pay for water by land area, not by volume, and use a shared system, for example with a rotation system determined by a users' committee. Thus they have little direct incentive to save water, and limited control over timing, depending on the flexibility of the system and politics within the users' committee (Meinzen-Dick and Zwarteveen 1998). Furthermore, in this region, water leaking from one field into another may contribute to other farmers' crops, to other water uses or the health of the whole agro-ecosystem, so more water than the actual crop requirement may be applied (Moya et al. 2004).

Research on AWD adoption has recognized and addressed some of these problems, primarily advocating changes to water governance mechanisms like stricter water pricing, and the creation of infrastructure such as storage ponds (Mushtaq et al. 2006; Li and Li 2010). It also focuses on education, dissemination and uptake at a variety of scales (Lampayan et al. 2015). However, context is important, and recommended policy, pricing or infrastructural fixes are rarely simple in practice, since farmers are rarely able to respond only to stable legal and economic factors (Berry 1993). Although agricultural practices may not be optimized due to rapid and externally driven change, they are usually contextually embedded. This means they are adapted to and implicated in a wide range of political, social, cultural and economic practices and institutions at different scales, as well as ecological and infrastructural conditions (see, for example, Berry 1993; Lansing 2006). Consequently, apparently simple changes to agricultural practices, such as AWD, may in fact have wider impacts and necessitate wider changes (Lansing 2006; Burnham et al. 2014). In Agyauli, for example, irrigation scheduling is socially and politically mediated, for example the management of water by *kulo* committees and of electricity by central government. It is linked to cultural and social factors including familiarity with particular techniques, labor availability and ease of weeding. As in previous field trials in Nepal, the impacts of AWD on soil properties take on new significance when part of a (rice-wheat) rotation system (Mandal et al. 2009). Further agronomic trials exploring the short- and long-term effects of AWD on pest and weed populations and soil properties would help ensure a holistic view of proposed changes to agricultural practices. Some of the features of AWD, such as its use of a simple and affordable field water tube, have promoted its large-scale uptake in other rice-growing areas such as the Philippines, Vietnam and Bangladesh (Lampayan et al. 2015). However, Nepal is a different context, especially in terms of water governance, irrigation infrastructure and agricultural extension services, and as such will require a different approach to promoting AWD.

Agricultural anthropologists such as Richards (1989) have long called for agricultural and agronomic research to consider farming as a performance: a lived process in which change is determined by dynamic environmental, cultural and social factors as well as economic costs and benefits. Additionally, the use of and changes to agricultural practices are dynamic. Richards argues that farming happens, unlike a predesigned field trial, "in time", with decisions and strategies made and developed in response to changing conditions, such as weather patterns and labor availability. Similarly, in Agyauli, irrigation scheduling happens *in time*. Although farmers generally use the same irrigation scheduling each year, the day-to-day practice of irrigation depends on various factors, such as weather, water availability, local water governance, neighbors' irrigation requirements and how level the field is. Carrying out an on-farm irrigation field trial, using the novel method of simultaneous experiment and participant observation, provided a lens onto day-to- day agricultural practice, enabling a better understanding of the potential suitability of AWD to this setting.

The contextual embeddedness of agricultural practices raises two interlinking, crucial but often neglected considerations for researchers working on new agricultural science and technologies: understanding a particular agricultural context through interdisciplinary research, and involving farmers, and their knowledge, opinions and needs, through participatory research and extension such as farmer to farmer training and farmer field schools (Fortmann 2008). There are growing concerns that the urgency and global focus of current food security narratives are marginalizing such approaches (Tomlinson 2011), despite the fact that locally specific, farmer-driven innovation may hold many of the answers to food security challenges (MacMillan and Benton 2014). Despite the wealth of detailed social science research into agricultural systems, and resilience and change within them, these understandings are rarely effectively integrated with conventional agricultural science research, with major implications for the relevance of research findings to the realities of farming communities (Cornwall et al. 1994; Giller et al. 2011). This paper offers an example of one form of interdisciplinary research which can help to incorporate social and physiological understanding of agronomic techniques. Major progress has been made in making AWD accessible and appropriate for farmers elsewhere in Asia (Lampayan et al. 2015); and more context-specific, interdisciplinary and participatory research could help adapt it for other rice- producing regions such as Nepal.

Conclusions

AWD proved to be an extremely effective water-saving irrigation technology in the context of *chaite* rice production in Agyauli VDC in lowland Nepal, greatly increasing the water use efficiency of locally important varieties without decreasing yields. Both irrigation treatment and cultivar affected early vigor, but had no effect on final yield. However, it is unlikely to be adopted in Agyauli and areas with similar socio-economic conditions unless local water management, and use (or both) is altered. Future extension work on AWD needs to consider ways of supporting farmers in changing water management and governance, but also ways in which the technology and the science underpinning it could be adapted for these contexts. Increasing the involvement of farmers in AWD research, through the use of participatory and interdisciplinary methods, could improve uptake of these or similar water-saving techniques.

Acknowledgements

KRH thanks LI-BIRD staff for assistance in running the field trial, the Agyauli farmers' committee for helping with social surveys and grain harvest; survey and interview respondents; and her hosts in Agyauli for their warm hospitality. ICD thanks BBSRC (BB/J003336/1) for continued support of research into AWD.

Conflict of Interest

None declared

References

Alcon, F., S. Tapsuwan, J. M. Martínez-Paz, R. Brouwer, and M. D. de Miguel. 2014. Forecasting deficit irrigation adoption using a mixed stakeholder assessment methodology. Technol. Forecast. Soc. Chang. 83:183–193.

Belder, P., B. Bouman, R. Cabangon, L. Guoan, E. Quilang, L. Yuanhua, et al. 2004. Effect of water-saving irrigation on rice yield and water use in typical lowland conditions in Asia. Agric. Water Manag. 65:193–210.

Berry, S. 1993. No condition is permanent: the social dynamics of agrarian change in Sub- Saharan Africa. Univ. of Wisconsin Press, Madison.

Blanke, A., S. Rozelle, B. Lohmar, J. Wang, and J. Huang. 2007. Water saving technology and saving water in China. Agric. Water Manag. 87:139–150.

Borzogi, H. R., A. Faraji, R. K. Danesh, A. Keshavarz, E. Azarpour, and F. Tarighi. 2011. Effect of plant density on yield and yield components of rice. World Applied Sciences Journal. 12: 2053–2057.

Bouman, B. A. M., and R. M. Lampayan. 2009. Rice fact sheet – alternate wetting drying (AWD). International Rice Research Institute, Los Baños (Philippines).

Bouman, B., R. Lampayan, and T. Tuong. 2007. Water management in irrigated rice: coping with water scarcity. International Rice Research Institute, Los Baños (Philippines).

Bouman, B., and T. Tuong. 2001. Field water management to save water and increase its productivity in irrigated lowland rice. Agric. Water Manag. 49:11–30.

Brouwer, R., T. Dekker, J. Rolfe, and J. Windle. 2009. Choice certainty and consistency in repeated choice experiments. Environ. Resource Econ. 46:93–109.

Budruk, M. 2010. Cross-language measurement equivalence of the Place Attachment Scale : a multigroup confirmatory factor analysis approach. J. Leis. Res. 42:25–42.

Burnham, M., Z. Ma, and D. Zhu. 2014. The human dimensions of water saving irrigation: lessons learned from Chinese smallholder farmers. Agric. Hum. Values 32:347–360 doi:10.1007/s10460-014-9565-8.

Cai, X., Y. Cui, J. Dai, and Y. Luo. 2012. Local storages: the impact on hydrology and implications for policy making in irrigation systems. Water Int. 37:395–407.

Casadesús, J., and D. Villegas. 2014. Conventional digital cameras as a tool for assessing leaf area index and biomass for cereal breeding. J. Integr. Plant Biol. 56:7–14.

Cornwall, A., I. Guijt, and A. Welbourn. 1994. Extending the horizons of agricultural research and extension: methodological challenges. Agric. Hum. Values 11:38–57.

Davies, W. J., J. Zhang, J. Yang, and I. C. Dodd. 2011. Novel crop science to improve yield and resource use efficiency in water-limited agriculture. J. Agricult. Sci. 149:123–131.

Dodd, I. C., J. Puértolas, K. Huber, J. G. Pérez-Pérez, H. R. Wright, and M. S. A. Blackwell. 2015. The importance of soil drying and re-wetting in crop phytohormonal and nutritional responses to deficit irrigation. J. Exp. Bot. 66: 2239–2252 doi:10.1093/jxb/eru532.

Fortmann, L. 2008. Participatory research in conservation and rural livelihoods: doing science together. Wiley-Blackwell, Oxford, U.K.

Gambín, B. L., and L. Borrás. 2010. Resource distribution and the trade-off between seed number and seed weight: a comparison across crop species. Ann. Appl. Biol. 156:91–102.

Gée, C., J. Bossu, G. Jones, and F. Truchetet. 2008. Crop/weed discrimination in perspective agronomic images. Comput. Electron. Agricult. 60:49–59.

Gill, G., E. Humphreys, S. S. Kukal, and U. S. Walia. 2011. Effect of water management on dry seeded and puddled transplanted rice. Part 1: Crop performance. Field. Crop. Res. 120:112–122.

Giller, K. E., M. Corbeels, J. Nyamangara, B. Triomphe, F. Affholder, E. Scopel, et al. 2011. A research agenda to explore the role of conservation agriculture in African smallholder farming systems. Field. Crop. Res. 124:468–472.

Gomez, K. 1972. Techniques for field experiments with rice. International Rice Research Institute, Los Baños (Philippines).

Kariali, E., and P. K. Mohapatra. 2007. Hormonal regulation of tiller dynamics in differentially- tillering rice cultivars. Plant Growth Regul. 53:215–223.

Koundouri, P., C. Nauges, and V. Tzouvelekas. 2006. Technology adoption under production uncertainty: theory and application to irrigation technology. Am. J. Agric. Econ. 88:657–670.

Kürschner, E., C. Henschel, T. Hildebrandt, E. Jülich, M. Leineweber, and C. Paul. 2010. Water saving in rice production – dissemination, adoption and short term impacts of alternate wetting and drying (AWD) in Bangladesh. Humboldt-Universitat Zu Berlin SLE Publication Series:S241.

Lampayan, R. M., R. M. Rejesus, G. R. Singleton, and B. A. M. Bouman. 2015. Adoption and economics of alternate wetting and drying water management for irrigated lowland rice. Field. Crop. Res. 170:95–108.

Lansing, J. S. 2006. Perfect order: recognizing complexity in Bali. Princeton Univ. Press, Princeton.

Lee, K.-J., and B.-W. Lee. 2013. Estimation of rice growth and nitrogen nutrition status using color digital camera image analysis. Eur. J. Agron. 48:57–65.

Li, H., and M. Li. 2010. Sub-group formation and the adoption of the alternate wetting and drying irrigation method for rice in China. Agric. Water Manag. 97:700–706.

Liu, Y., Q. Wang, Y. Ding, G. Li, J. Xu, and S. Wang. 2011. Effects of external ABA, GA3 and NAA on the tiller bud outgrowth of rice is related to changes in endogenous hormones. Plant Growth Regul. 65:247–254.

MacMillan, T., and T. G. Benton. 2014. Engage farmers in research. Nature 509:25–27.

Mandal, S. N., A. Regmi, J. Ladha, and T. Tuong. 2009. Crop establishment, tillage, and water management effects on crop and water productivity in the rice-wheat rotation in Nepal. Pp. 239–260 *in* Ladha, J.K., Y. Singh, O. Erenstein, and B. Hardy (eds) Integrated crop and resource management in the rice-wheat system of South Asia. International Rice Research Institute, Los Baños (Philippines).

Meinzen-Dick, R., and M. Zwarteveen. 1998. Gendered participation in water management: issues and illustrations from water users 'associations in South Asia. Agric. Hum. Values 15:337–345.

Meyer, G. E., and J. C. Neto. 2008. Verification of color vegetation indices for automated crop imaging applications. Comput. Elect. Agricult. 63:282–293.

Mishra, A., and V. M. Salokhe. 2010. The effects of planting pattern and water regime on root morphology, physiology and grain yield of rice. J. Agron. Crop Sci. 196:368–378.

Mishra, H. S., T. R. Rathore, and R. C. Pant. 1990. Effect of intermittent irrigation on groundwater table contribution, irrigation requirement and yield of rice in Mollisols of the Tarai region. Agric. Water Manag. 18:231–241.

Moya, P., L. Hong, D. Dawe, and C. Chongde. 2004. The impact of on-farm water saving irrigation techniques on rice productivity and profitability in Zhanghe Irrigation System, Hubei. China. Paddy Water Environ. 2:207–215.

Mullings, B. 1999. Insider or outsider, both or neither: some dilemmas of interviewing in a cross- cultural setting. Geoforum 30:337–350.

Mushtaq, S., D. Dawe, H. Lin, and P. Moya. 2006. An assessment of the role of ponds in the adoption of water-saving irrigation practices in the Zhanghe Irrigation System, China. Agricult. Water Manag. 83:100–110.

Nuijten, E. 2011. Combining research styles of the natural and social sciences in agricultural research. NJAS – Wageningen J. Life Sci. 57:197–205.

Panta, S. K., and B. P. Resurrección. 2014. Gender and caste relations amidst a changing political situation in Nepal: insights from a farmer-managed irrigation system. Gender Technol. Develop. 18:219–247.

Peng, S., G. Khush, and K. Cassman. 1994. Evolution of the new plant ideotype for increased yield potential. Pp. 5–20 *in* K. G. Cassman, ed. Breaking the yield barrier: proceedings of a workshop on rice yield potential in favorable environments. International Rice Research Institute, Manila (Philippines).

Price, A. H., G. J. Norton, D. E. Salt, O. Ebenhoeh, A. A. Meharg, C. Meharg, et al. 2013. Alternate wetting and drying irrigation for rice in Bangladesh: Is it sustainable and has plant breeding something to offer? Food Energy Sec. 2:120–129.

Rao, S. M., and K. Revanasiddappa. 2006. Influence of cyclic wetting drying on collapse behaviour of compacted residual soil. Geotech. Geol. Eng. 24:725–734.

Richards, M., and B. O. Sander. 2014. Alternate wetting and drying in irrigated rice: implementation guidance for policy makers and investors. CGIAR Research Program on Climate Change, Agriculture and Food Security, Montpellier.

Richards, P. 1989. Agriculture as a performance. Pp. 39–42 *in* R. Chambers, A. Pacey, and L. Thrupp, eds. Farmer first: farmer innovation and agricultural research. Intermediate Technology, London.

Rockström, J., M. Lannerstad, and M. Falkenmark. 2007. Assessing the water challenge of a new green revolution in developing countries. Proc. Natl Acad. Sci. USA 104:6253–6260.

Sanchez, P. 1973. Puddling tropical rice soils: 2. Effects of water losses. Soil Sci. 115:303–308.

Senthilkumar, K., P. S. Bindraban, W. de Boer, N. de Ridder, T. M. Thiyagarajan, and K. E. Giller. 2009. Characterising rice-based farming systems to identify opportunities for adopting water efficient cultivation methods in Tamil Nadu, India. Agricult. Water Manag. 96:1851–1860.

Tabbal, D. F., B. A. M. Bouman, S. I. Bhuiyan, E. B. Sibayan, and M. A. Sattar. 2002. On-farm strategies for reducing water input in irrigated rice; case studies in the Philippines. Agric. Water Manag. 56:93–112.

Tomlinson, I. 2011. Doubling food production to feed the 9 billion: a critical perspective on a key discourse of food security in the UK. J. Rural Stud. 29:81–90.

Uprety, R. 2011. Participatory learning for technology shaping and its dissemination: a case from Nepal. Extens. Farm. Syst. J. 7:37.

Van Nguyen, N., and A. Ferrero. 2006. Meeting the challenges of global rice production. Paddy Water Environ. 4:1–9.

Venkateswarlu, B., G. Prasad, and A. Prasad. 1981. Studies on the nature of relationships between grain size, spikelet number, grain yield and spikelet filling in late duration varieties of rice (*Oryza sativa* L.). Plant Soil 60:123–130.

Willis, P., and M. Trondman. 2002. Manifesto for ethnography. Cultural Studies ↔ Critical Methodologies 2:394–402.

Yang, J., D. Huang, H. Duan, G. Tan, and J. Zhang. 2009. Alternate wetting and moderate soil drying increases grain yield and reduces cadmium accumulation in rice grains. J. Sci. Food Agric. 89:1728–1736.

Yang, J., and J. Zhang. 2010. Crop management techniques to enhance harvest index in rice. J. Exp. Bot. 61:3177–3189.

Yoshida, S. 1981. Fundamentals of rice crop science. International Rice Research Institute, Los Baños (Philippines).

Zhang, H., T. Chen, Z. Wang, J. Yang, and J. Zhang. 2010. Involvement of cytokinins in the grain filling of rice under alternate wetting and drying irrigation. J. Exp. Bot. 61:3719–3733.

Zhang, H., H. Li, L. Yuan, Z. Wang, J. Yang, and J. Zhang. 2012. Post-anthesis alternate wetting and moderate soil drying enhances activities of key enzymes in sucrose-to-starch conversion in inferior spikelets of rice. J. Exp. Bot. 63:215–227.

Zhao, D. L., G. N. Atlin, L. Bastiaans, and J. H. J. Spiertz. 2006. Comparing rice germplasm groups for growth, grain yield and weed-suppressive ability under aerobic soil conditions. Weed Res. 46:444–452.

4

Feeding 11 billion on 0.5 billion hectare of area under cereal crops

Rattan Lal

Carbon Management and Sequestration Center, The Ohio State University, Columbus, Ohio 43210

Keywords
global cereal yield, global population, per capita cropland area, soil health, Sustainable intensification.

Correspondence
Rattan Lal, Carbon Management and Sequestration Center, The Ohio State University, Columbus, OH 43210.

E-mail: lal.1@osu.edu

Funding Information
No funding information provided.

Abstract

Despite impressive increase in global grain production since 1960s, there are 795 million food-insecure and ~2 billion people prone to malnutrition. Further, global population of 7.4 billion in 2016 is projected to increase to 9.7 billion by 2050, with almost all increase occurring in developing countries. Thus, it is recommended that global food production be increased by 60 to 70% between 2005 and 2050. Global crop production increased threefold between 1965 and 2015 with a net increase of only 67 million ha (Mha) of cropland area. Nonetheless, agronomic yield of food staples can still be tripled or quadrupled in Sub-Saharan Africa (SSA), South Asia (SA), and the Caribbean by a widespread adoption of site-specific best management practices of sustainable intensification (SI). Rather than expanding the area under cropland, agriculturally marginal and degraded soils can be set aside for nature conservancy. The global average cereal yield of 3.27 Mg ha^{-1} in 2005 can be increased to 5 Mg ha^{-1} by 2050, 6 Mg ha^{-1} by 2080 and 7 Mg ha^{-1} by 2100 through SI of agroecosystems in SSA, SA and elsewhere. The strategy of "producing more from less" necessitates restoration of soil health and increasing soil organic carbon (SOC) concentration to be more than 1.5–2.0% in the rootzone. The goal of SOC sequestration is in accord with the "4 per Thousand" initiative proposed at COP21 in 2015. Therefore, global food demands can be met despite the decreasing trends in the per capita cropland area by 2050 – 0.17 ha in the world and 0.15 ha in developing countries. While enhancing productivity by SI, the strategy is to simultaneously reduce food waste, increase access and distribution of food, and promote plant-based diet. The goal is to reconcile high production with better environmental quality, develop urban agriculture (aquaponics, aeroponics, and vertical farms), promote nutrition-sensitive farming, and restore degraded soils. Sustainable intensification of agroecosystems can produce enough food grains to feed one person for a year on 0.045 ha of arable land.

Introduction

Achieving food security, eliminating hunger, and removing poverty have been among important foci of the Agenda 21 (UN, 1992), Millennium Development Goals (UN, 2000), and the Sustainable Development Goals (UN 2015a,b). While there are strong criticisms of broad-brushed initiatives (Horton 2014), these are generic guidelines at all levels of governance. Yet, the insidious problem of food and nutritional insecurity plagues humanity even more now than ever before. While the number of food-insecure population has progressively decreased since 1990s, there remain some 795 million food-insecure people (FAO, 2015a; Fig. 1) and an additional ~2 billion are prone to hidden hunger or malnutrition caused by the deficiency of micronutrients and protein (Ruel-Bergeron et al. 2015).

The elusive problems of food and nutritional insecurity are intricately interconnected with those of the degradation of soil and water resources. The extent and severity of degradation may be related to the increasing rate of population, which have and continue to differ among different regions of the world (Fig. 2). The rate of population increase has and continue to be more in Asia, Africa, and Latin America (AAL) compared with that in

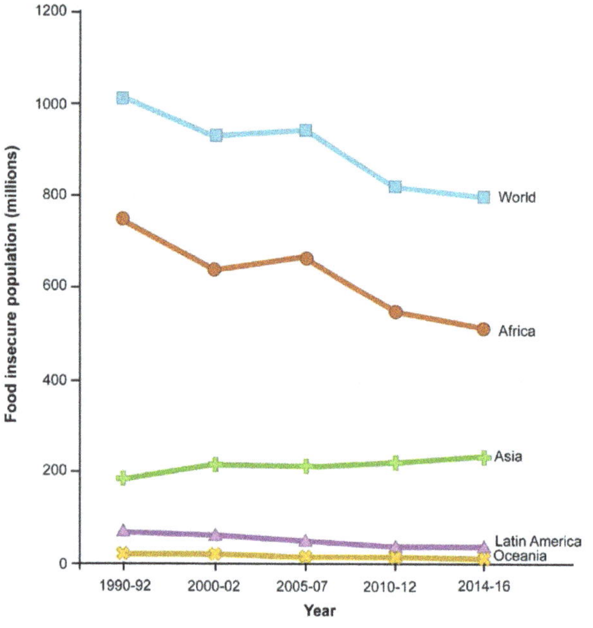

Figure 1. Global extent of food-insecurity and undernourishment (Redrawn from the data in FAO 2015b,c).

North America, Europe, and Oceania (NEO). Almost all of the future increase in world population will occur in the AAL region (Fig. 3), with a strong impact on soil, water, and the environment. These naturally reinforcing problems (population increase, food insecurity, and soil degradation) are affected both by biophysical factors and the serious issues pertaining to the human dimensions (e.g., social, economic, cultural, ethnical, and political). Indeed, there exists a strong interaction among processes (mechanisms), factors (environmental controls), and causes (anthropogenic activities) which mutually reinforce and aggravate the problems of soil/water degradation and perpetuation of food insecurity and hidden hunger (Fig. 4).

As a medium for crop and pasture production, soil management affects agronomic yield and nutritional quality. The goals of sustainable soil management are to maintain and restore the stock of soil organic carbon (SOC) at above the threshold/critical level of 1.5–2.0% in the root zone (Loveland and Web 2003; Patrick et al. 2013), especially for efficient management of nitrogen (Patrick et al. 2013) and water resources. It is widely documented that management of SOC is important to maintaining soil productivity, and especially the yield stability (Overstreet and DeJong-Huges 2009; Seremesic et al. 2011)).

Improving agronomic productivity and achieving yield stability under changing and uncertain climate are important to feeding the world population of 7.4 billion in 2016 and projected to be 9.7 billion by 2050 and 11.2 billion by 2100 (UN 2015a,b; Table 1, Fig. 3). Therefore, the objective of this article is to document that rather than bringing more land under agriculture, food demands can be met with soil and water resources lesser than those being used at present through judicious use and sustainable intensification (SI) of agroecosystems, so that extra land and water can be saved for nature conservancy?

Prevalence of Food Insecurity

Food insecurity remains an issue in Africa and Asia (Fig. 1), and will continue to do so for decades to come. Yet, the number of food-insecure population in the world has decreased from more >1 billion in 1990–92 to 795

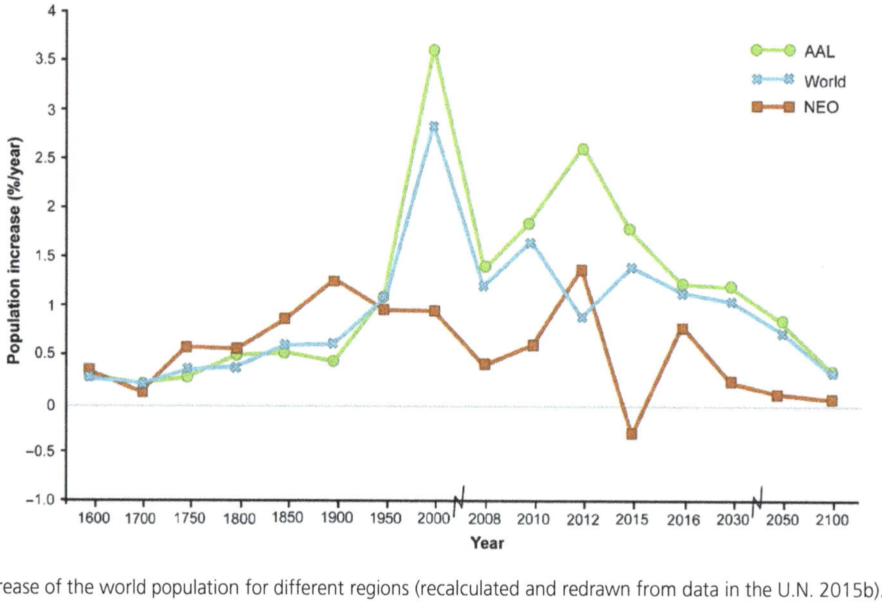

Figure 2. Rate of increase of the world population for different regions (recalculated and redrawn from data in the U.N. 2015b).

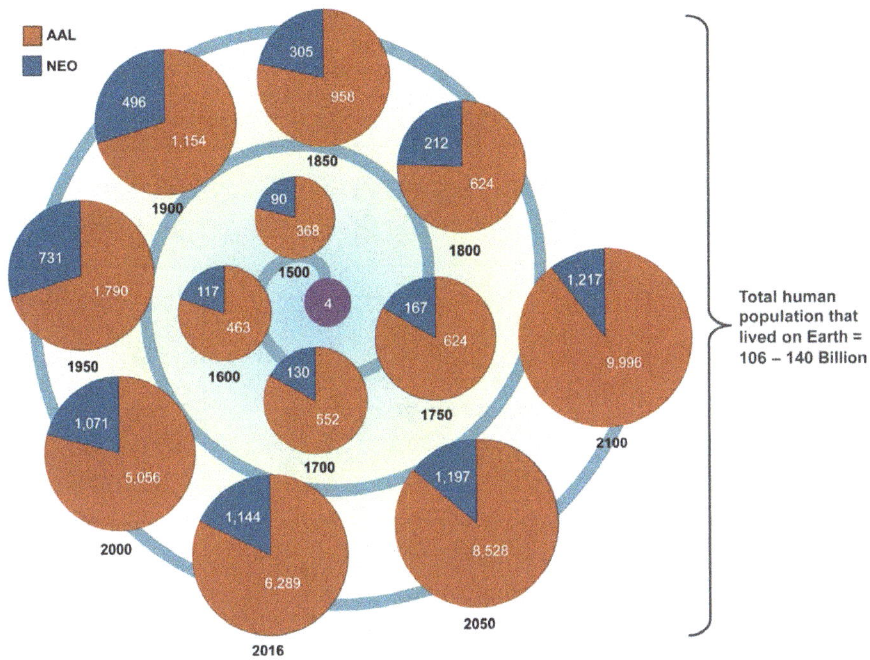

Figure 3. Trajectory of the growth of human population from ~4 million in 10,000 BCE and 300 million at 1 AD, to 11.2 billion by 2100, with almost all of future growth occurring in presently developing countries of Asia, Africa, Latin America, and the Caribbean (Plotted from the data by Fyfe 1997; Haub 2011; U.N. 2015a,b; Geohive 2016 and other sources). AAL = Africa, Asia, and Latin America; NEO = North America, Europe, and Oceania.

million in 2014–16 (Fig. 1). In this context, a considerable progress has also been made in developing countries of Asia and Africa. The absolute number of food-insecure population in developing countries was 810 million (19.7%) in 1990–92, 827 million (15.9%) in 2005–07, 687 million (11.7%) in 2015, and is projected to be 543 million (7.9%) in 2030 and 318 million (4.1%) in 2050 (Alexandratos and Bruinsma 2012). A high rate of population growth in developing countries, where almost all the future population growth will occur (Figs. 2, 3), along with slow adoption of the proven best management practices (BMPs) are among the reasons of persistence of hunger through 2050 and beyond. Thus, it has been suggested that feeding a population of 9.7 billion by 2050 will require overall increase in food production by 60–70% between 2005–07 and 2050 (Bruinsma 2009; FAO, 2015c). If this assumption is correct, it would require additional cereal production of 1 billion Mg, and meat production by over 200 million Mg to a total of about 500 million Mg by 2050 (FAO 2015c). Rather than producing more food, better utilization of the food produced and efficient use of soil and water resources would be a better option.

Global crop production has increased threefold over the 50-year period from 1965 to 2015, primarily through increase in yield per unit area. Increase in global cropland area during this period was only 67 Mha, with increase of 107 Mha in developing countries and decrease of 40 Mha in developed countries. Similarly, future increase in production must also come from existing land resources through the adoption of BMPs for site-specific situations. The trend of decreasing total agricultural land area must continue throughout the 21st century, so that agriculturally marginal and degraded lands can be set aside for nature conservancy, and SOC sequestration.

Land and Water Resources for Agroecosystems

Land area (million hectare, Mha) allocated for cropland and grazing land in 2000 was 1381 and 3357, respectively (FAO, 2015b,c). Of this, land area equipped for irrigation was 280 Mha. For a population of 6896 million in 2000, the per capita allocation of land resources in 2000 was 0.20 ha for cropland, 0.49 for grazing and pasture land, and 0.04 ha for irrigated land. In 2013, when the world population reached 7162 million, land resources (Mha) allocated were 1408 for cropland, 3354 for grazing land and pastures, and 325 equipped for irrigation. Thus, the per capita land resources in 2013 were 0.197 ha of cropland, 0.47 ha of grazing land and pastures, and 0.045 ha of land equipped for irrigation. However, the per capita arable land area will decline by 2050–0.18 ha in the world compared with 0.15 ha in developing and 0.42 ha in developed countries (Fig. 5). In many densely populated countries (e.g., Bangladesh), the per capita arable land area is already <0.04 ha. With adoption of BMPs, and

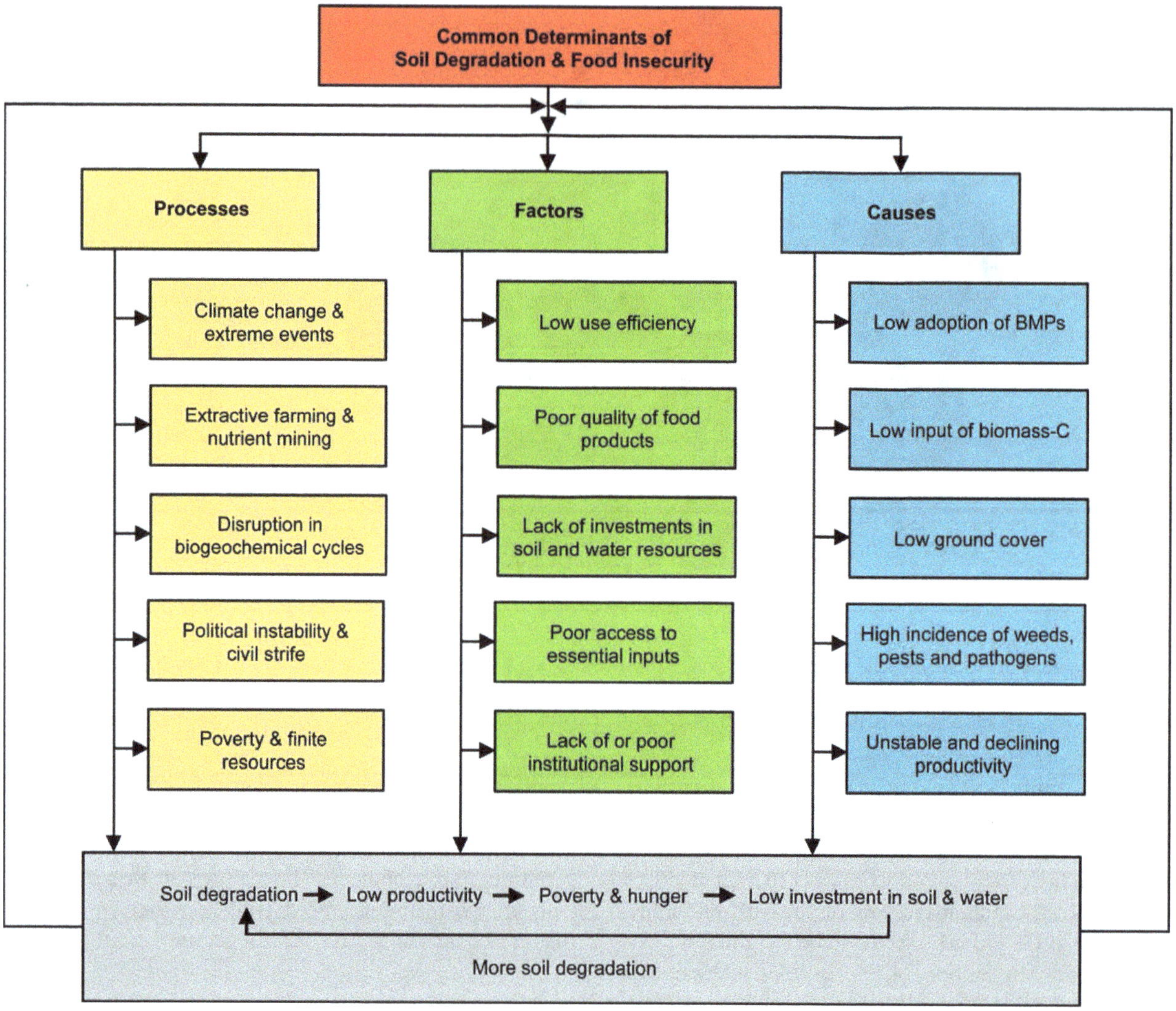

Figure 4. Common determinants of soil degradation and food/nutritional insecurity.

Table 1. Food demand during the 21st century (Data on population, per capita food consumption, and cereal production are from FAO 2009; Alexandratos and Bruinsma 2012; U.N. 2015a,b)

Parameter	Year			
	2005–07	2050	2080	2100
Population	6.4	9.7	10.6	11.2
Per capita food consumption (kcal)	2772	3070	3200	3300
Cereal production (10^6 Mg)	2012	3012	3350	3540
Land area needed for intensive agriculture (Mha)	613	600	560	500
The desired global average cereal yield (kg ha^{-1})	3280	5000	6000	7000

judicious use of soil and water resources, these land resources are more than sufficient to meet the expected food needs of the world population of 9.7 billion by 2050. The strategy is to manage the soil and water resources according to the site-specific BMPs (conservation agriculture, precision farming, drip sub-irrigation, integrated nutrient management, and improved varieties) so that agronomic yield can be increased through enhancement of the use efficiency of inputs and decreasing losses by erosion, leaching, and volatilization.

Yet, the finite soil resources must be protected for other ecosystem services. The land area allocated to agriculture in 2013 included 1408 Mha for cropland for a total cereal production of 2.82 billion Mg, 3354 Mha for grazing land for a total meat production of 290 million Mg, and 325 Mha of land area equipped for irrigation (FAOSTAT, 2016). The fertilizer use (10^6 Mg of nutrients per year) in 2012, 2020, and 2050 is projected to be 123, 135, and 236 for N, and 46.0, 47.6 and 83.7 for P (IFDC, 2004; FAOSTAT, 2016). Global pesticide use was estimated at 2 million Mg in 2014,

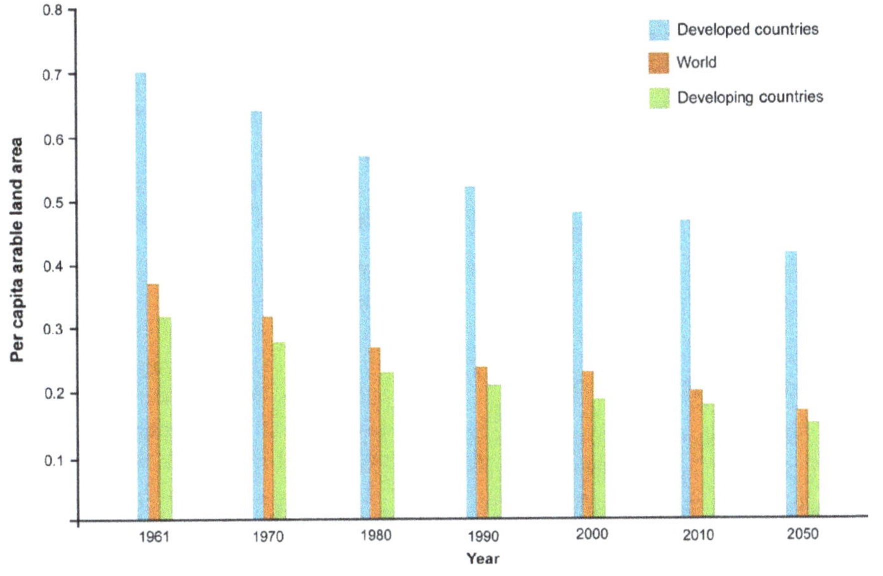

Figure 5. Estimates of temporal changes in per capita arable land area in the world, developing, and developed countries between 1961 and 2050 (Redrawn from Bruinsma 2009).

of which 25% was used in the U.S., 45% in Europe, and 30% in other countries of the world (De et al. 2014). Further, the use of fertilizers, pesticides, and chemicals, along with the irrigation water must be reduced to the minimum required.

Global Food Demand

Cereals presently account for only 37% of total caloric intake in developed countries compared with 71% in least developed, and 54% in other countries (OECD-FAO, 2015). Thus, rising income and growing affluence is changing food habits in preference of animal-based diet. *Per capita* caloric intake is projected to increase between 2002–04 and 2012–14 from 1750 to 2150 kcal per day per person in least developed countries, 2400–2800 in other developing countries, to 2750–2900 in developed countries. However, the proportional caloric intake from meat-based diet will increase more rapidly in developing than in developed countries (OECD-FAO, 2015). Thus, global meat consumption is expected to grow by an annual average of 1.4%, with additional consumption of 51 million Mg of meat by 2024 (OECD-FAO, 2015). The overall demand for agricultural products is projected to grow at 1.1% per year from 2005–07 to 2050 due to population growth, increase in per capita consumption, and changes in diet preferences toward animal-based food (Alexandratos and Bruinsma 2012). Regions with increase in food demands, especially the animal-based food, include sub-Saharan Africa (SSA) and South Asia (SA). This is an important issue that must be addressed through education at all levels.

With the business-as-usual (BAU), there are several estimates of the required cereal yield and total production (Table 1). Wild (2003) estimated that the global average cereal yield of 3.27 Mg ha^{-1} and total production of 2.24 billion Mg would have to be increased to 3.60 Mg ha^{-1} and 2.78 billion Mg in 2025, and 4.3 Mg ha^{-1} and 3.3 billion Mg by 2050. However, with the growing preference for animal-based diet, global average cereal yield and total cereal production would have to be 4.4 Mg ha^{-1} and 3.63 billion Mg by 2025, and 6.0 Mg ha^{-1} and 4.6 billion Mg by 2050, respectively (Wild 2003). Yet, a widespread adoption of proven BMPs can and must increase the global average cereal yield to 5, 6, 7 Mg ha^{-1} by 2050, 2080, and 2100, respectively. This is an achievable goal with already available and proven technology. That being the case, it is possible to decrease the cropland area to as low as 500 Mha by 2100 (Table 1).

While enhancing agronomic yields through adoption of BMPs and sustainable intensification, there are other constraints on available land resources which must also be addressed. Soil degradation remains to be a major issue, with accelerated soil erosion affecting 1.1 billion hectare (Bha) by water and 0.55 Bha by wind erosion (Oldeman 1994). Secondary salinization, a serious problem in arid and semiarid regions, affects 20% of the irrigated land area. Nonpoint source pollution and eutrophication of surface waters are causing algal blooms in the Great Lakes, Golf of Mexico, Chesapeake Bay, Lake Taihu in China, Baltic Sea, and other regions receiving intensive use of chemical fertilizers. Intensive irrigation is leading to rapid water depletion in many regions (Gleesen et al. 2012), including Ogallala (Alley et al. 1999; Amelinckx 2015), North China Plains (Yang et al. 2015), and the Indo-Gangetic Plains (Kerr 2009). Conversion of prime agricultural land to urban expansion and surface sealing by cement and asphalt to develop infrastructure are reducing land area available for agriculture. By 2030, global urban land cover will increase by 152 Mha, more than 10% of the present cropland area of the world. These degraded lands must be restored, and used for implementation of the "4 per Thousand" program (UNFCCC, 2015).

Strategies for Feeding the World

Resources (land, water, chemicals, energy) already used for agriculture must be reduced by enhancing the use efficiency and decreasing losses. For example, 38% of Earth's terrestrial surface is used for agriculture (cropland, grazing land, agroforestry etc.), and 70% of agricultural land (3354 Mha out of 4762 Mha) is allocated to raising animals for production of 290 million Mg of meat and dairy products. Further, 70% of the global freshwater withdrawals are used for irrigation of 325 Mha. In the context of anthropogenic emissions, 30–35% of global greenhouse gas (GHG) emissions are directly and indirectly contributed by agriculture and food-related processes (IPCC, 2014). Despite the disproportional utilization of all these resources, with severe ecological and environmental footprint, 1 in 9 persons (795 out of 7400 million) is prone to hunger, and 1 in 5 to malnutrition. Apparently, the BAU is not an option, and must be objectively and critically assessed.

It is widely believed that 2.8 billion Mg of cereals produced annually are adequate to feed 7.4 billion people, with per capita consumption of 378 kg per person per year. Indeed, the total amount of food produced (grains, meat, fish, poultry, fruits, vegetables, etc.) may be adequate to feed even a world population of 10 billion because almost 1 billion Mg of grains produced are wasted enroute from field to the fork (Lipinski et al. 2013; Nunley 2013). Therefore, a prudent strategy to advance food and nutritional security must consider the followings:

1. Reducing the food waste equivalent to 30% of total production;
2. Increasing access to food by addressing poverty, civil strife, inequity, wars, political instability, and political/ecological refugees;
3. Improving food distribution by strengthening infrastructure at global, regional, and local scale;
4. Enhancing use of pulses and plant-based diets and identifying alternative sources of proteins;
5. Accepting personal responsibilities of not taking food and natural resources (soil, water, vegetation) for granted; and
6. Raising agronomic productivity from existing lands, restoring degraded lands, sequestering SOC and enhancing biological nitrogen fixation, and setting aside some of the existing agricultural land for nature conservancy.

Sustainable Intensification for Saving Land and Water Resources

The strategy of SI implies producing "more from less," of less land area, less water used for irrigation and processing, less fertilizers and chemicals applied, less energy consumed in farm operations and processing/transport etc., and less gaseous emissions (CO_2, CH_4, N_2O) in farm-related activities, food processing, and waste disposal. In the context of soil and water management, it is essential to replace all the nutrients harvested in agricultural produce, respond rationally (based on scientific concepts) to anthropogenic perturbations (e.g., land use and land use change), and predict environmental impact of humanity (I = PAT, P is population, A is affluence and T is technology) (Holdren and Eicher 1974), and reduce it.

The importance of enhancing soil health through management of SOC cannot be overemphasized (Lal 2016; Fig. 6). The average increase in grain yield by increase in SOC pool in the rootzone (0–20 cm) by 1 MgC ha^{-1} may be as much as (kg ha^{-1}) 100–300 for corn, 20–50 for soybean 20–70 for wheat, 30–60 for beans, 10–50 for rice, 80–140 for sorghum, and 30–70 for millet (Lal 2006, 2010a). Residue retention is essential to maintaining SOC (Karlen et al. 2009), and the latter to improving soil health and human nutrition (Lal 2009). Judicious management of the coupled cycling of H_2O, C, N, P, and S (Lal 2010b) is critical to improving biogeochemical cycling of elements and enhancing the use efficiency of inputs (Lal 2010a,b,c). The strategy is to adopt site-specific BMPs which create a positive soil C budget. The latter involves land application of biomass-C (C_i = biomass input by crop residues, deposition and use of inputs such as manure, composting, cover cropping, biochar, etc.) to exceed its losses (C_l = erosion, decomposition, leaching). Thus, the removal of crop residues for cellulosic ethanol (Malcolm and Aillery 2009) and other competing uses must be done prudently and rationally because there may be no such thing as free biofuels from crop residues (Lal and Pimentel 2007). Biomass from other sources such as switchgrass (Schmer et al. 2007), miscanthus, and other C-4 grasses (van der Weijde et al. 2013) may be better alternatives than that of harvesting crop residues from arable lands.

Future Strategies

Agriculture, as we know it, is on a crossroads. Low grain yields obtained under extractive farming practiced widely in

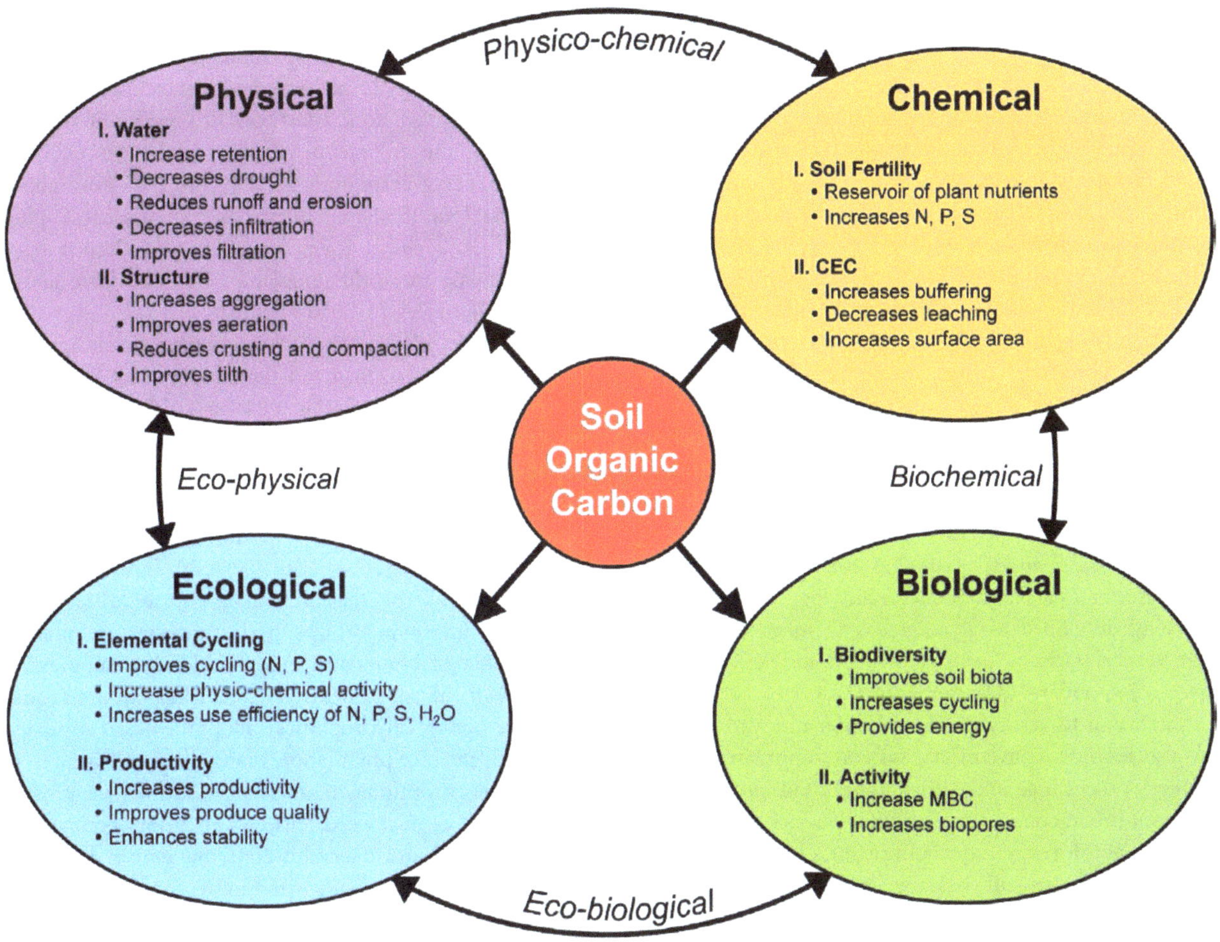

Figure 6. Impacts of soil organic carbon pool on improvements of soil physical, chemical, biological, and ecological properties.

SSA and SA, producing ~1 Mg ha^{-1} of cereals and 300–400 kg ha^{-1} of grain legumes, must be replaced by science-based practices of SI for producing 3–5 times the present crop yields. Yet, the ecological footprint of SI must be reduced. Any strategies of achieving high production must be reconciled with low environmental footprint. Indeed, the so-called C-civilization (Lal 2007) is also challenged by the issue of "peak soil" and "endangered or extinct soils" (Pimentel and Sparks 2000; Ditzler 2003; Tennesen 2015). In addition to its role as a medium for crop growth under traditional agriculture, soil is a source of numerous other ecosystem services including the source of antibiotics (Christina et al. 2013; Ling et al. 2015; Khapton 2016) and other pharmaceuticals.

Managing the carbon cycle

The severe disturbance of the global C cycle by anthropogenic activities is the root cause of numerous ecological issues of the 21st century, including the climate change (Fig. 4). The atmospheric C pool is approaching 840 Pg, in 2016, in comparison with 6000 Pg of the total pedological pool (SOC and SIC to 3-m depth including SOC pool in the Cryosols) (Batjes 2014; Le Quéré et al. 2015). Being a smaller pool compared with that of the terrestrial pool (pedologic plus biotic), the atmospheric pool is easily overwhelmed by anthropogenic perturbations. Thus, sustainable management of world soils (Lal 2004a,b) and biotic resources (Lal 2010c) is critical to mitigating the climate change through implementation of the initiative such as the "4 per Thousand" (UNFCCC, 2015; Chambers et al. 2016).

Urban agriculture (UA)

Global urban population is increasing rapidly, and about 70% of the 9.7 billion (6.8 billion) will live in urban centers by 2050 (U.N. 2015a,b). Land area under urban centers in the U.S. may be as much as 7–16 M ha by 2050 (USDA, 2016). Populations of mega cities has, and will continue to grow. For example, the expected increase in population (million) from 1950 to 2050 is 7.9 to 22.8 by a factor of 2.9 for New York, 2.9 to 14.3 by a factor of 5 for Rio de Janeiro compared with 0.1 to 14.3 by a factor of 158 for Nairobi, Kenya (Table 2). These megacities require a lot of food and other essentials for their sustenance. For example, it takes 30,000 ha of land to provide accommodation and infrastructure to 1 million people. The annual global increase of 75 million people would divert about 2.25 Mha of land mostly out of agricultural land use. Megacities with population of 10 million or more are 28 in 2015 and will be 41 in 2030 (U.N. 2015a,b). A city of 10 million requires 6000 Mg of food per day. Thus, a vast amount of plant nutrients (N, P, S etc.) are brought into the city and are not returned to the land. Therefore, urban agriculture (UA) must be developed to meet 20–25% of the food demands (e.g., fresh produce, vegetables, aquaculture) through production from within the urban and peri-urban areas. The strategy is to recycle nutrients and water, which are now a liability and cause of environmental pollution, through the strategies of soil-water-waste nexus (Lal 2016). Vertical farms, using aeroponics and aquaponics technology (without soil), are being developed to grow greens in mega cities. One such vertical farm is being developed in Newark, New Jersey (Despomier 2011; AeroFarms, 2016).

In addition to vertical farm and other innovative ideas, UA also involves conventional system comprising of cultivating, processing, and distributing food around and within an urban community including vegetables, livestock produce, agroforestry, aquaculture etc. It also includes backyard, rooftop and balcony gardening, community gardening in vacant lots and parks and on restored land from abandoned homesteads and parking lots. Among U.S. cities with well-established UA are Austin (TX), Boston (MA), Chicago (IL), Cleveland (OH), Minneapolis (MN), New York (NY), Portland (OR), San Francisco (CA), and Seattle (WA). UA must be expanded in all large, medium, and small urban communities.

Innovation in agriculture

Agriculture and food production have been a success story of the 21st century, especially since 1960s, and is capable of producing food and other victuals for the expected population of 9.7 billion by 2050 and 11.2 billion by 2100. The question is not whether the technology exists to produce enough food, but how to produce it without jeopardizing the environment? Indeed, the process thus far in enhancing agronomic yield in the U.S. from about 0.8 Mg ha^{-1} in 1750 to >10 Mg ha^{-1} at present is phenomenal. On a global scale, advances in agriculture have led to increase in world population (billion) from about 0.8 in 1750 to 7.4 in 2016 (Fig. 7).

Table 2. Temporal change in urban population of some mega cities (Adapted from UN, 2015b)

City	1950	1980	2015	2025	2050	Factor
New York	7.9	7.1	8.5	20.6	22.8	2.9
Rio de Janeiro	2.9	8.7	11.9	13.4	14.3	5
Los Angeles	2	3	18.7	13.7	16.4	8.3
Mexico City	3.1	8.8	21.2	21	24.3	8
Mumbai	2.9	8.1	26.1	26.4	42.4	14.6
Delhi	1.4	5.6	16.8	22	35.2	25.3
Lagos	0.3	4.4	23.2	24.2	32.6	112.5
Nairobi	0.1	0.9	3.8	5.9	14.3	158.3

Progressive innovations in agricultural technology include those based on: hand tools (1750–1850); draught animals and crop rotations (1850–1950); the Green Revolution technologies of input (fertilizer, irrigation) – responsive varieties (1975–2000); conservation agriculture, precision farming, perennial culture (Glover et al. 2010; Kane et al. 2016), genetically modified crops, and soil C sequestration (2000–2025) (Lal, 2004a, b). Innovative agricultural options for the immediate future (2015–2025) include SI, improving rhizospheric processes to enhance use efficiency of inputs and strengthen disease-suppressive attributes of the rootzone, develop soil-less culture (aquaponics, aeroponics, hydroponics), adopt the nexus approach (Lal 2016), enhance phytobiome management (Beattie et al. 2015), explore options of alternative sources of protein (e.g., algae, sea/duckweed, insects), develop agricultural robots (Astrand and Baerveldt 2002; Bangert et al. 2013), and strengthen smartphone extension/communication systems (Xue et al. 2010; Lantzos et al. 2013). Algae is a crop of the future, and microalgae is a potential source of proteins (Becker 2007). Microbial protein is another option for human and animal feed (Kuhad et al. 1994). Between 2025–2050, soil scientists must reach out and open new frontiers such as precision agriculture in a broader context, synthetic soils, soil-less farming, vertical farms, soil microbiomes, and space farming to understand pedological processes in a gravity-less environment (Fig. 7).

Saving soil

The overall strategy is to save the soil for ecosystem services other than agriculture and food production, produce more from less, and convert some of the extra land back to nature for sequestering C, strengthening biodiversity, improving water quality, and promoting nature conservation. The projected climate change will affect changes in soil-forming factors in terms of the input of biomass-C into soil, soil moisture and temperature regimes, hydrological processes (water retention and movement) and change in components of the hydrological cycle (rainfall, runoff, evapotranspiration) and erosivity of rainfall (intense rains with a high kinetic

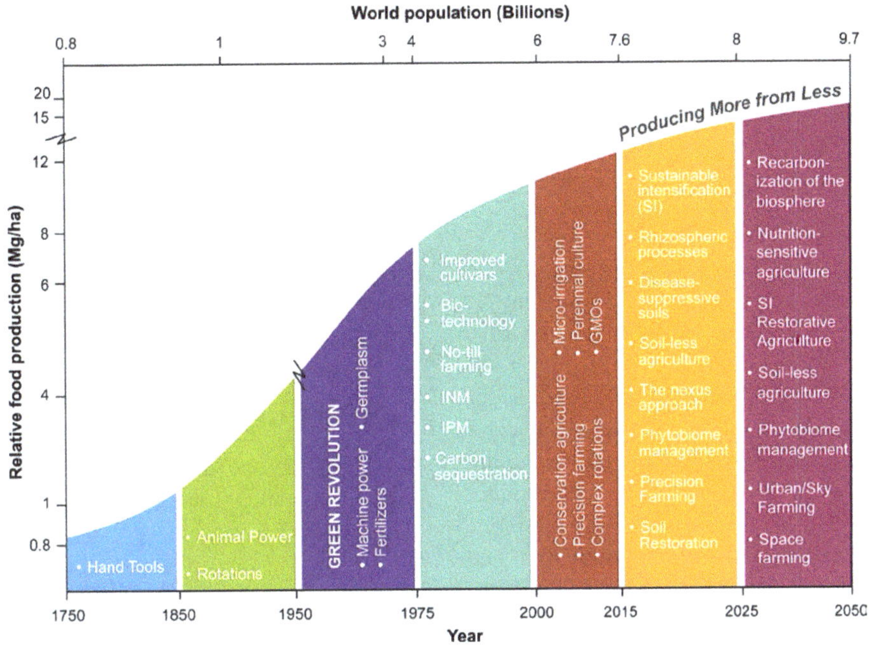

Figure 7. Technological evolutions in agriculture from 1750 to 2050. Revised and updated from Lal (2000).

energy and momentum). Thus, climate change would bring about changes in soil properties and processes (Buol et al. 1990; Goryachkin and Targulian 1990; Scharpenseel et al. 1990; Sombroek 1990; Brinkman and Sombroek 1996). Because of the shifts in biomes, agricultural activities will expand into the present boreal regions. Thus, understanding soil processes in cold environments (Olivia et al. 2014) under warmer climates (Brevik 2012) are important considerations. The temperature dependence of SOC decomposition (Kirschbaum 1995), feedback to climate change (Kirschbaum 2000; Davidson and Janssens 2006), and effects on microbial processes (Classen et al. 2015) are priority researchable issues. In the context of food security, it is pertinent to understand the site-specific impacts of climate change on soil properties and processes (Brevik 2013) in relation to agronomic production. Relation between climate variation and trends in crop yield (Vermeulen et al. 2012; Ray et al. 2015) will need to be established under a range of biophysical environments.

Space farming

Attempts toward human colonization of space (Moon and Mars) will be made between 2020 and 2050. Therefore, it would be pertinent to study space agriculture under micro- and hypo-gravity conditions. There is an increasing interest in soil-based agriculture involving bioregeneration support systems to produce food and generate oxygen (O_2). Yet, hydrodynamics and biogeochemical functioning of soil systems under low gravity environments of Moon (0.16 g), Mars (0.38 g), and the international space station (0 g) are not understood (Maggi and Pallud 2010). In conjunction with the study of soil processes, those of plant growth and morphogenesis must also be assessed under different gravity conditions (Hoson 2014; Kiss 2014).

Soil Organic Carbon as an Indicator of Soil Health and Environment Quality

There exists a strong need to enhance soil structure/soil health (Pagliai et al. 2002; Rojas et al. 2016) and restore environment quality for mitigating climate change and reducing risks of eutrophication. The strategy is to feed the soil to feed the plant, while restoring soil health and environment quality. The magnitude, quality, and dynamics of SOC are important indicators of soil health and environment quality (Ogle and Paustian 2005). Among specific attributes of SOC, microbial biomass C (MBC) is a pertinent indicator of changes in soil health and the attendant impact on environment quality (Hargreaves et al. 2003). Changes in particulate organic C (POC) are also sensitive to land use and land use change (Cambardella and Elliot 1991). Thus, SOC sequestration in agricultural soils is an important strategy to mitigate climate change (UNFCCC, 2015) and advance food security (Lal 2004a, 2010b). There are numerous technological options of withdrawing CO_2 from the atmosphere and storing it in the soil (Kane 2015). The potential of C sequestration in the terrestrial biosphere (soil and vegetation) is about 3.0 PgC yr^{-1} (Pacala and Socolow 2004; Le Quéré et al. 2015) to 3.8 PgC yr^{-1} (Lal 2010b). However, there are challenges to adapt technologies under site-specific conditions, such as conservation agriculture (Powlson et al. 2014), and issues of C saturation (Six et al. 2002).

Conclusion

The food demands of the growing world population, 7.4 billion in 2016 and projected to be 9.7 billion by 2050 and 11.2 billion by 2100, can be met by reducing the allocated land area for cereals and water resources for irrigation through adoption of site-specific practices of sustainable intensification and restoration of soil health. Sustainable intensification of agroecosystems can produce enough grains on 0.045 ha to feed one person for one year. Crop yields in Sub-Saharan Africa, South Asia, and elsewhere in the developing countries (e.g., the Caribbean) can be tripled or quadrupled despite the decrease in per capita cropland area to 0.15–0.17 ha by 2050. Soil health can be enhanced by improving the SOC concentration to 1.5–2.0% in the rootzone, as has also been proposed by the "4 per Thousand" program at COP21 in November 2015 in Paris. The global average cereal yield of 3.27 Mg ha^{-1} in 2005 can be increased to 5.0 Mg ha^{-1} by 2050, 6.0 Mg ha^{-1} by 2080, and 7.0 Mg ha^{-1} by 2100 through adoption of innovative technologies. In addition to sky farming based on soil-less culture and other systems of strengthening UA, understanding of rhizospheric processes can lead to creation of disease-suppressive soils, precision farming, and the nexus approach to enhancing use efficiency of resources. Innovative ideas include: adoption of SI, improvement of rhizospheric processes, creation of disease suppressive soils, use of soil-less culture based on the nexus approach, recarbonization of the biosphere, development of nutrition-sensitive agriculture, creating innovative technology for strengthening UA including space farming, and wide-spread use of precision agricultural practices. These innovations will launch the new era of agriculture in the 21st century (Fig. 7).

Soils will continue to play a pivotal role in feeding the world population of 9.7 billion by 2050, and beyond. Yet, soil-less agriculture (aeroponics, aquaponics, synthetic soils) will also become increasingly important through new innovations in sky farming and other forms of urban agriculture. Simultaneously, space agriculture based on bioregenerative systems will be an emerging field of soil science. Recycling biowaste to produce food and generate oxygen will necessitate understanding of biogeochemical and hydrological processes under low gravity and no-gravity conditions. Nonetheless, the human dimensions (political and social factors) will play an increasingly important role in achieving food security especially with regard to reducing waste, increasing food access, identifying new sources of protein (e.g., algae) and creating peace and political stability.

Although sustainable management of soil is crucial to mitigating climate change and advancing food security, stewardship of earthly resources and understanding of the biogeochemical and hydrological processes of soils of celestial bodies (Moon, Mars) will emerge important themes of future research. Soil science must evolve to meet challenges of the 21st century.

Conflict of Interest

The author is the director of the Carbon Management and Sequestration Center of the Ohio State University, Columbus, OH.

References

AeroFarms. 2016. World's largest vertical farm grows without soil, sunlight or water in Newark. The Guardian. Available at https://www.theguardian.com/environment/2016/aug/14/world-largest-vertical-farm-newark-green-revolution. (accessed 14 August 2016).

Alexandratos, N., and J. Bruinsma. 2012. World Agriculture Towards 2030/2050. 2012 Revision. ESA Working Paper No. 12-03. FAO, Rome, Italy.

Alley, W. M., T. E. Reilly, and O. L. Franke. 1999. Sustainability of groundwater resources. U.S. Geol. Surv. Circ., 1186:79.

Amelinckx, A. 2015. Even without drought, we are depleting groundwater at an alarming pace. Modern Farmers. Available at http://modernfarmer.com/2015/07/ogallala-aquifer-depletion/.

Astrand, B., and A. J. Baerveldt. 2002. An agricultural mobile robot with vision-based perception for mechanical weed control. Auton. Robots 13:21–35.

Bangert, W., A. Kielhorn, F. Raha, A. Albert, P. Biber, S. Grzonka, et al. 2013. Field-robot-based agriculture: "RemoteFarming.1" and "BoniRob-Apps". 71st conference LANDTECHNIK-AgEng 2013, Hannover, Germany.

Batjes, N. H. 2014. Total carbon and nitrogen in soils of the world. Eur. J. Soil Sci. 65:10–21.

Beattie, G. A., J. E. Leach, K. E. Eversole, et al. 2015. Phytobiomes: a roadmap for research and translation. 15 pp. American Phytopathological Society, St Paul, MN.

Becker, E. W. 2007. Micro-algae as a source of protein. Biotechnol. Adv. 25:207–210.

Brevik, E. C. 2012. Soil and climate change: gas fluxes and soil processes. Soil Horiz., 53:12–23. doi:10.2136/sh12-04-0012.

Brevik, E. C. 2013. The potential impact of climate change on soil properties and processes and corresponding influence on food security. Agriculture 3:398–417.

Brinkman, R., and W. G. Sombroek. 1996. The effect of global change on soil conditions in relation to plant growth and food production. Pp. 49–63 *in* F. Bazzaz and W. Sombroek, eds. Global climate change and agricultural production. Direct and indirect effects of changing hydrological, pedological and plant physiological processes. FAO, Rome, Italy. ISBN-13:9789251039878

Bruinsma, J. 2009. The resource outlook to 2050. FAO Expert Meeting, 24–26 June 2009.

Buol, S. W., P. A. Sanchez, J. M. Kimble, and S. B. Weed. 1990. Predicted impact of climatic warming on soil properties and use. Am. Soc. Agron. Spec. Publicaiton 53:71–82.

Cambardella, C. A., and E. T. Elliot. 1991. Particulate soil organic-matter changes across a grassland cultivation sequence. Soil Sci. Soc. Am. J. 56:777–783.

Chambers, A., R. Lal, and K. Paustian. 2016. Soil carbon sequestration potential of US croplands and grasslands: Implementing the 4 per Thousand Initiative. J. Soil Water Conserv. 71:68A–76A.

Christina, A., V. Christopher, and S. Bhore. 2013. Endophytic bacteria as a source of novel antibiotics: an overview. Pharmacogn. Rev. 7:11–16.

Classen, A. T., M. K. Sundquist, J. A. Henning, G. S. Newman, J. A. M. Moore, M. A. Cregger, et al. 2015. Direct and indirect effects of climate change on soil microbial and soil microbial-plant interactions: what lies ahead. Ecosphere 6:1–21.

Davidson, E. A., and I. A. Janssens. 2006. Temperature-sensitivity of soil carbon decomposition and feedbacks to climate change. Nature 440:165–173.

De, A., R. Bose, A. Kumar and S. Mojumdar. 2014. Worldwide pesticide use. Pp. 5–6 *in* De, A., R. Bose, A. Kumar and S. Mojumdar eds. Targeted delivery of pesticide using biodegradable polymeric nanoparticles. Springer, New Delhi, India.

Despomier, D. 2011. The vertical farm: feeding the world in the 21st century. 304 pp. Thomas Dunne Books, New York.

Ditzler, C. 2003. Endangered Soils. National Cooperative Soil Survey Newsletter, Issue 25, November 2003.

FAO. 2009. How to feed the world 2050. High Level Expert Forum, 12-13 October 2009. FAO, Rome, Italy.

FAO. 2015a. World hunger falls to under 800, eradication is next goal. FAO News 27, May 2015. Available at https://www.wfp.org/news/news-release/world-hunger-falls-under-800-million-eradication-next-goal-0.

FAO. 2015b. The state of food and agriculture. Social protection and agriculture: breaking the cycle of rural poverty. 129 pp. FAO, Rome, Italy.

FAO. 2015c. The state of food insecurity in the world. 62 pp. FAO/IFAD/WFP, Rome, Italy.

FAOSTAT. 2016. Global crop production. 17 August 2016. FAO, Rome, Italy.

Fyfe, W. S. 1997. Earth sciences for the 21st century, evolve or die. GFF 199:85–90.

Geohive. 2016. Population of the entire world, yearly 1950–2100. Available at http://geohive.com/earth/his_history3.aspx.

Gleesen, T., Y. Wada, M. F. P. Bierkens, and L. P. H. Van Beek. 2012. Water balance of global aquifers revealed by groundwater footprint. Nature 488:197–200.

Glover, J. D., J. P. Reganold, L. W. Bell, J. Borevitz, E. C. Brummer, E. S. Buckles, et al. 2010. Increased food and ecosystem security via perennial gains. Science 328:1638–1639.

Goryachkin, S. V., and V. O. Targulian. 1990. Climate-induced changes and subpolar soils. Dev. Soil Sci. 20:191–209.

Hargreaves, P. R., P. C. Brookes, G. J. S. Ross, and P. R. Poulton. 2003. Evaluating soil microbial biomass carbon as an indicator of long-term environmental change. Soil Biol. Biochem. 35:401–407.

Haub, C. 2011. How many people have ever lived on Earth. Population Reference Bureau. Available at http://www.prb.org/Publications/Articles/2002/HowManyPeopleHaveEverLivedonEarth.aspx.

Holdren, J. P., and P. R. Eicher. 1974. Human population and the global environment. Am. Sci. 62:282–292.

Horton, R. 2014. Why the sustainable development goals will fail. The Lancet 383:2196.

Hoson, T. 2014. Plant growth and morphogenesis under different gravity conditions: relevance to plant life in space. Life 4:205–216.

IFDC. 2004. Global and regional data on fertilizer production and consumption, 1961/62–2002/03. 73 pp. FODC, Muscle Shoals, AL.

IPCC. 2014. Climate change 2014, synthesis report. 151 pp. WMO, Geneva, Switzerland.

Kane, D. 2015. Carbon sequestration potential on agricultural lands: a review of current sciences and available practices. National Sustainable Agriculture Cooperation. Available at http://sustainableagriculture.net/publications

Kane, D. A., P. Rogé, and S. S. Snapp. 2016. A systematic review of perennial stable crops literature using topic modeling and bibliometric analysis. PLoS ONE 11:e0155788.

Karlen, D., R. Lal, R. F. Follett, J. M. Kimble, J. L. Hatfield, J. M. Miranowski, et al. 2009. Crop residues: The rest of the story. Environ. Sci. Technol. 43:8011–8015.

Kerr, R. A. 2009. Northern India's groundwater is going, going, going. Science 325:798–798.

Khapton, S. 2016. First new antibiotic in 30 years discovered in major breakthrough. The Telegraph, 7 January 2016.

Kirschbaum, M. U. F. 1995. The temperature-dependence of soil organic matter decomposition, and the effet of global warming on soil organic carbon storage. Soil Biol. Biochem., 27:753–760.

Kirschbaum, M. U. F. 2000. Will changes in soil organic carbon act as a positive or negative feedback on global warming. Biogeochemistry 48:21–51.

Kiss, J. Z. 2014. Plant biology in reduced gravity on the Moon and Mars. Plant Biol. 16:12–17.

Kuhad, R. C., A. Singh, K. K. Tripathi, R. K. Saxena, and K. E. L. Eriksson. 1994. Microorganisms as an alternative source of protein. Nutr. Rev. 55:65–75.

Lal, R. 2000. Controlling greenhouse gases and feeding the world through soil management. Distinguished University Lecture, 17 February 2000, OSU, Columbus, OH

Lal, R. 2004a. Soil carbon sequestration impacts on global climate change and food security. Science 304:1623–1627.

Lal, R. 2004b. Soil carbon sequestration to mitigate climate change. Geoderma 123:1–22.

Lal, R. 2006. Enhancing crop yields in the developing countries through restoration of soil organic carbon pool in agricultural lands. Land Degrad. Dev. 17:197–209.

Lal, R. 2007. Soil Science and the carbon civilization. Soil Sci. Soc. Am. J. 71:1095–1108.

Lal, R. 2009. Soil degradation as a reason for inadequate human nutrition. Food Secur. 1:45–57.

Lal, R. 2010a. Enhancing eco-efficiency in agroecosystems through soil C sequestration. Crop Sci. 50:S120–S131.

Lal, R. 2010b. Managing soils and ecosystems for mitigating anthropogenic carbon emissions and advancing global food security. Bioscience 60:708–721.

Lal, R. 2010c. Managing soils for a warming earth in a food-insecure and energy-staved world. J. Plant Nutr. Soil Sci. 173:4–15.

Lal, R. 2016. Global food security and nexus thinking. J. Soil Water Conserv. 71:85A–90A.

Lal, R., and D. Pimentel. 2007. Biofuels from crop residues. Soil Till. Res. 93:237–238.

Lantzos, T., G. Koykoyris, M. Salampasis, and A. Theodoridis. 2013. "FarmManager: an Android application for the management of small farms". 6th International Conference on Information and Communication Technologies in Agriculture, Food and Environment (Haicta 2013), 8, 587–592, doi: 10.1016/j.protcy.2013.11.084.

Le Quéré, C., R. Moriarty, R. M. Andrew, J. G. Canadell, S. Sitch, J. I. Korsbakken, et al. 2015. Global Carbon Budget 2015. Earth Syst. Sci. Data 7:349–396.

Ling, L. L., T. Scheider, A. J. Peoples, A. L. Spoering, I. Engles, B. P. Conlon, et al. 2015. A new antibiotic kills pathogens without detectable resistance. Nature, 517:455–459.

Lipinski, B., C. Hansen, J. Lomax, L. Kitinoja, R. Waite, and T. Searchinger. 2013. Reducing food loss and waste. Working Paper, Installment of creating a Sustainable Food Future. World Resources Institute, Washington, D.C.

Loveland, P., and J. Web. 2003. Is there a critical level of organic matter in the agricultural soils of the temperate regions? A review. Soil Till. Res. 70:1–18.

Maggi, F., and C. Pallud. 2010. Space agriculture in micro- and hypo-gravity: a comparative study of soil hydraulics and biogeochemistry in a cropping unit on Earth, Mars, the Moon and the space station. Planet. Space Sci. 58:1996–2007.

Malcolm, S., and M. Aillery. 2009. Cellulosic ethanol from crop residues is no free lunch? Farm Economy, 1 December 2009.

Nunley, P. 2013. From farm to fork to landfill: food waste and consumption in America. Pitzer College Senior Thesis Paper 37. Available at http://scholarship.claremont.edu/pitzer_theses/37.

OECD-FAO. 2015. Agriculture outlook 2015–2024. 143 pp. OECD Publishing, Paris, France.

Ogle, S. M., and K. Paustian. 2005. Soil organic carbon as an indicator of environmental quality at the national scale: Inventory monitoring methods and policy relevance. Can. J. Soil Sci. 85:531–540.

Oldeman, R. 1994. The global extent of soil degradation. Pp. 99–118 *in* D. J. Greenland and I. Szabolas, eds. Soil resilience and sustainable land use. CAB International, Wallingford, UK.

Olivia, M., P. Pereira, J. Bockheim, and A. Navas. 2014. Preface: Soil processes in cold-climate environments. Solid Earth 5:1205–1208.

Overstreet, L. F., and J. DeJong-Huges. 2009. The importance of soil organic matter in cropping systems of the northern Great Plains. University of Minnesota Extension, Minneapolis, MN.

Pacala, S., and R. Socolow. 2004. Stabilization wedges: solving the climate problem for the next 50 years with current technologies. Science 305:968–972.

Pagliai, M., N. Vignozzi, and R. Jones. 2002. The soil pore system as an indicator of soil quality. Sustain. Land Manag. Environ. Prot. 35:71–82.

Patrick, M., J. S. Tenywa, P. Fbanyat, M. M. Tenywa, D. N. Mubiru, T. A. Basamba, et al. 2013. Soil organic carbon thresholds and N management in tropical agroecosystems: concepts and prospects. J. Sustain. Dev., 6:31–43.

Pimentel, D., and D. Sparks. 2000. Soils as an endangered ecosystem. Bioscience 50:947–947.

Powlson, D. S., C. M. Stirling, M. L. Jat, B. G. Gerard, C. A. Palm, P. A. Sanchez, et al. 2014. Limited potential of no-till agriculture for climate change mitigation. Nat. Clim. Chang. 4:678–683.

Ray, D. K., J. S. Gerber, G. K. MacDonald, and P. C. West. 2015. Climate variation explains a third of global crop yield variability. Nat. Commun. 6:5989–5989.

Rojas, R. V., M. Achouri, J. Maroulis, and L. Caon. 2016. Healthy soils, a prerequisite for sustainable food security. Environ. Earth Sci. 75:108.

Ruel-Bergeron, J. C., G. A. Stevens, J. D. Sugimoto, F. F. Roos, M. Ezzati, R. E. Black, et al. 2015. Global update and trends of hidden hunger 1995-2011: The hidden hunger index. PLoS ONE 10:e0143497.

Scharpenseel, H. W., M. Schomaker, and A. Ayoub, eds. 1990. Soils on a warmer earth. Effects of expected climatic change on soil processes, with emphasis on the tropics and subtropics. Dev. Soil Sci., 20:1–274.

Schmer, M. R., K. P. Vogel, R. B. Mitchell, and R. K. Perrin. 2007. Net energy of cellulosic ethanol from switchgrass. Curr. Issue 105:464–469.

Seremesic, S., D. Milosev, I. Djalovic, T. Zeremski, and J. Ninkov. 2011. Management of soil organic carbon in maintaining soil productivity and yield stability of winter wheat. Plant Soil Environ. 57:216–211.

Six, J., R. Conant, E. Paul, and K. Paustian. 2002. Stabilization mechanisms of soil organic matter: Implications for C-saturated soils. Plant Soil 241: 155–176.

Sombroek, W. G. 1990. Soils on a warmer earth: the tropical regions. Dev. Soil Sci. 20:157–174.

Tennesen, M. 2015. Rare earth. Science 346:692–695.

U.N. 1992. Agenda 21, U.N. Conference on Environment and Development, 3-14 June 1992, Rio de Janeiro, Brazil.

U.N. 2000. Millennium development goals and beyond 2015. U.N. Department of Public Affairs, New York, NY.

U.N. 2015a. Transforming our world: The 2030 agenda for sustainable development. A/RES/70/1. U.N., New York, NY.

U.N. 2015b. World population prospects: key findings and advance table, 2015 revision. 59 pp. United Nations, New York, NY.

UNFCCC. 2015. The "4 per Thousand" Initiative: Soils for Food Security and Climate. Available at http://agriculture.gouv.fr/4-pour-1000-et-si-la-solution-climat-passait-par-les-sols-03

USDA. 2016. Second Biennial Report of the United States of America under the UNFCC. (https://unfccc.int/files/national_reports/biennial_reports_and_iar/submitted_biennial_reports/application/pdf/2016_second_biennial_report_of_the_united_states_.pdf)

Vermeulen, S. J., B. M. Campbell, and J. S. I. Ingram. 2012. Climate change and food systems. Environ. Resour. J. 37:195–222.

van der Weijde, T., C. L. A. Karnel, A. F. Torres, W. Vermerris, O. Dolstra, R. F. Visser, et al. 2013. The potential of C-4 grasses for cellulosic biofuel production. Front. Plant Sci., 107:1–18. doi:10.3389/fpls.2013.00107.

Wild, A. 2003. Soils, land and food: managing the land during the 21st century. Cambridge University Press, Cambridge, UK.

Xue, H., L. Gao, L. Wang, Y. Wang, and W. Li. 2010. The wireless automatic control system for orchard irrigation based on Smart-phone Platform. Hunan Agric. Sci. 1:1–19.

Yang, X., Y. Chan, S. Pacenka, W. Gao, P. Sui, and T. S. Steenhuis. 2015. Recharge and groundwater use in the North China Plain for six irrigated crops for an eleven year period. PLoS ONE 10:e0115269.

5

Agronomic evaluation of seedbed and mulching alternatives with plant spacing for dry-season fluted pumpkin in coarse-textured tropical soil

Sunday E. Obalum[1], Ifeoma G. Edeh[1], Odirichi N. Imoh[2], Ogechi M. Njoku[1], Ifeyinwa M. Uzoh[1], Vincent N. Onyia[2], Charles A. Igwe[1] & Jose M. Reichert[3]

[1]Department of Soil Science, University of Nigeria, Nsukka 410001, Enugu State, Nigeria
[2]Department of Crop Science, University of Nigeria, Nsukka 410001, Enugu State, Nigeria
[3]Soils Department, Federal University of Santa Maria, Av. Roraima, 1000, 97105–900 Santa Maria, Brazil

Keywords
Close spacing, fresh leaf yield, seedbed configuration, surface mulch, tropical dry season

Correspondence
Sunday E. Obalum, Department of Soil Science, University of Nigeria, Nsukka 410001, Enugu State, Nigeria.

E-mails: sunday.obalum@unn.edu.ng or ijewelle@yahoo.com

Abstract

Fluted pumpkin (*Telfairia occidentalis*) has high agronomic and food relevance in the tropics. Production of this vegetable in the dry season with irrigation is increasingly popular and is mostly on droughty soils, thus requiring research on appropriate combination of soil moisture conservation/management practices. This study evaluated the interactions among seedbed configuration (flat and raised), surface mulching (grass-mulched and no-mulch) and plant spacing (100^2, 50^2 and 25^2 cm^2, designated wide, intermediate and close spacing, respectively) on fluted pumpkin performance in a sandy-loam soil in southeastern Nigeria during 2014/2015 dry season. Seedling emergence (ratio of emerged to expected seedlings) was highest and lowest in flat/mulched/wide and flat/no-mulch/close combinations, respectively. Main effects showed higher seedling emergence in mulched (67.9%) than no-mulch plots (53.1%) and in wide (70.8%) and intermediate spacing (64.6%) compared to close spacing (46.0%). Fresh leaf yield assessed biweekly during 6–14 weeks after sowing was generally highest and lowest in flat/mulched/close and flat/no-mulch/wide combinations, respectively. Seedbed configuration had no distinct effect on leaf yield; hence, flat/mulched/close option was similar to raised/mulched/close option. Yields from mulched/wide and no-mulch/close options were similar with a mean (2735 kg/ha) approximating the overall mean (2868 kg/ha). Flat/mulched/close combination is the best management practice to grow fluted pumpkin with irrigation on droughty soils during the dry season of the tropics. Where this best practice is compromised due to mulch or seed scarcity, the ensuing no-mulch/close and mulched/wide, respectively can support dry-season production of pumpkin in droughty tropical soils at average levels.

Introduction

Among the vegetable crops that are indigenous to the tropical Africa, fluted pumpkin (*Telfairia occidentalis* Hook F.) is the most cultivated in the region mainly in southeastern Nigeria owing to its edible succulent shoots and leaves. This vegetable is widely eaten, serving as a vital vegetable supplement in the carbohydrate-dominated diet of the people. The tender vines, leaves and seeds are consumed by humans, whereas the mature vines are an important fiber source in animal diet. Today, fluted pumpkin is cultivated in almost all parts of tropical Africa not only as backyard crop but also as commercial crop during the rainy and dry seasons of the region.

Fluted pumpkin is a rich source of protein, oil, vitamins and minerals (Aregheore 2007; Nwite et al. 2013). This vegetable has also gained medicinal recognition as blood purifiers (Aletor et al. 2002), and as food supplement to maintain good health among resource-poor farmers in developing countries.

Research has shown that some production techniques could affect fluted pumpkin yield. These include time of sowing (Asiegbu 1985), tillage practice (Awodun 2007a), manure rate (Ogar and Asiegbu 2005; Awodun 2007b), plant staking (Egun 2007; Chukwudi and Agbo 2013), and cutting frequency (Ogar and Asiegbu 2005; Chukwudi and Agbo 2013). In Nigerian savanna where bulk of agriculture is done, low status of plant nutrients and poor soil physical condition constitute serious agronomic constraints (Salako 2003). For the low soil fertility, combined use of organic amendments and inorganic fertilizers to improve productivity of fluted pumpkin has been advocated (Ndor and Dauda 2013; Iren et al. 2014). The use of organic amendments (poultry droppings) alone has also been found to be effective for fluted pumpkin particularly in acid Ultisols of the Nigeria savanna (Ndor and Dauda 2013; Nwite et al. 2013).

The problem of poor soil physical condition for fluted pumpkin could also be ameliorated by right combination of organic and inorganic fertilizers (Nwite et al. 2012). One major factor still lowering the crop's productivity in tropical Africa is disregarding this vegetable as major crop during the rainy farming season in the region. Hence, it is often grown with residual soil moisture and manual irrigation in the dry season.

Considering the crop's role in food and nutrition security, its skeletal production in the rainy season needs to be expanded into the dry season of the tropics using viable resource management practices. Fluted pumpkin is known to prefer a loose, friable soil with ample humus and shade (Olaniyi and Odedere 2009). Although the crop is also known for drought tolerance, its productivity might be increased with appropriate seedbed-mulch combination to enhance moisture storage in the root zone while adopting the plant density that could provide adequate protective cover and optimize resource use.

Seedbed configuration plays an important role in conservation of soil water. Raised seedbed systems have been used since time immemorial by farmers in many parts of the world (Govaerts et al. 2007). Their use has traditionally been linked with water management issues; to drain the cropped farmland of excess water or to irrigate crops via furrows in semi-arid and arid regions (Sayre 2004). Flat seedbeds are also commonly used in crop production but they offer smaller inducement to rooting and drainage, and this often makes for poorly structured soil. Such conditions deter the establishment of adequate protective crop cover early in the season. Narkhede et al. (2015) reported that opening of furrows between rows increased the growth and yield attributes of cotton (*Gossypium* spp.) compared to flatbed sowing.

Mulching is also an important soil and water management practice in dry season farming. The many benefits of surface mulch to the soil and the crop grown therein are well known, including maintenance of uniformly high soil water status thus reducing irrigation frequency especially in dry season production (Obalum et al. 2011a). Odiaka et al. (2008) noted that in Nigeria fluted pumpkin is not commonly grown with mulch as the crop is assumed to adequately cover the soil. Nevertheless, this concept can be contested for the droughty soils of the savanna zones. In the present location at Nsukka, for instance, surface mulch was found to increase the yield of soybean (*Glycine max* M.) that provided cover to the soil, and this was even under rainfed conditions (Obalum et al. 2011b). There have been similar reports of mulch-induced increases in fresh and dry weights of vegetables including fluted pumpkin (Okugie and Ossom 1988; Mahadeen 2014; Aniekwe 2015).

Furthermore, the practice of adopting a plant spacing that gives the desired plant density for simultaneously conserving resources and optimizing crop yield is an important factor in agronomic production. Improper plant spacing may cause either too dense or too sparse a population resulting in reduced yield, whereas optimum plant density ensures efficient use of resources to give maximum yield. Some researchers have reported increased above-ground yield of vegetable crops at close spacing (Firoz et al. 2009; Uko et al. 2013). In Nigeria, fluted pumpkin is traditionally spaced 100 cm × 100 cm (Nwite et al. 2013), but this is deemed too wide especially under conditions of limiting soil water in the dry season.

The trio of seedbed configuration, surface mulch and plant spacing on a droughty soil in a tropical environment is thought to produce interaction effect on performance of dry-season fluted pumpkin, but this has yet been reported. This hypothesis was therefore tested in the dry season on a coarse-textured soil at Nsukka in southeastern Nigeria with the crop's performance indexed by seedling emergence and by periodic fresh leaf yield.

Materials and Methods

Study area

The experiment was conducted at the University of Nigeria Teaching & Research Farm, Nsukka, located at 06°52′N, 07°24′E. The site is on an altitude of approximately 400 m above sea level. Nsukka has a tropical humid climate characterized by the existence of two distinct seasons; the rainy season (April–October) and the dry season (November–March). The study was done during the 2014/2015 dry season. Mean annual total rainfall is about 1600 mm with the mean annual evapotranspiration often exceeding this value in recent years. Temperature is uniformly high all-year-round, with mean minimum and maximum annual values of 21°C and 31°C, respectively, though the latter can approach 35°C during the hottest months. Rarely does the relative humidity fall outside the range 55–80% throughout the year.

The soil at the experimental site, derived from false-bedded sandstones, is deeply weathered, brownish red in color, coarse-textured, excessively porous and well drained. The soil is classified as Acrisol by the World Reference Base for Soil Resources (IUSS Working Group 2006). Soil moisture is usually recharged with the onset of rains and sometimes there can be moisture surplus of about 260 mm at the peak of the rainy season, which depletes to a mean deficit of 650 mm in the dry season (Mbagwu 1987). Within the control section, the soil has an ustic moisture regime and an isohyperthermic thermal regime. The prevailing vegetation is grassland typifying the area as derived savanna in southeastern Nigeria.

Field study

The field study was sited some 40 m away from where a similar study with information on key properties of the soil of the area had been conducted (Obalum et al. 2011c). Just before setting up the field for this study, six replicate samples of the topsoil (0–20 cm) were randomly collected from the site. Analysis of these samples reveal that the topsoil has mean sand, silt and clay contents of 750, 70 and 180 g/kg soil, respectively thus belonging to the textural class of sandy loam. The soil was acidic with mean pH of 4.8 while having mean organic carbon concentration of 17.88 g/kg soil. The mean value of total nitrogen in the soil was 0.56 g/kg while the mean cation exchange capacity of the soil was 12.40 cmol/kg. Furthermore, the contents of the exchangeable bases were for K^+, Ca^{2+}, Mg^{2+} and Na^+ 0.04, 3.20, 0.80 and 0.03 cmol/kg soil, respectively.

The experimental field (35 m × 20 m; ~0.07 ha) was cleared, ploughed and harrowed with tractor, after which the soil was further manually harrowed to provide good tilth.

Three factors were considered including seedbed configuration, surface mulching and plant spacing. The experiment had 12 treatments derived from the factorial combination of two forms of seedbed configuration (flat seedbed and raised seedbed) with two status of surface mulching (no-mulch and mulched) and three levels of plant spacing (100 cm × 100 cm, 50 cm × 50 cm and 25 cm × 25 cm, designated wide, intermediate and close spacing, respectively; Table 1). Treatments were replicated four times in a randomized complete block design (RCBD) and were nested in a 2 × 2 × 3 split-split plot design; with seedbed configuration as main plots, surface mulching as sub-plots and plant spacing as sub-sub-plots. Blocks were spaced ~0.75 m apart. The treatment plots (sub-sub-plots) each measuring 3.00 m × 3.00 m were demarcated from one another using earthen bunds 0.30 m high and 0.30 m wide.

Poultry droppings was applied uniformly to all sub-sub-plots at the rate of 10 t/ha, adopted after Ndor and Dauda (2013) and Nwite et al. (2013) in studies of fluted pumpkin responses to soil amendments on taxonomically similar soils in the Nigerian savanna. Three weeks later, the sub-sub-plots were pulverized to make flat seedbeds from among which raised seedbeds (level-top beds ~0.25 m high) were prepared according to the field layout of the experiment. Mulch (comprising dry leaves of a grass species mixture) was applied at 10 t/ha on the plots meant for surface mulching; this was done on 9th Jan., 2015. Three days after mulching, fluted pumpkin seeds were sown with the various plant spacings at one seed per hill. The plant spacing of this study gave plant density of 9, 36 and 121 plants per sub-sub-plot in the wide, intermediate and close spacing, respectively, corresponding to 10,000, 40,000 and 160,000 plants/ha, respectively.

The plots were manually watered daily from the time of mulching through maturity until about 11 weeks after sowing (WAS) when rains set in. The watering was done to predetermined field capacity of the soil, assumed to be its moisture content at –33 kPa (Obalum et al. 2011d), down to 60-cm depth. Weeding was done at two-week intervals during the

Table 1. Summary of the various combinations of the factors and ensuing treatments of the study.

Seedbed configuration × Surface mulching		Plant spacing		
		Wide	Intermediate	Close
Flat	No-mulch	Flat/no-mulch/wide	Flat/no-mulch/ intermediate	Flat/no-mulch/close
	Mulched	Flat/mulched/wide	Flat/mulched/ intermediate	Flat/mulched/close
Raised	No-mulch	Raised/no-mulch/wide	Raised/no-mulch/ intermediate	Raised/no-mulch/close
	Mulched	Raised/mulched/wide	Raised/mulched/ intermediate	Raised/mulched/close

Wide – 100 cm × 100 cm; Intermediate – 50 cm × 50 cm; Close – 25 cm × 25 cm.

Table 2. Interaction of seedbed configuration, surface mulching and plant spacing on seedling emergence of fluted pumpkin.

Seedbed form	Mulch status	Plant spacing	Seedling emergence (%)	Number of emerged seedlings per plot
Flat	No-mulch	Wide	66.7	6.0
Flat	No-mulch	Intermediate	51.4	18.5
Flat	No-mulch	Close	27.7	30.5
Flat	Mulched	Wide	86.1	7.8
Flat	Mulched	Intermediate	63.9	23.0
Flat	Mulched	Close	49.3	54.2
Raised	No-mulch	Wide	58.3	5.2
Raised	No-mulch	Intermediate	70.8	25.5
Raised	No-mulch	Close	43.4	47.7
Raised	Mulched	Wide	72.2	6.5
Raised	Mulched	Intermediate	72.2	26.0
Raised	Mulched	Close	63.4	69.7
$LSD_{(0.05)}$			25.5	

Wide – 100 cm × 100 cm; Intermediate – 50 cm × 50 cm; Close – 25 cm × 25 cm.

study. Agronomic parameters assessed were seedling emergence and fresh leaf yield at two-week intervals. For each sub-sub-plot having different plant densities as per treatment and hence different number of expected plants (NEP) per plot, number of emerged seedlings (NES) was monitored and percentage seedling emergence (%SE) calculated as:

$$\frac{NES}{NEP} \times 100$$

To assess fresh leaf yield, harvestable fresh leaves were cut off and weighed. This was done biweekly as from 6 WAS; this first harvest corresponded to 23rd Feb., 2015. Harvest was stopped at 14 WAS when leaf yield of fluted pumpkin normally peaks (Nwite et al. 2013). Since dry season and manual irrigation lasted until 11 WAS, the pumpkin was sustained by irrigation in the bulk (roughly four-fifth) of its growth phase and by rainfall in the rest.

Data analysis

Statistical analyses of data was performed by analysis of variance (ANOVA) procedure for a split-split-plot experiment in an RCBD using the software GENSTAT 3, 7.2 Edition (VSN International Ltd., Hemel Hempstead, UK). Significant treatment means were separated and compared using Least Significant Difference (LSD), and all inferences were made at 5% level of probability.

Results

Seedling emergence

The factors under investigation (seedbed configuration, surface mulching and plant spacing) showed interactions in seedling emergence with the highest and lowest percentage emergence due to treatments flat/mulched/wide and flat/no-mulch/close, respectively (Table 2). These results show that, under limited soil moisture in the dry season, achieving maximum germination and crop stand for fluted pumpkin depends not on seedbed configuration but on pre-sowing soil mulch status and plant spacing.

Hence, the interaction of surface mulching and plant spacing is also presented (Table 3), showing highest and lowest percentage emergence under mulched/wide and no-mulch/close plots, respectively. Notably, seedling emergence tended to decrease with closeness of planting spacing, and surface mulch appeared to compensate for this effect. This role of mulch could be said to be pronounced with closeness of plant spacing, as evident in the fact that no-mulch and mulched plots showed similar values for seedling emergence under both wide and intermediate spacing but not under close spacing (Table 3). The weak role of seedbed configuration vis-à-vis the importance of pre-sowing mulch status and plant spacing in seedling emergence is evident in their main effects shown in Figure 1.

Fresh leaf yield of fluted pumpkin

Seedbed configuration, surface mulching and plant spacing produced significant interaction on fresh leaf yield of fluted pumpkin, with consistently highest values under treatment flat/mulched/close; the only exception was at 6 WAS when the highest values were recorded under raised/mulched/close (Table 4). The data also show that the lowest yields were obtained under treatments flat/no-mulch/wide and raised/no-mulch/wide, except at 10 WAS. Yield increases due to treatment flat/mulched/close relative to flat/no-mulch/wide averaged over 250% (i.e., three and a half-fold increase); that due to raised/mulched/close relative to raised/no-mulch/wide was about 242%. Again, these results

Table 3. Interaction of surface mulching and plant spacing on seedling emergence of fluted pumpkin.

Mulch status	Plant spacing	Seedling emergence (%)	Number of emerged seedlings per plot
No-Mulch	Wide	62.5	5.6
No-Mulch	Intermediate	61.1	22.0
No-Mulch	Close	35.6	39.1
Mulched	Wide	79.2	7.1
Mulched	Intermediate	68.1	24.5
Mulched	Close	56.4	62.0
$LSD_{(0.05)}$		17.2	

Wide – 100 cm × 100 cm; Intermediate – 50 cm × 50 cm; Close – 25 cm × 25 cm.

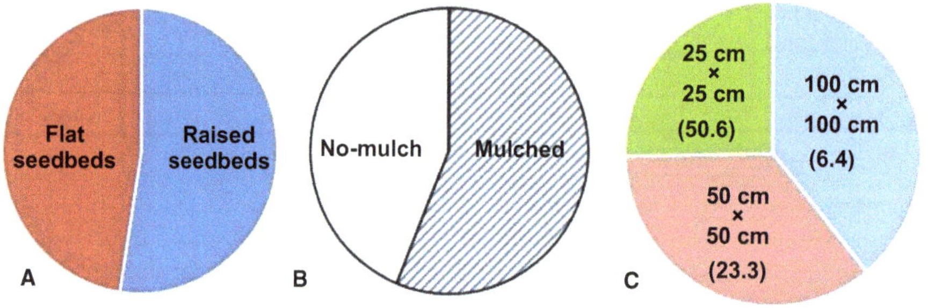

Figure 1. Main effects of (A) seedbed configuration, (B) surface mulching and (C) plant spacing on seedling emergence of fluted pumpkin (numbers in parenthesis represent emerged seedlings). Different annotated colors define distribution.

show that the fresh leaf yield of fluted pumpkin under the dry season and droughty soil condition of this study depended less on seedbed configuration and more on soil mulch status and plant spacing.

The interaction of surface mulching and plant spacing is thus shown (Fig. 2). Mulched/close plots consistently produced higher fresh leaf yields compared to the rest while no-mulch/wide plots generally gave the lowest leaf yields. No-mulch/close plots showed rather comparable values with mulched/wide plots. The yields from these two plots averaged 2735 kg/ha while the overall mean yield was 2868 kg/ha. The mulched/intermediate plots out-yielded no-mulch/close plots on some of the harvest dates (Fig. 2). Also the mulched/wide plots out-yielded the no-mulch/intermediate plots on most of the harvest dates. These observations further underscore the role of surface mulch in the interaction presented.

To validate the inferences made on the relative contributions of the three factors investigated, their main effects on fresh leaf yield of fluted pumpkin are shown (Figs. 3–5). Seedbed configuration affected yield with higher values from raised and flatbed seedbeds at 6 and 8 WAS, respectively, beyond which both forms of seedbed produced similar yields with a trend of the former tending to out-yield the latter (Fig. 3). Yield was always higher in mulched plots compared to their no-mulch counterparts, with this mulch effect being more pronounced during 6–10 WAS and being insignificant only at 12 WAS (Fig. 4). On the average, fresh leaf yield increased in the mulched plots relative to the no-mulch plots by about 74%. The main effect of plant spacing shows that, on most of the assessment dates, leaf yield was higher due to close spacing compared with intermediate and wide spacings for which values were similar on most of the dates (Fig. 5). On the average, leaf yield due to close spacing represented increases over intermediate and wide spacings by about 41% and 86%, respectively.

Discussion

Although raised seedbeds provide both better drainage and greater rooting space compared to flat ones, the major aim of factoring in seedbed configuration in this dry season trial was to enhance not drainage but rooting space, the effect on pumpkin root growth of which was evaluated with leaf yield as proxy. Thus, the weak role of seedbed configuration in seedling emergence was reasonably due to the prevailing dry season condition of the study whereby there was no need to drain the seedbeds of excess water. The higher seedling emergence in mulched plots than no-mulch plots could be attributed to the effect of surface mulch in moderating and in making uniform the soil hydrothermal regime (El-Shaikh and Fouda 2008; Hochmuth et al. 2015), deemed crucial for seed germination. Wide spacing promoted seedling emergence probably because of the lesser competition to access environmental

Table 4. Interaction of seedbed configuration, surface mulching and plant spacing on fresh leaf yield (kg/ha) of fluted pumpkin.

Seedbed form	Mulch status	Plant spacing	6	8	10	12	14	Mean
			Weeks after sowing					
Flat	No-mulch	Wide	396	463	741	2778	3037	1483.0
Flat	No-mulch	Intermediate	519	500	870	2000	6074	1992.6
Flat	No-mulch	Close	1185	1148	1815	3444	7889	3096.2
Flat	Mulched	Wide	1000	1556	2667	2037	6741	2800.2
Flat	Mulched	Intermediate	2741	1630	3315	2222	8667	3715.0
Flat	Mulched	Close	4926	4148	3630	3907	9444	5211.0
Raised	No-mulch	Wide	500	630	889	1519	2667	1241.0
Raised	No-mulch	Intermediate	1204	981	1278	2667	4000	2026.0
Raised	No-mulch	Close	1833	1222	1630	2556	5222	2492.6
Raised	Mulched	Wide	1093	981	1833	2185	6667	2551.8
Raised	Mulched	Intermediate	2593	1667	290	2630	4852	2406.4
Raised	Mulched	Close	5704	2333	3304	3093	6778	4242.4
$LSD_{(0.05)}$			1088	887	1057	1989	3496	1037

Wide – 100 cm × 100 cm; Intermediate – 50 cm × 50 cm; Close – 25 cm × 25 cm.

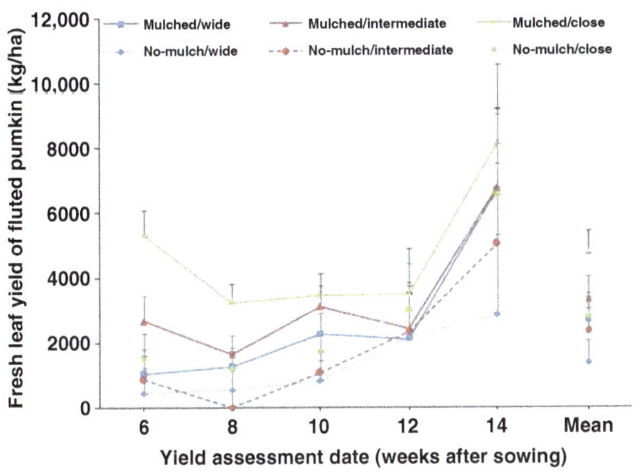

Figure 2. Interaction of surface mulching and plant spacing on fresh leaf yield of fluted pumpkin. Error bars represent least significant difference at $P \leq 5\%$.

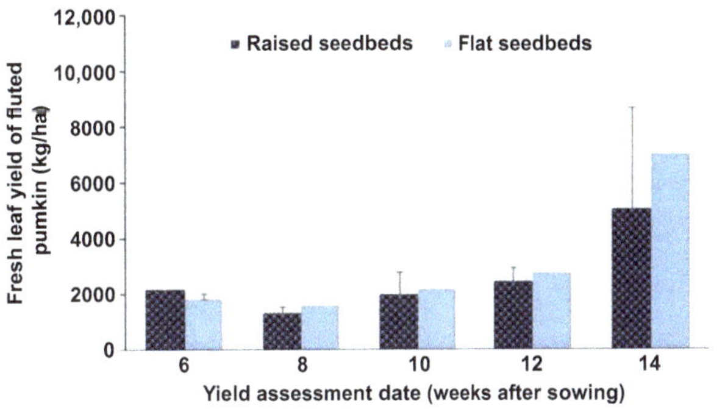

Figure 3. Main effect of seedbed configuration on fresh leaf yield of fluted pumpkin. Error bars represent least significant difference at $P \leq 5\%$.

resources needed for germination. Elhag and Osman (2013) similarly reported that wider spacing had positive effect on germination and percentage emergence of onion (*Alluim cepa* L.) at northern Sudan. However, Moniruzzaman et al. (2007), Firoz et al. (2009) and Asaduzzaman et al. (2012) reported that plant spacing had no effect on germination percentage of some vegetable crops in Bangladesh.

The fresh leaf yield of fluted pumpkin was generally insensitive to seedbed configuration. It could be that fluted pumpkin, being a shallow-rooting crop, does not need enhanced rooting space offered by raised seedbeds for increased productivity. Contrary to these results, Ugale et al. (1995) and Deshmuh et al. (2013) reported that ridges and furrows produced higher yields of deeper-rooting pearlmillet (*Pennisetum glaucum* L.) compared to flatbeds under rainfed and irrigation systems, respectively in India. Hence, the fact that the present study thrived on manual irrigation in the dry season was only to partially blame.

The effect of seedbed configuration on fresh leaf yield of fluted pumpkin was inconsistent early in the assessment period, being higher on raised seedbeds than flat seedbeds at 6 WAS and vice versa at 8 WAS. The initially higher yield observed for raised seedbeds might be due to greater availability of soil moisture when the structure of the tilled, pulverized soil was yet to fully reform. Conversely, the subsequent higher yield on flat seedbeds and tendency for such during 10–14 WAS suggests that, after the soil structure reformation, these flat seedbeds sustained more favorable soil moisture than the raised seedbeds that likely had part of the irrigation water percolating down below the reach of the plants. Because of its coarse texture, the soil can be excessively macroporous (Obalum and Obi 2014).

The positive effect of mulch on fluted pumpkin yield was pronounced early in the assessment period, but this effect was insignificant at 12 WAS. These results are

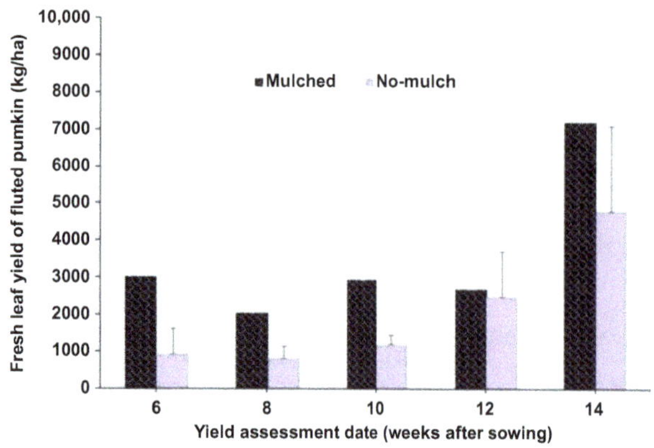

Figure 4. Main effect of surface mulching on fresh leaf yield of fluted pumpkin. Error bars represent least significant difference at $P \leq 5\%$.

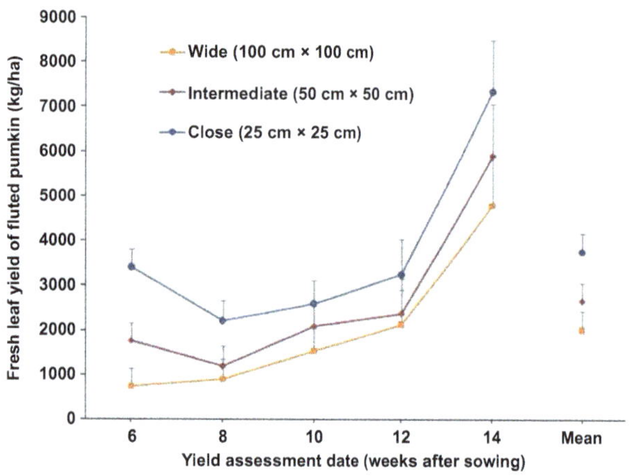

Figure 5. Main effect of plant spacing on fresh leaf yield of fluted pumpkin. Error bars represent least significant difference at $P \leq 5\%$.

explained by the degree of soil coverage due to the crop before and after attaining maturity. The onset of rains about a week earlier might have also contributed to the diminished role of mulch at 12 WAS, to a lesser extent though, by the reason of the higher yield due to mulch later at 14 WAS. This effect of mulching likely resulted from temperature modification, evaporation reduction and hence more uniform soil moisture regime in the mulched compared to the no-mulch plots (Mbagwu 1991; Qin et al. 2015). Our findings agree with the results of Okugie and Ossom (1988) that mulch increased the yield of fluted pumpkin in southern Nigeria. Hudu et al. (2008) and Nwokwu and Aniekwe (2014) reported similar results for other vegetables including watermelon (*Citrullus lanatus*) and tomato (*Lycopersicon esculentum* Mill.) under irrigated conditions in Nigeria. Increases in fresh and dry yields of vegetable crops have been attributed to mulch-induced high soil moisture and the associated optimal transpiration, nutrient uptake and increased photosynthesis required for plant growth (Mahadeen 2014; Aniekwe 2015).

The progressive increase in yield of fluted pumpkin with reduction in plant spacing are explained by the increase in number of plants per plot with reduction in plant spacing, represented by number of emerged seedlings highlighted in Figure 1. Similar results have been reported for other leafy and fruit vegetables (Firoz et al. 2009; Uko et al. 2013). Contrary reports of higher crop yields at low plant density than at high plant density are usually for tuber crops (Arsenault and Malone 1999; Mauromicale et al. 2003).

In the mulching-×-spacing interaction, seedling emergence was highest and lowest under mulched/wide and no-mulch/close, respectively, yet these two plots gave similar yields. This similarity was despite the gap in their plant densities – number of plants per plot (see Table 3), thus highlighting the rather equal strengths of mulching and spacing in the interaction. With regard to conserving soil moisture, close spacing plausibly compensated for the absence of mulch by providing coverage just as surface mulch did for wide spacing.

In a fertility trial in a similar environment in southeastern Nigeria, Nwite et al. (2013) showed that the mean fresh leaf yield of fluted pumpkin from plots amended with poultry droppings at 10 t/ha and grown without mulch at wide spacing was 3575 kg/ha between 6 and 14 WAS. Considering the dry season condition of the present study, the duo of mulched/wide and no-mulch/close with a mean yield of 2735 kg/ha can not only be said to represent the overall mean yield (2868 kg/ha) but to also represent average yields under conventional practice of no-mulch and wide spacing in the rainy, growing season. The observation that highest yield came from mulched/close plots is attributed to the droughtiness of the coarse-textured soil of this study, exacerbated by the prevailing dry season. Even under high plant density, therefore, mulch can enhance fluted pumpkin yield under conditions of limited soil moisture and rainfall.

The insensitivity of fresh leaf yield to seedbed configuration notwithstanding, the effect of surface mulching on leaf yield was more pronounced on flat seedbeds compared to raised ones (see Appendix 1). Also, of the three plant spacings, narrow spacing showed the greatest tendency for higher yield from flat seedbeds than raised ones (see Appendix 2). It is therefore unsurprising that the interaction of the three factors investigated showed that mulched/close combination did better on flat seedbeds than on raised seedbeds.

Conclusion

The data presented support the production of fluted pumpkin in droughty tropical soils with irrigation in the dry season using the combined soil and agronomic management practice of growing on tilled flat seedbeds with surface mulch (dry grass at 10 t/ha) and at a close spacing of 25 cm × 25 cm. When grown for up to 14 weeks, an average of three and a half-fold increase in mean fresh leaf yield could be achieved with this best practice relative to the conventional practice of growing on flat seedbeds without mulch and at a wide spacing of 100 cm × 100 cm. A close substitute would be to grow the crop on raised seedbeds still with surface mulch and close spacing, if raising the seedbeds rather than leaving them flat subsists the farmer beyond the dry season. In either case of cultivating on flat or raised seedbeds, surface mulching can potentially compensate for wide spacing just as growing without mulch can be compensated for by close spacing. Under conditions of limited mulch material or pumpkin seeds, therefore, irrigated plots without mulch but at close spacing and those with mulch but at wide spacing could have similar agronomic value in terms of fresh leaf yield of fluted pumpkin; a value somewhat in-between the best and the conventional practices.

Acknowledgments

The assistance of the batch of students in their 4th year of the B.Agric. Degree Programme of the Faculty of Agriculture of the University of Nigeria Nsukka who offered the course SSC441 (Agricultural Meteorology, Irrigation & Drainage) in the 2014/2015 Academic Session in setting up the field experiment and in the manual irrigation and data collection is gratefully acknowledged. Deserving of special mention is the class leader, Mr Odinaka C. Ezenwanne, who co-ordinated the activities of the students that participated in the project. The support of the Farm Operations Unit of the University of Nigeria Teaching & Research Farm in providing the tractor used for ploughing/harrowing is also gratefully acknowledged.

Conflict of Interest

None declared.

References

Aletor, O., A. Oshodi, and K. Ipinmoroti. 2002. Chemical composition of common leafy vegetables and functional properties of their leaf protein concentrates. Food Chem. 78:63–68.

Aniekwe, N. 2015. Comparative effects of organic and plastic mulches on the environment, growth and yield of okra in a Derived Savanna Zone of Nigeria. Int. J. Sci. Res. 4:1860–1864.

Aregheore, E. 2007. Voluntary intake, nutrient digestibility and nutritive value of foliage of fluted pumpkin (*Telfairia occidentialis*) – haylage mixtures by goats. Livestock Res. Rural Dev. 19, Article No. 56.

Arsenault, W., and A. Malone. 1999. Effects of nitrogen fertilization and in row see piece spacing on yield of three potato cultivars in Prince Edward Island. Am. J. Potato Res. 6:227–229.

Asaduzzaman, M., M. M. Hasan, M. M. Hasan, M. Moniruzzaman, and M. H. K. Howlander. 2012. Effect of bulb size and plant spacing on seed production of onion (*Allium cepa* L.). Bangladesh J. Agric. Res. 37:405–414.

Asiegbu, J. 1985. Influence of planting date on growth and productive life of fluted pumpkin *Telfairia occidentalis* Hook F. Trop. Agric. 64:281–284.

Awodun, M. 2007a. Effect of tillage practices on the growth and yield of fluted pumpkin (*Telfaria occidentalis* Hook F). Afr. Crop Sci. Conf. 8:463–465.

Awodun, M. 2007b. Effect of poultry manure on the growth, yield and nutrient content of fluted pumpkin (*Telfaria occidentalis* Hook F.). Asian J. Agric. Res. 1:67–73.

Chukwudi, U., and C. Agbo. 2013. Effect of trellis height and cutting frequency on leaf and fruit yield of fluted pumpkin (*Telfairia occidentalis* Hook F.) over two seasons in Nsukka. Crop Sci. Soc. Nigeria Conf. 1:348–357.

Deshmuh, S., J. Patel, and A. Patel. 2013. Ensuing economic gains from summer pearlmillet (*Pennisetum glaucum* L.) due to different dates of sowing and land configuration. Afr. J. Agric. Res. 8:6337–6343.

Egun, A. 2007. Comparative marketable leaf yield of staked and unstaked pumpkin (*Telfairia occidentalis*) in a tropical ultisols. Stud. Home Comm. Sci. 1:27–29.

Elhag, A., and H. Osman. 2013. Effect of plant spacing on onion (*Allium cepa* L.) seeds quality. Univ. J. Appl. Sci. 1:52–55.

El-Shaikh, A., and T. Fouda. 2008. Effect of different mulching types on soil temperature and cucumber production under Libyan conditions. Misr. J. Ag. Eng. 25:160–175.

Firoz, Z., M. Aslam, M. Uddin, and S. Khatun. 2009. Effect of sowing time and spacing on lettuce seed production in hilly region. Bangladesh J. Agric. Res. 34:531–536.

Govaerts, B., K. D. Sayre, K. Lichter, L. Dendooven, and J. Deckers. 2007. Influence of permanent raised bed planting and residue management on physical and chemical soil quality in rainfed maize/wheat systems. Plant Soil 291:39–54.

Hochmuth, G., R. Hochmuth, and S. Olson. 2015. Polyethylene Mulching for Early Vegetable Production in North Florida. Horticultural Sciences Department, UF/IFAS Extension.

Hudu, A., K. Futuless, and N. Gworgwor. 2008. Effect of mulching intensity on the growth and yield of irrigated tomato (*Lycopersicon esculentum* Mill) and weed infestation in Semi-Arid Zone of Nigeria. J. Sustain. Agric. 21:37–45.

Iren, O. B., N. M. John, and E. A. Imuk. 2014. Effects of sole and combined applications of organic manures and urea on soil properties and yield of fluted pumpkin (*Telfairia occidentalis* Hook F). Nigerian J. Soil Sci. 24:125–133.

IUSS Working Group. 2006. World Reference Base for Soil Resources. IUSS-ISRIC-FAO. World Soil Resources Report, 103. Food and Agricultural Organization (FAO), Rome.

Mahadeen, A. 2014. Effect of polyethylene black plastic mulch on growth and yield of two summer vegetable crops under rain-fed conditions under semi-arid region conditions. Am. J. Agric. Biol. Sci. 9:202–207.

Mauromicale, G., P. Signorelli, A. Lema, and S. Foti. 2003. Effects of intraspecific competition on yield of early potato grown in Mediterranean environment. Am. J. Potatoes Res. 80:281–288.

Mbagwu, J. S. C. 1987. Some hydrological characteristics of soil formed on a toposequence in southeastern Nigeria and consequences for dry-season irrigation. Beitr. Trop. Landwirtsch. Veterinarmed. 25:25–31.

Mbagwu, J. S. C. 1991. Mulching an ultisol in southern Nigeria: effects on physical properties and maize and cowpea yields. J. Sci. Food Agric. 57:517–526.

Moniruzzaman, M., M. Uddin, and A. Choudhury. 2007. Response of okra seed crop to sowing time and plant spacing in south eastern hilly region of Bangladesh. Bangladesh J. Agric. Res. 32:393–402.

Narkhede, W. N., S. K. Nayak, G. S. Khazi, and B. H. Jaware. 2015. Effect of land configuration and nutrient management on productivity of Bt cotton. Int. J. Trop. Agric. Ser. Publ. 33:2293–2297.

Ndor, E., and N. Dauda. 2013. Growth and yield performances of fluted pumpkins (*Telferia occidentalis* Hook F.) Under organic and inorganic fertilizer on ultisols of North Central Nigeria. Glob. J. Plant Ecophysiol. 3:7–11.

Nwite, J. C., E. N. Ogbodo, S. E. Obalum, V. C. Igbo, and C. A. Igwe. 2012. Short-term response of soil physical properties of an Ultisol, and nutrient composition of fluted pumpkin to organic and inorganic fertilizer mixtures. J. Biol. Agric. Healthcare 2:195–204.

Nwite, J. C., C. I. Keke, S. E. Obalum, J. B. Essien, M. U. Anaele, and C. A. Igwe. 2013. Organo-mineral amendment options for enhancing soil fertility and nutrient composition and yield of fluted pumpkin. Int. J. Vegetable Sci. 19:188–199.

Nwokwu, G., and L. Aniekwe. 2014. Impact of different mulching materials on the growth and yield of watermelon (*Citrullus lanatus*) in Abakaliki, Southeastern Nigeria. J. Biol. Agric. Healthcare 4:22–30.

Obalum, S. E., and M. E. Obi. 2014. Measured versus estimated total porosity along structure-stability gradients of coarse-textured tropical soils with low-activity clay. Environ. Earth. Sci. 72:1953–1963.

Obalum, S. E., G. I. Ezenne, Y. Watanabe, and T. Wakatsuki. 2011a. Contemporary global issue of rising water scarcity for agriculture: the quest for effective and feasible soil moisture and free-water surface conservation strategies. J. Water Resource Prot. 3: 166–175.

Obalum, S. E., C. A. Igwe, M. E. Obi, and T. Wakatsuki. 2011b. Water use and grain yield response of rainfed soybean to tillage-mulch practices in southeastern Nigeria. Sci. Agricola 68:554–561.

Obalum, S. E., I. M. Okpara, M. E. Obi, and T. Wakatsuki. 2011c. Short-term effects of tillage-mulch practices under sorghum and soybean on organic carbon and eutrophic status of a degraded Ultisol in southeastern Nigeria. Trop. Subtrop. Agroecosys. 14:393–403.

Obalum, S. E., C. A. Igwe, Hermansah, M. E. Obi, and T. Wakatsuki. 2011d. Using selected structural indices to pinpoint the field moisture capacity of some coarse-textured agricultural soils in southeastern Nigeria. J. Trop. Soils 16:151–159.

Odiaka, N. I., M. O. Akoroda, and E. C. Odiaka. 2008. Diversity and production methods of fluted pumpkin (*Telfairia occidentalis* Hook F.); Experience with vegetable farmers in Makurdi, Nigeria. Afr. J. Biotechnol. 7:944–954.

Ogar, E., and J. Asiegbu. 2005. Effect of fertilizer rates and cutting frequency on the marketable vegetable and fruit yields of in fluted pumpkin in Southern Nigeria. Agro-Sci. 4:66–69.

Okugie, D. N., and E. M. Ossom. 1988. Effect of mulch on the yield, nutrient concentration and weed infestation of the fluted pumpkin, *Telfairia occidentalis* Hook F. Trop. Agric. (Trinidad) 65:202–204.

Olaniyi, J., and M. Odedere. 2009. The effects of mineral N and compost fertilizers on the growth, yield and nutritional values of fluted pumpkin (*Telfairia occidentalis*) in Southwestern Nigeria. J. Anim. Plant Sci. 5:443–449.

Qin, W., C. Hu, and O. Omenema. 2015. Soil mulching significantly yields and water and nitrogen use effeciencies of maize and wheat: a meta-analysis. Sci. Rep. 5:162–170.

Salako, F. 2003. Soil physical conditions in Nigerian savannas and biomass production. Lecture given at the College on Soil Physics, Trieste, March 3–21, 2003.

Sayre, K. D. 2004. Raised bed cultivation. Pp. 1433–1436 *in* R. Lal, ed. Encyclopedia of soil science. Marcel Dekker Inc, New York.

Ugale, S., A. Wani, B. Patil, and M. Bhingrde. 1995. Performance of different sowing methods of pearl milliet on lighter type of soils under rainfed conditions. J. Maharashtra Agric. Univ. 20:311–312.

Uko, A. E., I. A. Udo, and J. O. Shiyam. 2013. Effects of poultry manure and plant spacing on the growth and yield of water leaf. J. Agron. 12:146–152.

6

Flood and drought tolerance in rice: opposite but may coexist

A. N. M. Rubaiyath Bin Rahman & Jianhua Zhang

School of Life Sciences and State Key Laboratory of Agrobiotechnology, The Chinese University of Hong Kong, Hong Kong, China

Keywords
aus, drought, ethylene signaling, flood, rice, stress tolerance.

Correspondence
Jianhua Zhang, School of Life Sciences and State Key Laboratory of Agrobiotechnology, The Chinese University of Hong Kong, Hong Kong, China.

E-mail: jhzhang@cuhk.edu.hk

Funding Information
We thank the National Basic Research Program of China (2012CB114300), Shenzhen Overseas Talents Innovation and Entrepreneurship Funding Scheme (The Peacock Scheme) and Hong Kong Research Grant Council (AoE/M-05/12, CUHK 473512, 474313, and 14122415) for their grant support.

Abstract

Flood and drought are the two most prevalent abiotic stresses causing major yield reduction globally. In the last decade, molecular mechanisms of flood tolerance in rice have been revealed with successful release of flash flood-tolerant varieties to farmers. However, despite extensive research, the breakthrough of drought tolerance is still to come. In this review, we have examined the distribution and population types of drought- and flood-tolerant rice accessions, synthesized recent progresses of flood and drought tolerance research, and proposed a hypothesis that the molecular mechanisms of both drought and flood tolerance may be regulated by cross-talked pathways and coexist in *aus* subpopulation. We conclude that it is the time to mine the key regulator(s) of drought tolerance through *de novo* assembly of drought-tolerant *aus* landrace(s) with other molecular approaches and develop drought-tolerant rice using genome manipulation weaponry.

Introduction

Flood and drought are the two most prevalent climate disasters that hamper growth and development of all crop plants, causing major yield reduction globally. These two abiotic stresses can completely destroy crop production in extreme conditions, and consequently both of stresses are considered as key determinants of global food security. Moreover, climate change is also projected to undermine global food security (IPCC, 2014) as climate change will exaggerate drought and flood occurrence and severity globally. For an instance, just 2 years ago California experienced most severe drought in record history where a recent report predicts more severe and more frequent droughts and floods may occur in California near future (Yoon et al. 2015).

Another growing global concern is fresh water limitation for agriculture. Agriculture accounts most share of freshwater withdrawals. In some dry areas it is even more than 90% where globally 70% (WWDR, 2014). Increasing demand of food for a still growing population may lead to severity of freshwater scarcity for agriculture. Rice is usually grown in areas with hot and wet climate. However, it is also projected that 15–20 million hectares of irrigated rice may experience some degree of water scarcity by 2025 (Bouman et al. 2007). Thus, the motto "*more crop per drop*" has become a trend of development of water-saving technologies either through agricultural practices or varietal development programs for virtually all the crops (Marris 2008).

Poaceae is the single angiosperm family having the most important sources of food crops like rice, maize, wheat, sorghum, oat, rye, millet, etc. All these crops account for the most calorie consumption of human and raised animals of the entire globe (Paterson et al. 2005). Among these, half of the world population consumes rice where rice accounts over 75% calorie demand in some developing countries (Maclean et al. 2002). Moreover, rice is one of the unique crops that has tolerant genotypes of both flash and prolonged floods. In the last decade, molecular mechanisms of flash (Xu et al. 2006) and prolonged (Hattori et al. 2009) flood adaptation of rice have been revealed with successful release of flash flood-tolerant varieties to farmers in several countries in Asia (Ismail et al. 2013). Unfortunately, despite extensive efforts of past decades, development and release of drought-tolerant varieties is largely scarce, although large number of genes has been reported to gain drought tolerance in rice (Todaka et al. 2015). Moreover, recent technological advancement, that is, sequencing of thousands of rice accessions, etc., has generated giga-data (Alexandrov et al. 2014) but breakthrough of drought tolerance research is still to come.

During the course of adaptation, some ecotype can efficiently sense particular stress and activate downstream adaptive mechanisms, hence becoming tolerant to certain extent, thus known as stress-tolerant genotype/accession. Genetic resources are the raw material of any crop improvement program (McCouch 2013) where success of the improvement program largely depends on the identification of true stress tolerance genotypes (Ismail et al. 2012) and gene mining strategy. So screening of hundreds of thousands genotypes from genebank accessions of rice has been done for differ types of biotic (Jackson 1995) and abiotic stresses (Angaji et al. 2010; Platten et al. 2013; Torres et al. 2013).

Drought and flood are considered as opposite abiotic stresses based on water status, either shortage or excess. So generally drought and flood are studied, analyzed and even reviewed separately. However, we have found that growing evidences may suggest that these two apparent opposite stresses may be regulated by some shared signaling pathways. So in this review, we analyze drought and flood tolerance together in different angles. Since molecular mechanisms of different types of flood tolerances are already known, whereas large-scale field level drought (Torres et al. 2013) and anaerobic germination tolerance (Angaji et al. 2010) screening results are available, so we first examine the distribution and population type of drought and flood-tolerant rice accessions. Then we review recent developments of flood and drought tolerance and related players and synthesize a hypothesis that the molecular mechanisms of both drought and flood tolerance may be regulated by cross-talked pathways and coexist in *aus* subpopulation of rice. Finally, we describe possible ways to mine the key regulator(s) of drought tolerance and development of drought-tolerant rice.

Distribution of Flood- and Drought-Tolerant Genotypes

Anaerobic germination tolerance

Flood may occur at any growth and developmental stages of rice. The destructive effect of flood first appears on seed germination and seedling establishment stage. Except rice, none of the major crops is flood tolerant during seed germination and seedling establishment (Ismail et al. 2012). Rice can efficiently germinate even in anaerobic condition but the tolerance is only limited to coleoptile development. Neither shoot nor root develops in anaerobic condition, and hence completely fail to establish a seedling (Ismail et al. 2012).

Hundreds of germplasms have been screened earlier for AG (anaerobic germination) tolerance in different studies (Yamauchi et al. 1993; Ling et al. 2004). However, these attempts largely failed mainly due to lack of true stress-tolerant genotypes (Ismail et al. 2012). Then scientists at IRRI (International Rice Research Institute) screened over 8000 genotypes for AG tolerance and identified only a few varieties with over 70% survival (Angaji et al. 2010). After repeated experiments, numbers of tolerant genotypes were limited to only 6. It includes Khaiyan and Kalonchi (Bangladesh), Nanhi (India), Khao Hlan On (Myanmar), Ma-Zhan Red (China), and Cody (USA) (Angaji et al. 2010). Cody was an improved variety developed in Missouri in 1944 from cross between Lady Wright and Colusa. Colusa was released in Louisiana in 1917 by selection from Chinese variety introduced from Italy in 1909 (Johnston 1958). Colusa was temperate *japonica* type as reveled in SNPs and population structure analysis (Kim and Tai 2013). Parentage of Lady Wright was unknown but may have similarity with temperate *japonica* as showing similarity with Smooth No. 3, a glabrous accession of also unknown origin (Kim and Tai 2013). Since Cody, an improved variety developed through breeding process where source of AG tolerance is obscure, so we focus on distribution of the rest 5 genotypes. Interestingly, out of these five, four of the tolerant genotypes are from three adjacent countries (India, Bangladesh, and Myanmar) where Bangladesh only shares borders with both India and Myanmar. Moreover, these countries also share culture and thousand years of rice cultivation history together.

Flash flood tolerance

During seedling stage, sudden heavy rainfall can lead to flash flood in rainfed lowland rice ecosystem. Around 20 million hectares of rice growing area, mostly in Bangladesh and India is at risk of being submerged (Ismail et al. 2012). Most of the rice varieties are highly sensitive to submergence and so they die within a week of complete submergence. FR13A, an *aus* variety was developed from pure-line selection of the flash flood-tolerant landrace, Dhalputtia (HilleRisLambers and Vergara 1982), can survive more than 2 weeks of complete submergence (Xu et al. 2006). QTL studies identified that the submergence tolerance trait is located on chromosome 9 where single QTL explains nearly 70% of the phenotypic variation. Further fine mapping and subsequent positional cloning identified SUB1 locus having three ethylene response-like transcription factors. Complementation of SUB1A in intolerant *japonica* rice confirmed the involvement of SUB1A in submergence tolerance (Xu et al. 2006).

The presence of SUB1A allele both in landraces/traditional varieties and in wild rice (both in *Oryza nivara* and *O. rufipogon*) was examined in different studies (Fukao et al. 2009; Li et al. 2011; Niroula et al. 2012; Iftekharuddaula et al. 2015). Interestingly, the presence of both wild and SUB1A rice together is limited to Bangladesh, India, and Nepal (Pucciariello and Perata 2013). Recently a study using 300 germplasm accessions of Bangladesh has identified eight accessions with survival ranging from 71.2% to 95.4% after 16 days complete submergence (Iftekharuddaula et al. 2015). Interestingly, five accessions of Bangladesh (DG1-349, Kalojoma, DSL-78-8, Damsi and Putidepa) showed higher submergence tolerance (83–95% survival) even better than FR13A although they do not possess same resistance allele as FR13A (Iftekharuddaula et al. 2015). Aside from these areas, landraces having SUB1A allele were also reported in Sri Lanka, Thailand, and Vietnam. So Pucciariello and Perata (2013) hypothesized that SUB1A-1 might have been introgressed into domesticated rice from wild rice where introgression probably occurred in the Ganges Basin, and subsequent spread of the SUB1A-1 to other parts of South Asia due to human migration.

Prolonged flood tolerance

Rice is an unique crop that has ecotypes that are well-adapted in prolonged (1–5 months) deepwater (1–5 meters) flood; hence they are called deepwater rice (Cattling 1992). Deepwater rice adapts with flood using a completely opposite strategy from SUB1A rice. After rising flood water, deepwater rice continuously elongates internode and leaf sheath as its top leaves always keep above the water (Cattling 1992). Molecular mechanisms of prolonged flood tolerance was identified from Bangladeshi deepwater rice, Gowai 38-9 (Hattori et al. 2009). QTL studies and subsequent fine mapping and complementation in intolerant genotypes identified involvement SNORKEL genes in deepwater adaptation. SNORKEL (SK1 and SK2) allows rice to adapt in deepwater condition by enhancing GA responsiveness (Hattori et al. 2009).

Three wild rice species were examined for the presence of SNORKEL. Both *O. rufipogon* and *O. nivara* possess SK1 and SK2. However, SK2 of *O. nivara* is truncated. Amazonian wild rice, *O. glumaepatula* also possesses SK2 and an SK2-like gene, but lack of SK1 (Hattori et al. 2009). Several deepwater rice were examined for the presence of SNORKEL genes and interestingly, all of them possess both SK1and SK2 (Hattori et al. 2009). However, other than deepwater rice, there is no report of presence of SNORKEL genes in other ecotypes or subspecies. So the rice reference genome, Nipponbare completely lacks such gene or allele.

Distribution of deepwater rice cultivation is mainly confined to the depression or river basin of the Ganges-Brahmaputra of Bangladesh (11.42%), India (3.2%) and Nepal (7.9%), Irrawady of Myanmar (5.75), the Mekong of Vietnam (2.77%), Cambodia (8.0%), and Chao Phraya of Thailand (3.54) (Cattling 1992; Huke and Huke 1997). Country-wise percentage of deepwater rice cultivation area is shown in parenthesis above. Some deepwater rice is also cultivated in upper and middle basins of the Niger River in West Africa but it is only 5% of total deepwater rice cultivated area (Cattling 1992). However, African deepwater rice is not *O. sativa*, rather another species, *O. glaberrima*. The majority of deepwater rice cultivation areas in Bangladesh and Indian states are adjacent to Bangladesh (West Bengal, Assam, Bihar, Orissa, Uttar Pradesh). Bangladesh is rich in deepwater water rice diversity. Three major types of deepwater rice are cultivated in Bangladesh, that is, typical deepwater, Bhadoia/Ashina, and Rayada (Bin Rahman and Zhang 2013). Among these, Rayada is the most primitive having year-long life cycle, and still shares some features of wild rice (Bin Rahman and Zhang 2013).

Drought tolerance

Drought is considered as one of the most complex and toughest stresses to overcome (Reyes 2009). Around 23 million hectares of rainfed rice growing area are estimated to be drought prone in South and Southeast Asia (Pandey et al. 2007) costing 17–40% yield reduction in severe drought years. Numerous attempts have been taken to develop drought-tolerant rice, but resulted in poor success. Earlier molecular genetic analyses identified numerous QTL (quantitative trait loci) of secondary traits such as root architecture, leaf water status, panicle water potential, osmotic adjustment and relative water content, etc., but failed to identify any major genes that regulate these traits; mainly due to low mapping resolution and weak phenotypic effect. However, huge numbers of genes are differentially expressed after drought stress in rice, that is, over 5000 genes upregulated and 6000 downregulated (Maruyama et al. 2014).

The major limitation of drought stress breeding is the lack of suitable screening methods along with its environment specific nature. Earlier studies tried to link secondary traits with drought tolerance. However, the recent and most effective selection strategy, for example, the direct selection for yield under drought and well-watered condition, is increasingly accepted for drought tolerance in rice (Kumar et al. 2008; Venuprasad et al. 2008; Torres et al. 2013). Recently scientists at IRRI screened nearly 1000 Genebank accessions originated from 47 countries for yield under drought and well-watered conditions during 2004–2009. Among the 988 accessions, total 65 accessions (originated from 13 different countries) were considered drought-tolerant accessions based on highest grain yield in any crop season (Torres et al. 2013). Interestingly, distributions of these tolerant accessions (country of origin) clearly show that most drought-tolerant accessions (48 accessions, i.e., 73% of total tolerant accessions) originated from Bangladesh and its adjacent India (particularly east India). The rest 17 tolerant accessions were originated from 11 different countries like Sri Lanka (4), Philippines (3), Vietnam (2), Afghanistan (1), China (1), Liberia (1), Malaysia (1), Nepal (1), Pakistan (1), Thailand (1), and USA (1). Interestingly, around 20% of Bangladeshi accessions are drought tolerant, whereas it is only 12% in the case of Indian accessions (Torres et al. 2013). Remarkably, all of the recommended varieties from drought stress breeding, that is, Kataktara Da2, Dular, Shada Shaita, and DA 28 (Torres et al. 2013), are still being cultivated in Bangladesh in *aus* growing season in considerable areas (Hossain et al. 2013).

Population Type of Drought- and Flood-Tolerant Accessions

Rice is categorized as *indica* and *japonica* types/subspecies based on morphological features since long (Kato et al. 1928), but population studies using molecular markers to recent sequencing and SNP genotyping have confirmed several more categories (Glaszmann 1987; Garris et al. 2005; Huang et al. 2011; Alexandrov et al. 2014; Wang et al. 2014). From isozyme to SNP genotyping, *aus* is always identified as a distinct category. However, few studies (isozyme and SSR genotyping) also identified separate categories for deepwater rice (Ashina and Rayada) (Glaszmann 1987; Wang et al. 2014). Recently over 4000 global rice accessions are sequenced and SNP genotyped (Huang et al. 2012; Alexandrov et al. 2014). So we have analyzed population types of both flood and drought-tolerant accessions.

Since AG tolerance accessions are very limited, that is, only 5, hence we extensively examined their population types in previous studies. Among the five AG-tolerant accessions, three of them, Nanhi (Baltazar et al. 2014), Khaiyan (Angaji et al. 2010) and Kalonchi (Platten et al. 2013) are reported as *aus* type where Ma-Zhan Red is *indica* (Septiningsih et al. 2013). Interestingly, Khao Hlan On was reported as aromatic rice of *japonica* type (Angaji et al. 2010) grown in the Shan state of Myanmar but morphological features categorized as *indica* (www.genesys-pgr.org), whereas recent sequencing and SNP analysis showed it is typically *aus* (Dwiyanti et al. 2014). So out of five AG-tolerant accessions, four of them are *aus* and mostly originated from Bangladesh and adjoining areas.

Among 65 drought-tolerant accessions, most of them are from two subpopulations, either *aus* (29) or *indica* (29), with the rest few from others subpopulations or admixture (Torres et al. 2013). Highest number of drought-tolerant *aus* accessions (19) are originated from Bangladesh, followed by India (7), whereas highest number of drought-tolerant *indica* accessions were originated from India (16) followed by Bangladesh (3) and Sri Lanka (3). Among all *indica* accessions (601), only 4.8% are drought tolerant, whereas remarkably 45.3% of *aus* accessions are drought tolerant among all *aus* accessions. So it clearly represents that drought-tolerant accessions are predominately originated from *aus* accessions. Similarly, flash-tolerant rice accessions (Sub1A) used in different molecular genotyping studies clearly showed that most of the tolerant accessions are from either *aus* or *indica* subpopulations.

From the above discussion, it is clear that all water stress, both drought and flood, tolerant accessions are predominately originated from *aus* type. What about deepwater rice? Several population studies also identified separate categories for deepwater rice (Glaszmann 1987; Wang et al. 2014). However, most population studies either exclude special type deepwater rice like Rayada/Ashina or include only a few accessions, hence separate category may not be identified in their studies. Since recently several thousand accessions were sequenced and SNP genotyped (Huang et al. 2012; Alexandrov et al. 2014), so we can check population type of deepwater rice from these database. It is noteworthy that *aus* and *indica* varieties cannot be differentiated by morphological features. Interestingly, after examining 3000 rice genome SNP databases (Alexandrov et al. 2014), we identified numerous deepwater rice of Bangladesh such as BORA DIGA, HIJOL DIGA, LAKHSMI DIGHA, JHUL DIGA, BAZAIL 975, BAZAIL 980, BHADOIA 303, DHARIAL, etc. as typical *aus*.

Among different drought adaptive traits, features of root (root angle, length, density, etc.) are largely considered as traits of drought tolerance (Blum 2011). Recently, Uga et al. (2013) identified DEEPER ROOTING 1 (DRO1) QTL from breeding populations of IR64 and Kinandang Patong, a deep-rooting landrace from the Philippines. DRO1 significantly increases yield performance under drought conditions due to root angle and longer root system. Among 20 OryzaSNP genotypes, *aus* accessions showed longest root length density in drought condition (Gowda et al. 2012). Although literatures identified Kinandang Patong as *tropical japonica* variety (Uga et al. 2008), we were bit curious. So we checked its subpopulation type in all recent sequencing-based population and genetic studies whether any study used Kinandang Patong. Interestingly, we identified that Huang et al. (2012) used Kinandang Patong (GP75) in their studies. Interestingly, sequencing and SNP genotyping data clearly showed Kinandang Patong is also typical *aus* type. So from the above analyses of distribution and population type of drought- and flood-tolerant rice accessions, it is evident that all water-related stress (either drought or flood) tolerant accessions are predominantly *aus* type and mostly originated from Bangladesh and neighboring areas.

Bangladesh: Highly Flood and Drought Prone & Center of Diversity of *aus* and Deepwater Rice

Bangladesh is one of the most flood and drought prone countries in the world (Paul 1997, 1998) where rice is cultivated throughout the year (Khush 1997). Recurrent occurrence of drought and flood in Bangladesh is common where devastating effect of drought and flood of some particular years (e.g.,1865, 1943–44, 1974) led to famines in past (Habiba et al. 2011) since rice accounts over 75% calorie demand in Bangladesh (Maclean et al. 2002). Both flood and rice is indispensable feature of Bangladesh. Almost every year, nearly half of rice cultivating areas are at risk of flood and drought of varied intensities (Dasgupta et al. 2011; Habiba et al. 2011).

Aside from flood and drought, Bangladesh has wide variations in land, topography, seasons, rainfall, etc. So genetic diversity of rice particularly drought and flood adaptive genotypes (*aus* and deepwater rice) are highly enriched (Bin Rahman and Zhang 2013). Nationwide survey identified over 3500 *aus* (Nasiruddin and Miah 1982) and 3820 deepwater rice (Bashar et al. 2004) varieties in Bangladesh. Center of diversity for both *aus* and deepwater rice is considered for Bangladesh and adjoining Indian states of West Bengal and Assam (Khush et al. 2003). Interesting, a separate domestication event was proposed earlier for *aus* (Londo et al. 2006) and a more recent reanalysis of large-scale global sequencing database of rice also supported the independent domestication of *aus* besides *indica* and *japonica* (Civáň et al. 2015). Fortunately, numerous varieties of *aus* are still being cultivated in considerable areas in Bangladesh (Hossain et al. 2013) although areas of these rice cultivation are continuously decreasing due to adoption of modern rice varieties with improved irrigation and agronomic practices.

Flood and Drought Tolerance Together

Since both drought and flood-tolerant accessions are predominantly originated from *aus* accessions of Bangladesh and adjacent areas, so it would be interesting to check drought tolerance of these flood-tolerant accessions. Submergence-tolerant *aus* genotype, FR13A, from which Sub1A was identified (Xu et al. 2006), was also reported as drought tolerant in seedling stage (Ito et al. 1999). Similarly some other *aus* accessions having SUB1A was also reported to be drought tolerant. Moreover, recently it was identified that SUB1A enhances drought recovery, whereas its overexpression increased ABA responsiveness (Fukao et al. 2011). Likewise many deepwater rice varieties (SHAIL KOTA, MANIKDIGHA, HABIGONJ AMAN II, GOIROL, etc.) showed pronounced drought tolerance comparing to some well-accepted drought-tolerant upland varieties (Cattling 1992). Excellent recovery capacity from drought stress was also reported in some Bangladeshi deepwater rice varieties (SARSARI, BHABANI, HIJALDIGHA, and TILBAZAL (De Datta and O'Toole 1977). Due to yearlong life cycle, Rayada rice generally exposes to cold, drought and flood in its lifecycle. Rayada rice varieties were reported as multiple stresses tolerant, that is, cold, flood, submergence, drought (Bin Rahman and Zhang 2013) even salt (Platten et al. 2013). Distribution of Rayada rice was completely confined to certain area in Bangladesh. However, deep exploration of Rayada varieties for stress tolerance gene mining has always been neglected (Bin Rahman and Zhang 2013).

Interestingly, excellent recovery capacity is one of the common features of both flood and drought-tolerant accessions/genotypes. Recovery capacity of submergence and drought (at vegetative stage) are evaluated on the basis of emergence and growth of new leaves and tillers several days to weeks after withdrawal of respective stresses (Fukao et al. 2011; Singh et al. 2014). SUB1A enhances faster recovery manifested by prompt emergence of tillers and new leaves, whereas almost all of the leaves of submergence intolerant lines withered within days after submergence (Xu et al. 2006; Singh et al. 2014). Remaining sugar and starch concentration in stem after submergence is highly correlated with survival percentage. Interestingly, both FR13A and traditional rice landraces adapted to deepwater and floating rice possess high stem carbohydrates. FR13A is higher submergence tolerant than Sub1A lines where FR13A possess high carbohydrate before submergence than Sub1A lines (Singh et al. 2014). Several studies also suggest the presence of other QTLs for submergence tolerance. It is more likely that cytokinin may play an important role in the prompt emergence of tillers and new leaves during recovery stage; however, direct evidence requires further experimentation. Three genes of group VII AP2/ERF showed upregulated in Sub1A introgressed line (but downregulated in submergence-intolerant line), interestingly, the same genes were also upregulated by cytokinins (Jung et al. 2010). Recently, it was shown that downregulation of cytokinin degradation enzyme, cytokinin oxidase 2 (OsCKX2) increases tiller number and improves rice yield (Yeh et al. 2015). Similarly, expression of isopentenyltransferase (IPT) (enzyme that mediates the rate-limiting step in cytokinin synthesis) driven by the promoter of senescence- and stress-activated receptor kinase (SARK) showed excellent drought tolerance comparing to wild-type plants (Peleg et al. 2011).

Hydrosensor: Still Unknown for Both Drought and Flood

Although over five thousand genes were reported as drought-responsive genes (Wang et al. 2011), master regulator of drought tolerance is still unknown. However, physiological regulation of root growth after soil drying is known since long. Several decades ago it was reported that mild drought stress (soil drying) can stimulate root growth (Sharp and Davies 1979). Moreover, it was also shown that root growth is less inhibited (than shoot) after water stress (Sharp et al. 1988). However, several decades have passed but still we do not know the answer of simple and fundamental question: how do plants sense the soil drying and regulate their root growth in response to it? Surely the answer of this question is critically important for our understanding as well as the development of drought-tolerant plants.

Sensing water or moisture gradient and differential growth (bending) of root toward increasing moisture is classically defined as hydrotropism. Despite known over centuries, very few studies were done on hydrotropic response due to the complexity of dissecting hydrotropism (Eapen et al. 2005; Cassab et al. 2013). Moreover, both gravitropism and thigmotropism interact with hydrotropism while numerous factors of soil such as soil compactness, aeration, nutrient richness or deficiencies, etc., can interfere root growth pattern. However, growing evidences suggest that root cap may sense moisture gradient as decaped root unable to respond hydrotropically (Eapen et al. 2005; Cassab et al. 2013). However, hydrosensing ability of root cap is not an unique feature of root cap since it is also involved in sensing gravity, touch, and light (Cassab et al. 2013).

Only four mutants (*miz1, miz2, ahr1,* and *nhr1*) were characterized for hydrotropic responses in Arabidopsis. However, none of them are strictly specific for hydrotropism. MIZU-KUSSEI 1 (MIZ1) encodes protein of domain of unknown function (DUF617) mainly expressed in ER membrane (Yamazaki et al. 2012) and root tip and hydathodes (Kobayashi et al. 2007). Recent research on MIZ1 showed it is involved in lateral root development (Moriwaki et al. 2011). Similarly *miz2* is an ahydrotropic mutant caused by single-base alteration unable to hydrotropic bending (Miyazawa et al. 2008). It was proposed that MIZ1 and MIZ1 like genes may regulate root architecture through modulation of auxin and cytokinins content or signaling (Moriwaki et al. 2011; Cassab et al. 2013). Rest two hydrotropic mutants, *altered hydrotropic response 1* (*ahr1*) and *no hydrotropic response 1* (*nhr1*) (Eapen et al. 2003) mutants were reported having involvement in hydrotropism although specific genes were not identified (Cassab et al. 2013).

Identification of tissue that first sense water stress is critically important for identification of molecular regulator of stress tolerance. Despite unknown nature of hydrosensor, increasing evidences suggest the root tip may sense differential hydration. Interestingly, recently identified DRO1 is also preferentially expressed in root tip (Uga et al. 2013). However, initial perception of flood certainly not in root tip as rice is generally cultivated in flooded conditions (of few centimeters). SUB1A, master regulator of flash flood tolerance, was shown preferentially expressed in node (Singh et al. 2010), whereas deepwater adaptation is governed by elongation of internode by SNORKEL genes (Cattling 1992; Hattori et al. 2009). So collectively it can be concluded that root tip may initially perceive or sense water shortage (drought) where node or internode may sense flood. Since all cellular or physiological functions are regulated by phytohormones, so it seems necessary to analyze the hormonal changes in these initial perception sites of both drought and flood.

Ethylene: Common Player in Both Flood and Drought Tolerance

Hormones play critical role in all growth and developmental regulations as well as stress adaptations. ABA (Abscisic acid) and ethylene are generally considered hormones of drought (Xiong 2007) and flooding adaptations (Bailey-Serres et al. 2012), respectively. Patterns of these hormone level changes after flood and drought in initial stress perception sites are interesting. After flooding, ABA level tends to decrease in shoot whereas ethylene increases sharply (25-fold increase within 24 h) (Hattori et al. 2009). Oppositely, after drought ABA level increases in root (Shi et al. 2015a), although change of ethylene level in rice root is not clearly evident. Moreover, there are inconsistencies in the literatures of ethylene changes in shoot after water stress (Wilkinson and Davies 2010). Earlier studies reported increases of ethylene in both water logging and drought. However, experiments using detached leaves suggest increasing ethylene level after water stress whereas intact plants showed decreased ethylene level (Morgan and Drew 1997). Speed of desiccation (slow or rapid) also showed opposite pattern (Morgan and Drew 1997). However, experiments using split root techniques in tomato after water deficit in one half showed increased ethylene synthesis (Sobeih 2004).

Soil microorganisms can play important role in ethylene homeostasis (Gamalero and Glick 2015). However, changes in ethylene in soil after gradual soil drying (since drought occurrence is slow in process in nature) are also not clearly evident. Interestingly, inoculation of ACC deaminase-producing bacteria, *Achromobacter piechaudii* ARV8, isolated from the rhizosphere of desert-thorn (*Lycium shawii*) of Negev desert, showed significant reduction in ethylene level after drought stress (Mayak et al. 2004). Remarkably, ARV8 inoculated tomato and pepper accumulated fourfold more biomass than noninoculated control. Similar experiments using different ACC deaminase-producing bacteria also showed similar pattern of drought tolerance in different plants such as pea, maize, wheat, and mung bean both in laboratory and field experiments (Gamalero and Glick 2015). After field trials throughout the USA, scientists of DuPont Pioneer recently revealed reduction in ethylene through downregulated ACSs (ACC synthases) leading to higher yielding corn under drought conditions (Habben et al. 2014). Growing evidences from research on modifying ethylene biosynthesis and signaling suggest critical involvement of ethylene in drought tolerance (Shi et al. 2015b).

Ethylene Response Factor (ERF): Participation in Drought and Flood Tolerance

APETALA2 (AP2)/Ethylene Response Factor (ERF), a large transcription factor superfamily, plays important roles in different biotic and abiotic stress tolerance. It consists of three families: (1) AP2 (two AP2/ERF domain), (2) ERF (one AP2/ERF domain), and (3) RAV (one AP2/ERF and B3 domain). AP2/ERF domain consists of 60–70 amino acids and involves in DNA binding (Nakano et al. 2006). Interestingly, both flash and prolonged flood adaption are regulated by same type ERF (group VII) transcription factor, SUB1A (Xu et al. 2006), and SNORKEL (Hattori et al. 2009), respectively. Recently, group VII ERF transcription factors of Arabidopsis have showed having function in oxygen sensing mechanism (Gibbs et al. 2011; Licausi et al. 2011). However, SUB1A (possibly SNORKEL-explained bellow) seems different than Arabidopsis group VII ERF members (Gibbs et al. 2011). Remarkably, rice has three times more VII ERF transcription factors than Arabidopsis (Nakano et al. 2006). Generally, SUB1A restricts GA responsiveness, whereas SNORKEL enhances thus results opposite adaptive phenotypes. Interestingly, both of the genes are completely absent in reference genome, Nipponbare, although intolerant allele of SUB1 is present.

Manipulation of several AP2/ERF transcription factors results in enhanced drought tolerance in rice. For example, knockdown of OsDERF1 increases ethylene production and also confers enhanced drought tolerance at seedling and tillering stages (Wan et al. 2011). Likewise, modulation of number of AP2/ERF transcription factors (either native or heterologous systems), such as DREB1A (Oh et al. 2005; Ito et al. 2006; Datta et al. 2012), OsDREB1B (Ito et al. 2006), DREB1C (Ishizaki et al. 2013), OsDREB2B (Matsukura et al. 2010), OsDREB2A (Cui et al. 2011), AP37 (Oh et al. 2009), OsERF3/OsERF4a (Joo et al. 2013), JERF3 (Zhang et al. 2010) showed enhanced drought tolerance. Moreover, OsETOL1, ETHYLENE OVERPRODUCER 1-like gene showed differentially drought and submergence tolerance in rice (Du et al. 2014).

SUB1A (and Plausibly SNORKEL): More Than Flood Tolerance

Recently simultaneously two groups scientists (Gibbs et al. 2011; Licausi et al. 2011) showed that group VII ERFs play critical role in oxygen sensing mechanism. More specifically, N-end rule pathway of targeted proteolysis acts as a homeostatic sensor of severe low oxygen levels in Arabidopsis (Gibbs et al. 2011). All members of Arabidopsis group VII ERFs were shown as substrate of N-end rule pathway as showed *in vitro* heterologous rabbit reticulocyte lysate assay (Gibbs et al. 2011). Despite being group VII member and presence of similar N-terminal motif (MC), SUB1A is not the substrate for the N-end rule pathway. The N-end rule pathway is a proteolytic system where N-terminal residues of proteins determine half-life or stability of protein. Increasing evidences on N-end rule pathway suggest it as a major cellular proteolytic system. It was hypothesized that lack of optimally positioned lysine downstream of the N degron may be the reason why SUB1A escape the N-end rule pathway (Gibbs et al. 2011).

Apart initial MCC, rest N terminal amino acid sequences of SNORKEL2 are quite different from consensus sequences of group VII ERFs of Arabidopsis (Gibbs et al. 2011).

Moreover, phenotypic effect of SNORKEL2 is more pronounced than SNORKEL1 where Amazonian wild rice possesses two SNORKEL2 genes (Hattori et al. 2009). In addition, SNORKEL2 expression in T65 background showed high internode elongation even in aerobic conditions (nonflooded) (Hattori et al. 2009). Thus, it clearly indicates, plausibly alike SUB1A, SNORKEL (particularly SNORKEL2) might evade the N-end rule pathway. Enhanced stability of SUB1A may explain superior performance of SUB1A rice to multiple abiotic stresses as SUB1A mediates cross talk with drought tolerance (Fukao et al. 2011) and delayed leaf senescence under prolonged darkness (Fukao et al. 2012). However, alike SUB1A, multiple abiotic stress tolerance capacity of SNORKEL was never been explored.

Rationale of Our Hypothesis and Future Perspectives

Although we aware the complex nature of drought and flood stresses, we also know that the destination is much simple (Blum 2011). Rice type, *aus*, is evolved and domesticated in the areas where drought and flood are alternatively imposed. Sensing and subsequent regulations of responses to the stresses must be initiated frequently and cross talking in their signaling pathways may exist to avoid confusion and self-inflicted damages. From the above discussions, it is clear that

(1) Drought (and flood) tolerance accessions are predominately originated from *aus* subpopulation where
(2) Bangladesh and adjacent areas are considered center of diversity of *aus* and deepwater rice varieties (Khush et al. 2003).
(3) Ethylene may play a central role of activation of adaptive traits both in flood and drought tolerance.
(4) Despite opposite phenotypes, flash and prolonged flood tolerance are regulated by similar group VII ERF transcription factors (Xu et al. 2006; Hattori et al. 2009) where
(5) recent studies have shown that modulation of ethylene synthesis and signaling either through transgenesis or inoculation ACC deaminase containing bacteria (Gamalero and Glick 2015) result in enhanced drought tolerance.
(6) In addition, master regulators of flood tolerances, SUB1A (and plausibly SNORKEL2) evade the N-end rule pathway (Gibbs et al. 2011) where
(7) expression of SUB1A is also highly upregulated after water shortage (Fukao et al. 2011) and delay leaf senescence under prolonged darkness (Fukao et al. 2012).
(8) Some *aus* varieties also possess excellent recovery capacity after drought stress.
(9) Finally domestication of *aus* is also considered as separate event beside *indica* and *japonica* (Civáň et al. 2015).

After considering these altogether, we propose a hypothesis that both drought and flood tolerance may be regulated by cross-talked pathways and coexist in *aus* subpopulation where master regulators of flood adaptations have already been identified (i.e., SUB1A and SNORKEL) but that of drought tolerance still to be identified, which may be achieved through deep exploration of *aus* accessions of Bangladesh.

Our counterintuitive hypothesis consists of two parts. Interestingly, first part (*both drought and flood tolerance may be regulated by cross-talked pathways*) is partially already been established as Fukao et al. (2011) experimentally showed that the master regulator of submergence tolerance, SUB1A mediates cross talk between submergence and drought tolerance in rice. In addition, numerous deepwater rice cultivars have shown earlier having drought tolerance and excellent drought recovery capacity. We argue that plausibly SNORKEL2, master regulator of deepwater adaptation may also evade the N-end rule pathways. Thus, like SUB1A, SNORKEL may mediate cross talk between deepwater adaptation and drought tolerance. So, in our hypothesis we have broadened the horizon of cross talk as shown by Fukao et al. (2011), that is, not only limiting to submergence tolerance but extended to both form of flood (flash and prolonged) tolerance. In the cross-talked pathways concern, we mainly emphasize on ethylene and its mediated pathways. The second part of our hypothesis (*coexist in aus subpopulation*) is merely the outcome of our critical analysis and synthesis of distribution and population types of different types flood and drought-tolerant accessions/genotypes. Finally, we suggest ways to mine the key regulator(s) of drought tolerance through deep exploration of *aus* accessions of Bangladesh.

Recently, several novel genes such as SUB1A (Xu et al. 2006), SNORKEL (Hattori et al. 2009), PSTOL1 (Gamuyao et al. 2012), DRO1 (Uga et al. 2013) have been identified from *aus* genotypes, FR13A, Gowai 38-9, Kasalath, Kinandang Patong, respectively. Remarkably, the rice reference genome, Nipponbare either completely lacks such tolerant alleles (SNORKEL, PSTOL1) or only possesses intolerant alleles (SUB1A, DRO1). So it may not be unlikely to anticipate the presence of novel gene(s) (possibly transcription factors; more likely ERF) in drought-tolerant *aus* genotypes that may be completely absent in both Nipponbare and 93-11, the reference genomes of *japonica* and *indica*, respectively. Thus, to overcome such limitations, *de novo* assembly of drought-tolerant *aus* genotype(s) using long read sequencing technology such as Pacific Biosciences' SMRT sequencing (Roberts et al.

2013) may reveal such novel genes. Recommended drought stress-tolerant traditional *aus* varieties such as Shada Shaita, Kataktara, etc. (Torres et al. 2013) should be selected for *de novo* assembly rather than breeding lines developed through *aus* varieties.

It should be noted that translation of Arabidopsis-based drought research to field crops is relatively poor (Passioura 2006; Blum 2014) where agronomically important novel genes, such as SUB1A, SNORKEL, PSTOL1, DRO1 identified from *aus* accessions, are literally absent in Arabidopsis. In addition, separate domestication of *aus*, along with presence of drought-tolerant accessions among *aus* varieties suggest that deep exploration of the *aus* accessions for drought tolerance is needed. Apart *de novo* assembly of *aus* variety, concerted research approach such as large-scale genetic and genome wide association studies with routine molecular and physiological data may also resolve the genetic basis of drought tolerance.

Concluding Remark

To overcome the most complex and toughest stress and alleviate the threat of drought on global food security, we need alternative idea/hypothesis along with concerted research approach by utilizing genetic resources and modern technologies. We think it is high time to mine the key regulator(s) of drought tolerance from *aus* accessions through *de novo* assembly with large scale genetic studies using genome manipulation weaponry.

Acknowledgments

We thank the National Basic Research Program of China (2012CB114300), Shenzhen Overseas Talents Innovation and Entrepreneurship Funding Scheme (The Peacock Scheme),and Hong Kong Research Grant Council (AoE/M-05/12, CUHK 473512, 474313 and 14122415) for their grant support.

Conflict of Interest

None declared.

References

Alexandrov, N., S. Tai, W. Wang, L. Mansueto, K. Palis, R. R. Fuentes, et al. 2014. SNP-Seek database of SNPs derived from 3000 rice genomes. Nucleic Acids Res. 43:D1023–D1027.

Angaji, S. A., E. M. Septiningsih, D. J. Mackill, and A. M. Ismail. 2010. QTLs associated with tolerance of flooding during germination in rice (*Oryza sativa* L.). Euphytica 172:159–168.

Bailey-Serres, J., S. C. Lee, and E. Brinton. 2012. Waterproofing crops: effective flooding survival strategies. Plant Physiol. 160:1698–1709.

Baltazar, M. D., J. C. I. Ignacio, M. J. Thomson, A. M. Ismail, M. S. Mendioro, and E. M. Septiningsih. 2014. QTL mapping for tolerance of anaerobic germination from IR64 and the aus landrace Nanhi using SNP genotyping. Euphytica 197:251–260.

Bashar, M. K., M. M. Haque, and S. M. H. Zaman. 2004. Rice biodiversity and genetic wealth of flood-prone environment of Bangladesh. Rice Res. Dev. flood-prone Ecosyst. Proc. Int. Work. flood-prone rice Syst. held Gazipur, Bangladesh,9-11 January 2001. p 289

Bin Rahman, A. N. M. R., and J. Zhang. 2013. Rayada specialty: the forgotten resource of elite features of rice. Rice 6:41.

Blum, A. 2011. Drought resistance is it really a complex trait? Funct. Plant Biol. 38:753–757.

Blum, A. 2014. Genomics for drought resistance – getting down to earth. Funct. Plant Biol. 41:1191–1198.

Bouman, B. A. M., E. Humphreys, T. P. Tuong, and R. Barker. 2007. Rice and water. Adv. Agron. 92:187–237.

Cassab, G. I., D. Eapen, and M. E. Campos. 2013. Root hydrotropism: an update. Am. J. Bot. 100:14–24.

Cattling, D. 1992. Rice in deep water. Macmillan, London.

Civáň, P., H. Craig, C. J. Cox, and T. A. Brown. 2015. Three geographically separate domestications of Asian rice. Nat. Plants 1:1–5.

Cui, M., W. Zhang, Q. Zhang, Z. Xu, Z. Zhu, F. Duan, et al. 2011. Induced over-expression of the transcription factor OsDREB2A improves drought tolerance in rice. Plant Physiol. Biochem. 49:1384–1391.

Dasgupta, S., M. Huq, Z. H. Khan, M. S. Masud, M. M. Z. Ahmed, N. Mukherjee, et al. 2011. Climate proofing infrastructure in Bangladesh: the incremental cost of limiting future flood damage. J. Environ. Dev. 20:167–190.

Datta, K., N. Baisakh, M. Ganguly, S. Krishnan, K. Y. Shinozaki, and S. K. Datta. 2012. Overexpression of Arabidopsis and Rice stress genes' inducible transcription factor confers drought and salinity tolerance to rice. Plant Biotechnol. J. 10:579–586.

De Datta, S. K., and J. C. O'Toole. 1977. Screening deep-water rices for drought tolerance,. Proceeding Work. Deep. rice. International Rice Research Institute, Laguna, Philippines, p 239.

Du, H., N. Wu, F. Cui, L. You, X. Li, and L. Xiong. 2014. A homolog of ETHYLENE OVERPRODUCER, OsETOL1, differentially modulates drought and submergence tolerance in rice. Plant J. 78:834–849.

Dwiyanti, M. S., V. J. Ulat, H. Koh, R. Mauleon, J. Hyoun, and M. Thomson. 2014. Analysis of genome sequence data of 13 rice varieties to determine variants within genomic regions related to abiotic and biotic stress tolerance. Poster P 096 4th Int. Rice Congr. p 1

Eapen, D., M. Barroso, M. Campos, G. Ponce, G. Corkidi, J. G. Dubrovsky, et al. 2003. A no hydrotropic response Root Mutant that Responds Positively to Gravitropism in Arabidopsis. Plant Physiol. 131:536–546.

Eapen, D., M. L. Barroso, G. Ponce, M. E. Campos, and G. I. Cassab. 2005. Hydrotropism: root growth responses to water. Trends Plant Sci. 10:44–50.

Fukao, T., T. Harris, and J. Bailey-Serres. 2009. Evolutionary analysis of the Sub1 gene cluster that confers submergence tolerance to domesticated rice. Ann. Bot. 103:143–150.

Fukao, T., E. Yeung, and J. Bailey-Serres. 2011. The submergence tolerance regulator SUB1A mediates crosstalk between submergence and drought tolerance in rice. Plant Cell 23:412–427.

Fukao, T., E. Yeung, and J. Bailey-Serres. 2012. The submergence tolerance gene, SUB1A, delays leaf senescence under prolonged darkness through hormonal regulation in rice. Plant Physiol. 160:1795–1807.

Gamalero, E., and B. R. Glick. 2015. Bacterial modulation of plant ethylene levels. Plant Physiol. 169:13–22.

Gamuyao, R., J. H. Chin, J. Pariasca-Tanaka, P. Pesaresi, S. Catausan, C. Dalid, et al. 2012. The protein kinase Pstol1 from traditional rice confers tolerance of phosphorus deficiency. Nature 488:535–539.

Garris, A. J., T. H. Tai, J. Coburn, S. Kresovich, and S. McCouch. 2005. Genetic structure and diversity in *Oryza sativa* L. Genetics 169:1631–1638.

Gibbs, D. J., S. C. Lee, N. Md Isa, S. Gramuglia, T. Fukao, G. W. Bassel, et al. 2011. Homeostatic response to hypoxia is regulated by the N-end rule pathway in plants. Nature 479:415–418.

Glaszmann, J. C. 1987. Isozymes and classification of Asian rice varieties. Theor. Appl. Genet. 74:21–30.

Gowda, V. R. P., A. Henry, V. Vadez, H. E. Shashidhar, and R. Serraj. 2012. Water uptake dynamics under progressive drought stress in diverse accessions of the OryzaSNP panel of rice (*Oryza sativa*). Funct. Plant Biol. 39:402–411.

Habben, J. E., X. Bao, N. J. Bate, J. L. Debruin, D. Dolan, D. Hasegawa, et al. 2014. Transgenic alteration of ethylene biosynthesis increases grain yield in maize under field drought-stress conditions. Plant Biotechnol. J. 12:685–693.

Habiba, U., R. Shaw, and Y. Takeuchi. 2011. Chapter 2 Socioeconomic Impact of Droughts in Bangladesh. Community, Environ. Disaster Risk Manag. Vol. 8 - Droughts Asian Monsoon Reg. pp 25–48

Hattori, Y., K. Nagai, S. Furukawa, X.-J. Song, R. Kawano, H. Sakakibara, et al. 2009. The ethylene response factors SNORKEL1 and SNORKEL2 allow rice to adapt to deep water. Nature 460:1026–1030.

HilleRisLambers, D., and B. S. Vergara 1982. Summary results of an international collaboration on screening methods for flood tolerance. Proc. 1981 Int. Deep. Rice Work. International Rice Research Institute, Manila, Philippines, pp 347–353

Hossain, M., W. Jaim, M. S. Alam, and A. M. Rahman. 2013. Rice biodiversity in Bangladesh?: adoption, diffusion and disappearance of varieties a statistical report from farm survey in 2005. Research and Evaluation Division BRAC, Dhaka, Bangladesh.

Huang, X., Y. Zhao, X. Wei, C. Li, A. Wang, Q. Zhao, et al. 2011. Genome-wide association study of flowering time and grain yield traits in a worldwide collection of rice germplasm. Nat. Genet. 44:32–39.

Huang, X., N. Kurata, X. Wei, Z.-X. Wang, A. Wang, Q. Zhao, et al. 2012. A map of rice genome variation reveals the origin of cultivated rice. Nature 490:497–501.

Huke, R. E., and E. H. Huke. 1997. Rice area by type of culture: South, Southeast, and East Asia a revised and updated database. International Rice Research Institute, Los Banos, Philippines.

Iftekharuddaula, K. M., S. Ghosal, Z. J. Gonzaga, A. Amin, H. N. Barman, R. Yasmeen, et al. 2015. Allelic diversity of newly characterized submergence-tolerant rice (*Oryza sativa* L.) germplasm from Bangladesh. Genet. Resour. Crop Evol. 1–9. doi:10.1007/s10722-015-0289-4.

IPCC. 2014. Climate Change 2014: Impacts, Adaptation, and Vulnerability. Summaries, Frequently Asked Questions, and Cross-Chapter Boxes. A Contribution of Working Group II to the Fifth Assessment Report of the Intergovernmental Panel on Climate Change. Geneva, Switzerland

Ishizaki, T., K. Maruyama, M. Obara, A. Fukutani, K. Yamaguchi-Shinozaki, Y. Ito, et al. 2013. Expression of Arabidopsis DREB1C improves survival, growth, and yield of upland New Rice for Africa (NERICA) under drought. Mol. Breed. 31:255–264.

Ismail, A. M., D. E. Johnson, E. S. Ella, G. V. Vergara, and A. M. Baltazar. 2012. Adaptation to flooding during emergence and seedling growth in rice and weeds, and implications for crop establishment. AoB Plants 2012:pls019.

Ismail, A. M., U. S. Singh, S. Singh, M. H. Dar, and D. J. Mackill. 2013. The contribution of submergence-tolerant (Sub1) rice varieties to food security in flood-prone rainfed lowland areas in Asia. Field Crops Res 152:83–93.

Ito, O., J. O'Toole, and B. Hardy. 1999. Genetic Improvement of Rice for Water-limited Environments. Proc. Work. Genet. Improv. Rice Water-Limited Environ. 1-3 December 1998, Los Baños. International Rice Research Institute.Los Baños, Philippines., p 353

Ito, Y., K. Katsura, K. Maruyama, T. Taji, M. Kobayashi, M. Seki, et al. 2006. Functional analysis of rice DREB1/CBF-type transcription factors involved in cold-responsive gene expression in transgenic rice. Plant Cell Physiol. 47:141–153.

Jackson, M. T. 1995. Protecting the heritage of rice biodiversity. GeoJournal 35:267–274.

Johnston, T. H. 1958. Registration of rice varieties. Agron. J. 50:694–700.

Joo, J., H. J. Choi, Y. H. Lee, Y. K. Kim, and S. I. Song. 2013. A transcriptional repressor of the ERF family confers drought tolerance to rice and regulates genes preferentially located on chromosome 11. Planta 238:155–170.

Jung, K.-H., Y.-S. Seo, H. Walia, P. Cao, T. Fukao, P. E. Canlas, et al. 2010. The submergence tolerance regulator Sub1A mediates stress-responsive expression of AP2/ERF transcription factors. Plant Physiol. 152:1674–1692.

Kato, S., H. Kosaka, and S. Hara. 1928. On the affinity of rice varieties as shown by the fertility of rice plants. Cent. Agric. Inst. Kyushu Imp. Univ. 2:241–276.

Khush, G. S. 1997. Origin, dispersal, cultivation and variation of rice. Plant Mol. Biol. 35:25–34.

Khush, G. S., D. S. Brar, P. S. Virk, S. X. Tang, S. S. Malik, G. Busto, et al. 2003. Classifying Rice Germplasm by Isozyme Polymorphism and Origin of Cultivated Rice Germplasm by Isozyme Polymorphism and Origin of Cultivated Rice. IRRI Discussion Paper Series No. 46. International Rice Research Institute, Los Banos

Kim, S.-I., and T. H. Tai. 2013. Identification of SNPs in closely related Temperate Japonica rice cultivars using restriction enzyme-phased sequencing. PLoS ONE 8:e60176.

Kobayashi, A., A. Takahashi, Y. Kakimoto, Y. Miyazawa, N. Fujii, A. Higashitani, et al. 2007. A gene essential for hydrotropism in roots. Proc. Natl Acad. Sci. USA 104:4724–4729.

Kumar, A., J. Bernier, S. Verulkar, H. R. Lafitte, and G. N. Atlin. 2008. Breeding for drought tolerance: direct selection for yield, response to selection and use of drought-tolerant donors in upland and lowland-adapted populations. Field Crops Res 107:221–231.

Li, Z.-X., E. M. Septiningsih, S. M. Quilloy-Mercado, K. L. McNally, and D. J. Mackill. 2011. Identification of SUB1A alleles from wild rice *Oryza rufipogon* Griff. Genet. Resour. Crop Evol. 58:1237–1242.

Licausi, F., M. Kosmacz, D. A. Weits, B. Giuntoli, F. M. Giorgi, L. A. C. J. Voesenek, et al. 2011. Oxygen sensing in plants is mediated by an N-end rule pathway for protein destabilization. Nature 479:419–422.

Ling, J., H. MingYu, W. ChunMing, and W. JianMin. 2004. Quantitative trait loci and epistatic analysis of seed anoxia germinability in rice (*Oryza sativa*). Rice Sci. 11:238–244.

Londo, J. P., Y.-C. Chiang, K.-H. Hung, T.-Y. Chiang, and B. A. Schaal. 2006. Phylogeography of Asian wild rice, *Oryza rufipogon*, reveals multiple independent domestications of cultivated rice, *Oryza sativa*. Proc. Natl Acad. Sci. USA 103:9578–9583.

Maclean, J. L., D. C. Dawe, B. Hardy, and G. P. Hettel. 2002. Rice almanac source book for the most important economic activity on earth, 3rd ed. CABI Publishing A, Wallingford, U.K.

Marris, E. 2008. Water: more crop per drop. Nature 452:273–277.

Maruyama, K., K. Urano, K. Yoshiwara, Y. Morishita, N. Sakurai, H. Suzuki, et al. 2014. Integrated analysis of the effects of cold and dehydration on rice metabolites, phytohormones, and gene transcripts. Plant Physiol. 164:1759–1771.

Matsukura, S., J. Mizoi, T. Yoshida, D. Todaka, Y. Ito, K. Maruyama, et al. 2010. Comprehensive analysis of rice DREB2-type genes that encode transcription factors involved in the expression of abiotic stress-responsive genes. Mol. Genet. Genomics 283:185–196.

Mayak, S., T. Tirosh, and B. R. Glick. 2004. Plant growth-promoting bacteria that confer resistance to water stress in tomatoes and peppers. Plant Sci. 166:525–530.

McCouch, S. 2013. Feeding the future. Nature 499:23–24.

Miyazawa, Y., A. Takahashi, A. Kobayashi, T. Kaneyasu, N. Fujii, and H. Takahashi. 2008. GNOM-mediated vesicular trafficking plays an essential role in hydrotropism of Arabidopsis roots. Plant Physiol. 149:835–840.

Morgan, P. W., and M. C. Drew. 1997. Ethylene and plant responses to stress. Physiol. Plant. 100:620–630.

Moriwaki, T., Y. Miyazawa, A. Kobayashi, M. Uchida, C. Watanabe, N. Fujii, et al. 2011. Hormonal regulation of lateral root development in arabidopsis modulated by MIZ1 and requirement of GNOM activity for MIZ1 function. Plant Physiol. 157:1209–1220.

Nakano, T., K. Suzuki, T. Fujimura, and H. Shinshi. 2006. Genome-wide analysis of the ERF gene family. Plant Physiol. 140:411–432.

Nasiruddin, M., and N. M. Miah. 1982. Survey of conservation activities in Asian countries and proposals for Future action: Bangladesh. 1983 Rice germplasm Conserv. Work. International Rice Research Institute, Laguna, Philippines, p 109

Niroula, R. K., C. Pucciariello, V. T. Ho, G. Novi, T. Fukao, and P. Perata. 2012. SUB1A-dependent and -independent mechanisms are involved in the flooding tolerance of wild rice species. Plant J. 72:282–293.

Oh, S.-J., S. I. Song, Y. S. Kim, H.-J. Jang, S. Y. Kim, M. Kim, et al. 2005. Arabidopsis CBF3/DREB1A and ABF3 in transgenic rice increased tolerance to abiotic stress without stunting growth. Plant Physiol. 138:341–351.

Oh, S.-J., Y. S. Kim, C.-W. Kwon, H. K. Park, J. S. Jeong, and J.-K. Kim. 2009. Overexpression of the transcription factor AP37 in rice improves grain yield under drought conditions. Plant Physiol. 150:1368–1379.

Pandey, S., H. Bhandari, and B. Hardy. 2007. Economic costs of drought and rice farmers' coping mechanisms: a cross-country comparative analysis. International Rice Research Institute, Los Banos.

Passioura, J. 2006. The drought environment: physical, biological and agricultural perspectives. J. Exp. Bot. 58:113–117.

Paterson, A. H., M. Freeling, and T. Sasaki. 2005. Grains of knowledge: genomics of model cereals. Genome Res. 15:1643–1650.

Paul, B. K. 1997. Flood research in Bangladesh in retrospect and prospect: a review. Geoforum 28:121–131.

Paul, B. K. 1998. Coping mechanisms practised by drought victims (1994/5) in North Bengal, Bangladesh. Appl. Geogr. 18:355–373.

Peleg, Z., M. Reguera, E. Tumimbang, H. Walia, and E. Blumwald. 2011. Cytokinin-mediated source/sink modifications improve drought tolerance and increase grain yield in rice under water-stress. Plant Biotechnol. J. 9:747–758.

Platten, J. D., J. A. Egdane, and A. M. Ismail. 2013. Salinity tolerance, Na+ exclusion and allele mining of HKT1;5 in *Oryza sativa* and *O. glaberrima*: many sources, many genes, one mechanism? BMC Plant Biol. 13:32.

Pucciariello, C., and P. Perata. 2013. Quiescence in rice submergence tolerance: an evolutionary hypothesis. Trends Plant Sci. 18:377–381.

Reyes, L. C. 2009. Overcoming the toughest stress in rice: drought. Rice Today 8:30–32.

Roberts, R. J., M. O. Carneiro, and M. C. Schatz. 2013. The advantages of SMRT sequencing. Genome Biol. 14:405.

Septiningsih, E. M., J. C. I. Ignacio, P. M. D. Sendon, D. L. Sanchez, A. M. Ismail, and D. J. Mackill. 2013. QTL mapping and confirmation for tolerance of anaerobic conditions during germination derived from the rice landrace Ma-Zhan Red. Theor. Appl. Genet. 126:1357–1366.

Sharp, R. E., and W. J. Davies. 1979. Solute regulation and growth by roots and shoots of water-stressed maize plants. Planta 147:43–49.

Sharp, R. E., W. K. Silk, and T. C. Hsiao. 1988. Growth of the maize primary root at low water potentials. I. Spatial distribution of expansive growth. Plant Physiol. 87:50–57.

Shi, L., M. Guo, N. Ye, Y. Liu, R. Liu, Y. Xia, et al. 2015a. Reduced ABA accumulation in the root system is caused by ABA exudation in upland rice (*Oryza sativa* L. var. Gaoshan1) and this enhanced drought adaptation. Plant Cell Physiol. 56:951–964.

Shi, J., J. E. Habben, R. L. Archilbald, B. Drummond, M. A. Chamberlin, R. Williams, et al. 2015b. Over-expression of ARGOS genes modifies plant sensitivity to ethylene, leading to improved drought tolerance in both Arabidopsis and maize. Plant Physiol. 169:266–282.

Singh, N., T. T. M. Dang, G. V. Vergara, D. M. Pandey, D. Sanchez, C. N. Neeraja, et al. 2010. Molecular marker survey and expression analyses of the rice submergence-tolerance gene SUB1A. Theor. Appl. Genet. 121:1441–1453.

Singh, S., D. J. Mackill, and A. M. Ismail. 2014. Physiological basis of tolerance to complete submergence in rice involves genetic factors in addition to the SUB1 gene. AoB Plants 6:plu060.

Sobeih, W. Y. 2004. Long-distance signals regulating stomatal conductance and leaf growth in tomato (*Lycopersicon esculentum*) plants subjected to partial root-zone drying. J. Exp. Bot. 55:2353–2363.

Todaka, D., K. Shinozaki, and K. Yamaguchi-Shinozaki. 2015. Recent advances in the dissection of drought-stress regulatory networks and strategies for development of drought-tolerant transgenic rice plants. Front Plant Sci. 6:1–20.

Torres, R. O., K. L. McNally, C. V. Cruz, R. Serraj, and A. Henry. 2013. Screening of rice Genebank germplasm for yield and selection of new drought tolerance donors. Field Crops Res 147:12–22.

Uga, Y., K. Okuno, and M. Yano. 2008. QTLs underlying natural variation in stele and xylem structures of rice root. Breed. Sci. 58:7–14.

Uga, Y., K. Sugimoto, S. Ogawa, J. Rane, M. Ishitani, N. Hara, et al. 2013. Control of root system architecture by DEEPER ROOTING 1 increases rice yield under drought conditions. Nat. Genet. 45:1097–1102.

Venuprasad, R., M. T. Sta Cruz, M. Amante, R. Magbanua, A. Kumar, and G. N. Atlin. 2008. Response to two cycles of divergent selection for grain yield under drought stress in four rice breeding populations. Field Crops Res 107:232–244.

Wan, L., J. Zhang, H. Zhang, Z. Zhang, R. Quan, S. Zhou, et al. 2011. Transcriptional activation of OsDERF1 in OsERF3 and OsAP2-39 negatively modulates ethylene synthesis and drought tolerance in rice. PLoS ONE 6(9):e25216. doi:10.1371/journal.pone.0025216.

Wang, D., Y. Pan, X. Zhao, L. Zhu, B. Fu, and Z. Li. 2011. Genome-wide temporal-spatial gene expression profiling of drought responsiveness in rice. BMC Genom. 12:149.

Wang, C.-H., X.-M. Zheng, Q. Xu, X.-P. Yuan, L. Huang, H.-F. Zhou, et al. 2014. Genetic diversity and classification of *Oryza sativa* with emphasis on Chinese rice germplasm. Heredity (Edinb) 112:489–496.

Wilkinson, S., and W. J. Davies. 2010. Drought, ozone, ABA and ethylene: new insights from cell to plant to community. Plant, Cell Environ. 33:510–525.

WWDR. 2014. Water and energy: Facts and Figures in The United Nations World Water Development Report 2014. Perugia, Italy.

Xiong, L. 2007. Abscisic acid in plant response and adaptation to drought and salt stress. Adv. Mol. Breed.

Towar. Drought Salt Toler. Crop. Springer Netherlands, Dordrecht, the Netherlands, pp 193–221.

Xu, K., X. Xu, T. Fukao, P. Canlas, R. Maghirang-Rodriguez, S. Heuer, et al. 2006. Sub1A is an ethylene-response-factor-like gene that confers submergence tolerance to rice. Nature 442:705–708.

Yamauchi, M., A. M. Aguilar, D. A. Vaughan, and D. V. Seshu. 1993. Rice (*Oryza sativa* L) germplasm suitable for direct sowing under flooded soil surface. Euphytica 67:177–184.

Yamazaki, T., Y. Miyazawa, A. Kobayashi, T. Moriwaki, N. Fujii, and H. Takahashi. 2012. MIZ1, an essential protein for root hydrotropism, is associated with the cytoplasmic face of the endoplasmic reticulum membrane in Arabidopsis root cells. FEBS Lett. 586:398–402.

Yeh, S.-Y., H.-W. Chen, C.-Y. Ng, C.-Y. Lin, T.-H. Tseng, W.-H. Li, et al. 2015. Down-regulation of cytokinin oxidase 2 expression increases tiller number and improves rice yield. Rice 8:36.

Yoon, J.-H., S.-Y. S. Wang, R. R. Gillies, B. Kravitz, L. Hipps, and P. J. Rasch. 2015. Increasing water cycle extremes in California and in relation to ENSO cycle under global warming. Nat. Commun. 6:8657.

Zhang, H., W. Liu, L. Wan, F. Li, L. Dai, D. Li, et al. 2010. Functional analyses of ethylene response factor JERF3 with the aim of improving tolerance to drought and osmotic stress in transgenic rice. Transgenic Res. 19:809–818.

7

Common bean in highly variable weather conditions, on sandy soils and food security in a subtropical environment

José Miguel Reichert[1], Miriam Fernanda Rodrigues[2], Gabriel Oladele Awe[3], Ursino Federico Barreto Riquelme[4], Douglas Rodrigo Kaiser[5] & Dalvan José Reinert[1]

[1]Soils Department, Federal University of Santa Maria Av. Roraima 1000, CEP, 97105-900 Santa Maria, RS, Brazil
[2]Forest Engineering, Federal University of Santa Maria Av. Roraima 1000, CEP, 97105-900 Santa Maria, RS, Brazil
[3]Department of Crop, Soil and Environmental Sciences, Faculty of Agricultural Sciences, Ekiti State University, Ado Ekiti, Nigeria
[4]Universidad Privada del Guairá, Assunción, Paraguay
[5]Federal University of South Frontier Av. Major Antônio Cardoso 590, 97900-000 Cerro Largo, RS, Brazil

Keywords
Air-filled porosity, black bean, crop water stress, soil water storage, water availability to plants

Correspondence
Gabriel Oladele Awe, Department of Crop, Soil and Environmental Sciences, Faculty of Agricultural Sciences, Ekiti State University, Ado Ekiti, Nigeria.
E-mail: gabrielolaawe@yahoo.com

Funding Information
CNPq.

Abstract

Common bean is a vital component of food security and reduction in malnutrition in developing nations of the world. In a year of low rainfall and high temperature during flowering and high rainfall during grain filling, we studied the effects of (1) water availability and tillage on soil structural formation and on water retention, growth and productivity of common bean, and (2) time series analysis of soil available water and its relationship with related atmospheric variables. Soil management systems consisted of no-tillage (NT), conventional tillage (CT), chiseling 2 days before sowing (Ch_0), and 1 year before sowing (Ch_1), in complete randomized block design and four replications. Air temperature, rainfall, soil available water, and air capacity were monitored throughout the 78-days bean cycle, and time series analysis of data was performed. Water availability in topsoil reached limiting values, close to the permanent wilting point, limiting proper development of bean at flowering. Water availability was higher in periods of limited rainfall in untilled soils (NT and Ch_1). Low soil availability of water and high air temperatures during flowering and grain filling stages affected grain production but grain yield did not differ statistically among tillage systems, which was similar to the average yield obtained in the state of Rio Grande do Sul. Temporal distribution of soil available water did not correlate with daily rainfall but correlated with air temperature. Time series analysis was adequate for estimating soil available water from its previous and related variables. Conservation agriculture (no-tillage and old-chiseling) improves water conditions in soil, but with long periods of water deficit there is no guarantee for high food production; thus, supplementary irrigation may still be necessary to reduce the combined effects of abiotic factors of air temperature and rainfall.

Introduction

Because of their proven nutritional and therapeutic properties, beans are highly desirable as a component for fighting hunger (EMBRAPA 2003). Beans are a major element of the Brazilian diet, being an excellent protein source (Faria et al. 2014). The combination of beans and cereals (1:2), especially rice, provides a complementary protein, since these foods together provide eight amino acids essential for the human body (EMBRAPA 2003). In addition to protein content, beans have high content of alimentary fiber and contain vitamins B complex, iron, calcium and other minerals, bioactive compounds (protease inhibitors), carbohydrate complexes (e.g., amide), and small amounts of lipids, almost all unsaturated types (Faria et al. 2014). This crop, therefore, is a vital component for food security and reduction in malnutrition, especially in developing countries, such as Brazil.

Brazil and Mexico are the two top consumers of dry beans (including common beans), with a per capita consumption, respectively, of 16.4 and 7.5 kg year^{-1} in 2011. The top five producers of dry beans in 2013, in decreasing order, were Myanmar, India, Brazil, Mexico, and Tanzania. The highest yields, also in decreasing order (from about 7 to 5 t ha^{-1}), are obtained in Syrian Arab Republic, Iraq, Barbados, Ireland, and Tajikistan, that is, in countries with dryer weather conditions (FAO 2015). Globally, the total land area cropped to dry beans in 2013 was 29.05 million ha (FAO 2015). Dry beans production in Brazil, in 2013/2014, was 3.43 million tons from 3.35 million ha, with an average yield of 1024 kg ha^{-1} (CONAB 2014), which corresponds to 9.7 and 12.7%, respectively, of world's production area and production of dry beans (FAO 2015).

The variability in rainfall distribution and El Niño and La Niña events had significantly affected agricultural productivity in southern Brazil (Fontana and Almeida 2002). To mitigate the risk of agricultural production losses due to climatic reasons, the agricultural zoning of the Ministry of Agriculture, Livestock and Supply ("Ministério da Agricultura, Pecuária e Abastecimento") rejects soybean crop (*Glycine max* (L.) Merr.) but not common bean (*Phaseolus vulgaris* L.) in soils with a clay content greater than 10% and less than or equal to 15%, with soil depth greater than or equal to 50 cm; or clay content between 15 and 35%, with less than 70% sand, characterizing texture differences over the first 50 cm of soil and depth equal to or greater than 50 cm (Brasil 2008).

Notwithstanding the possibility of crop cultivation on sandy soils, bean root system is concentrated (62–87%) primarily in the first 0.10 m soil depth (Freddi et al. 2005) and shows strong sensitivity to drought and compaction (Montanari et al. 2010). The sensitivity is high because the water retention and availability depends mainly on particle size and soil structural quality. This can be expressed by the number, size, and continuity of pores (Portela et al. 2001), bulk density (Reichert et al. 2009a) and soil resistance to root penetration (Lima et al. 2010).

Intense traffic by agricultural machinery (Silva et al. 2004) under inadequate soil moisture conditions in soils under no-tillage causes soil compaction with increase in bulk density, especially the 0–20 cm soil layer, and high penetration resistance to root growth (Lima et al. 2010; Reichert et al. 2009a) and reducing crop root growth (Collares et al. 2006; Kaiser et al. 2009), thus hindering plant access to water stored in the soil matrix. During the period between flowering and grain filling, water deficit affects bean yield components (Collares et al. 2006), especially during periods of high temperature (Silva et al. 2006).

Soil chiseling has proven effective in reducing bulk density and increasing water infiltration into the soil (Camara and Klein 2005; Rosa et al. 2008). However, with subsequent traffic, alternate wetting and drying cycles, and crop growth, soil reconsolidation occurs (Veiga et al. 2008; Reichert et al. 2014), with negative effects on water infiltration, storage and availability to plants. The permanence of the positive effects of soil chiseling or subsoiling, however, is usually short (Abreu et al. 2004), but could be expanded if included cover crops without (Abreu et al. 2004) or after (Nicoloso et al. 2008) subsequent chiseling.

Spatio-temporal analysis of soil water status in the surface or the entire soil profile, as well as its relationship with other related soil and atmospheric variables has received research attention (Timm et al. 2006, 2011; Jia et al. 2013; Sur et al. 2013; Awe et al. 2015) and temporal relationship between soil water storage and atmospheric variables, for example, rainfall (Timm et al. 2011; Awe et al. 2015) has been established. However, there is still dearth of information on the temporal relationship between soil water storage and air temperature, especially for food crops such as edible beans. This information is vital for better understanding of the structure of spatial and temporal distribution of soil-plant-atmospheric processes at different scales and how the relationship influences the variability of crop production.

We hypothesized that (1) soil tillage significantly affect soil water retention and availability and structural quality as well as productivity of common bean and (2) temporal distribution of soil water available water correlates with other atmospheric variables. The objectives of this study were to evaluate, in a year characterized by low rainfall and high temperature during flowering and high rainfall during grain filling, the effect of (1) water availability and tillage practices on structural formation on soil water retention, and on growth and productivity of common bean; and (2) management practices on temporal variability of soil available water and its relationship with related atmospheric variables.

Materials and Methods

Study site

The study was conducted in the experimental research station of the Department of Soils, Federal University of Santa Maria, located on geographic coordinates latitude 29°41′S and longitude 53°48′W in the Central Depression physiographic region of Rio Grande do Sul State, Brazil. According to the Köppen climate classification, the climate is humid subtropical without dry season (Cfa), with mean temperature of the warmest month above 22°C (Moreno 1961; Álvares et al. 2013). The soil is classified as Argissolo Vermelho-Amarelo Distrófico according to the Brazilian System of Soil Classification (Santos et al. 2013) and Typic

Hapludult according to Soil Taxonomy (USDA 1999), with sandy loam texture in the surface horizons. The surface horizons (Ap and A) are homogeneous and, on average, consisted of 610 g kg^{-1} sand, 300 g kg^{-1} silt and 120 g kg^{-1} clay, whereas the Bt horizon has 360 g kg^{-1} clay, 210 g kg^{-1} of silt and 430 g kg^{-1} of sand.

The area had been cultivated for 10 years under black oat (*Avena strigosa* Schieb)/soybean (*Glycine max* L.) crop succession in no-tillage system. The site is characterized by high degree-of-compactness (DC), about 98.4% (Reichert et al. 2009a) and least limiting water range of 0.00 m^3 m^{-3} in the 0.10–0.15 m soil layer (Collares et al. 2006).

Weather data

Air temperature measured three times daily (at 9:00 h, 15:00 h, and 20:00 h) and rainfall were recorded on a daily basis whit an automatic weather station located about 1 km from the experimental site. A rain gauge was also installed at the center of the field to confirm the rainfall data from the weather station.

Experimental design and systems

The experiment was laid out in a randomized complete block design with four replications. The tillage systems consisted of long-term no-tillage (NT); conventional tillage of previously kept no-tillage (CT); chisel tillage just before bean sowing in 2002 (Ch_0); and chisel tillage about 1 year ago (Ch_1).

The no-tillage system (NT) consisted of an area not tilled for 10 years. A portion of the whole NT area was chiseled in 2001 (Ch_1), and another part was conventionally tilled. In the following year (2002), the conventional tillage (CT) and, in part of Ch_1, chiseling (Ch_0) operations were repeated. Conventional tillage system was done by plowing until about 0.20 m depth using three-disk plow, followed by harrowing. Soil chiseling was performed to a depth of 0.30 m using a ripper consisting of three slanted rods, spaced at 0.40 m, followed by harrowing for leveling. Sowing of bean (cultivar FT Bio nobre) was done in 16/12/2002, after establishment of the tillage systems.

Soil physical properties

The effect of tillage on soil physical and hydraulic properties was evaluated consisting of soil water retention, available and unavailable water capacity, air capacity, and void ratio.

Undisturbed soil samples collected from the soil layers of 0–0.05 and 0.1–0.15 m were used to evaluate soil properties. After preparation in the laboratory, the samples were saturated by capillary action for 48 h, and then subjected to 1, 6, and 10 kPa water tension applied in a sand column (Reinert and Reichert 2006), and 33 and 100 kPa water tension in Richard's pressure plates (Klute 1986). After equilibration at 100 kPa, the samples were placed in an oven and dried at 105°C until constant weight to determine soil bulk density (Blake and Hartge 1986).

Disturbed samples were also collected in the same 1500 kPa layers to determine soil moisture at water tension at 1500 kPa using a psychrometer (WP4 – Dew Point PotentiaMeter), following procedures of Klein et al. (2006) and Gubiani et al. (2013). Soil volumetric water content corresponds was obtained by multiplying the gravimetric water content by the corresponding soil bulk density.

Soil total porosity (TP) corresponds to the soil water content at saturation (θv at 0 kPa), with the assumption that all pores are filled with water. Field capacity (FC) and permanent wilting point (PWP) were determined as the moisture retained in the tensions of 10 and 1500 kPa, respectively. Available water capacity was calculated as the difference between the soil moisture at water tensions of 10 and 1500 kPa. Unavailable water was regarded as the soil moisture content at tension of 1500 kPa. Void ratio was calculated as the ratio between the TP and (1-TP). Air capacity was determined as the difference between the TP and the water content at each tension (1, 6, 10, 33, and 100 kPa).

Soil moisture content monitoring

To evaluate the temporal variability of soil moisture and water storage, soil volumetric moisture (θv) of soil layers of 0–0.10, 0.10–0.20, and 0.20–0.30 m was monitored throughout the bean cycle on 19 dates (1, 12, 22, 26, 29, 32, 35, 39, 42, 46, 50, 53, 57, 60, 63, 67, 71, 74, and 78 DAE) using time domain reflectometer (TDR) sensors. The TDR sensors, made of metal rods 0.2 m long, were installed and maintained in an angle in the soil to cover the entire thickness of each layer and integrate the measured layers. Soil θv was recorded manually using TDR datalogger (TDR 100, Campbell Scientific Inc., USA). The water available in each soil layer was calculated by the product of measured θv and layer thickness.

Temporal analysis of soil available water, rainfall, and average daily air temperature

Prior to normalization of data for state-time analysis, autocorrelation and cross-correlation of soil available water, rainfall, and average daily air temperature were determined using the Applied Statistical Times Series Analysis (ASTSA) software (Shumway 1988).

According to Hui et al. (1998), when the X_t data are scaled with respect to their mean (m) and standard deviation (σ) as:

$$X_t = \frac{[X_t - (m - 2\sigma)]}{4\sigma} \quad (1)$$

the normalized values of the X_t become dimensionless with mean of $\mu = 0.5$ and standard deviation of $\delta = 0.25$. This transformation allows the state coefficients of matrix ø have magnitudes proportional to their contribution to each state variables used.

The coefficient matrix (ø) of the state-time approach is described by the state equation (2):

$$X_t = \text{ø}X_{t-1} + \omega_1 \quad (2)$$

where: X_t and X_{t-1} are the state vectors (a set of p variables A_i, B_i, C_i,...) at times t and $t-h$; ø is a pxp transition matrix of state coefficients indicating the measure of temporal linear regression; ω_1 is the uncorrelated zero mean model errors with a common qxq covariance matrix Q. The detail description of the formulation of the state equation (2) can be found in Dourado-Neto et al. (1999).

Because soil available water data were not quantified on a daily basis, and for the dataset to appear in time series manner as the rainfall and average daily air temperature, zero (0) was used for days when AW observations were not made. Thus, state-time series analysis of soil available water, rainfall, and average daily air temperature was performed using the special state-algorithm (svar.exe) for the time series data.

Evaluation of bean growth and yield

Bean performance parameters evaluated include plant emergence, plant height, leaf area, and grain yield. Plant emergence was evaluated from 6 to 9 days after sowing (DAS), in an area, 1 m^2 in each plot. We considered complete seedling emergence with cotyledon at 10 mm above the ground surface.

Plant height and leaf area were evaluated at 16, 21, and 26 day after emergence (DAE). Plant height was determined by the vertical distance between the ground surface and the tip of the last leaf, visible on the main stem of the plant. For the leaf area, forty-four (44) leaves were selected from plants from each experimental plot and a regression equation was generated (eq. 3) using measured length and width of the leaves as:

$$A_L = 1.4602\text{CL}^{1.0334}, \quad r^2 = 0.98 \quad (3)$$

where: A_L is compound area of the three oval-shaped leaflets (cm^2), CL is the product of the length, and width of the central leaflet.

Grain yield was assessed at physiological maturity (84 DAE) from an area of 7.2 m^2, demarcated at the center of each plot. Grain yield was converted to kg ha^{-1} after moisture correction to 13%.

Statistical analysis

Data on soil physical and hydraulic properties, bean plant emergence, height, leaf area, and yield were subjected to analysis of variance (ANOVA) and means were compared by Duncan Multiple Range test at $P < 0.05$ using statistical package SAS (SAS Institute 1991).

Results and Discussion

Temporal variability of rainfall and air temperature

Temporal distributions of daily rainfall (P) and of average daily air temperature (Ta) during black bean growing season are shown in Figure 1A and B, respectively. The temporal variability of the AW strictly followed the course of rainfall distribution. Therewas virtually no rainfall between 3 and 18 days after common bean emergence (DAE), resulting in marked reduction in total soil available water capacity (from about 50 mm to 30 mm) in the 0–30 cm soil profile. Between 33 and 41 DAE, there was no rainfall, resulting in more intense reduction in AW, less than 20 mm in the soil profile. Between 42 and 60 DAE, water recharge from rainfall was not enough to increase the AW even as further reduction in AW was observed between 50 and 60 DAE as a result of very low recharge by rainfall (less than 10 mm). This shows that the rainfall amount may not be enough in meeting evaporative demand of the atmosphere and crop water requirement. It should also be noted that not all the rainfall amount contributes to soil wetting as some fractional amount is returned to the atmosphere by evaporation from the soil and plant surfaces. On 60 DAE, there was a sharp increase in AW (from <10 mm to >50 mm) due to high recharge by rainfall (>60 mm). Between 60 and 80 DAE, the AW was high as a result of adequate recharge from rainfall (Fig. 1C).

During the growing period, average daily air temperature ranged between 20 and 32°C (Fig. 1B). Forthis period, in most days (64 days) during the growing period, average daily air temperature was greater than 24°C (Fig. 1B), the optimum temperature for common beans (Balardin et al. 2000). Critical air temperature (30°C) (Hoffmann Júnior et al. 2007) occurred in

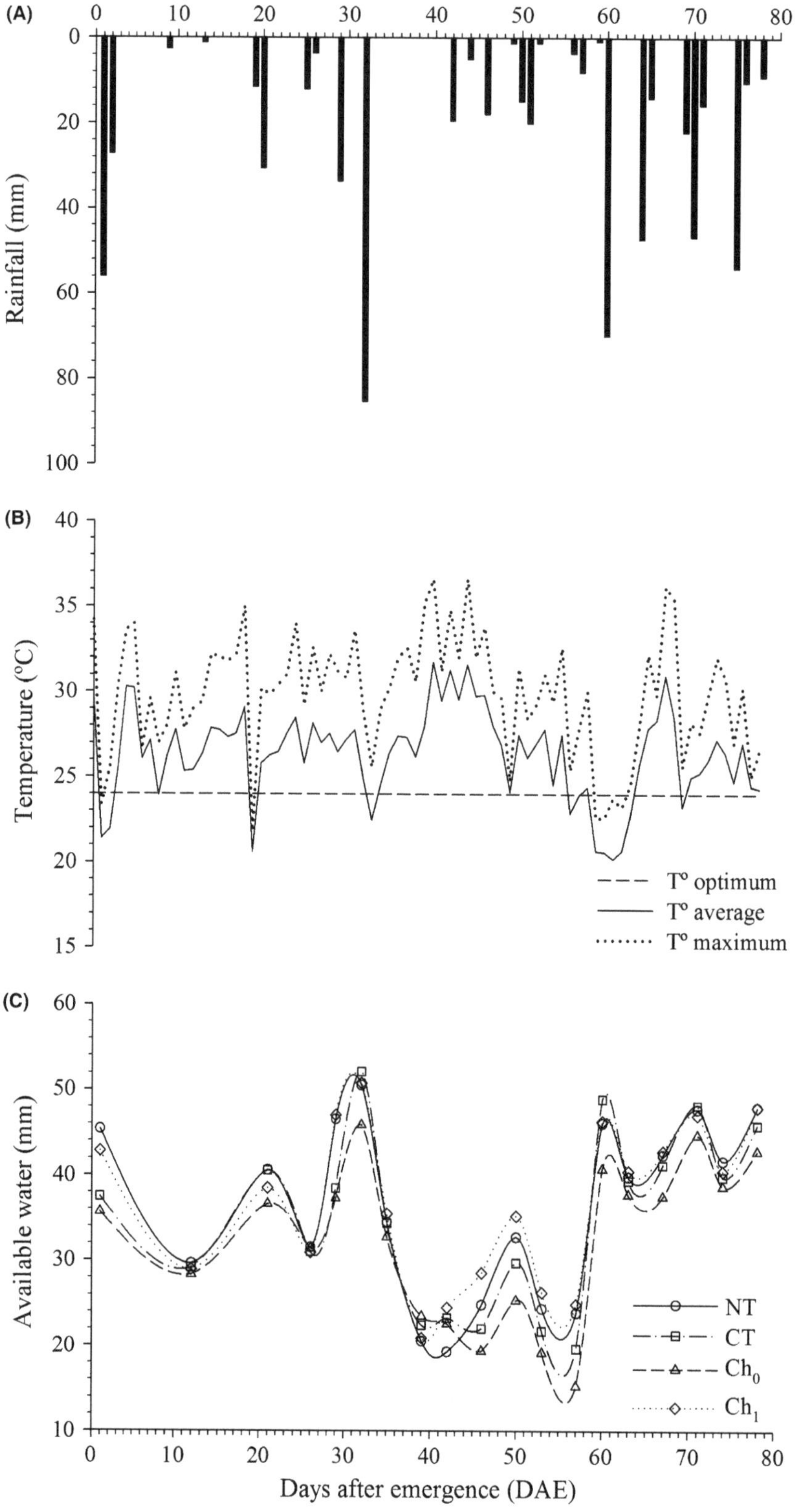

Figure 1. Temporal distribution of daily (A) rainfall, (B) optimum air temperature (T° optimum = 24°C), average (T° average) and maximum (T° maximum) air temperature (°C), and (C) total available water in the sandy loam Ultisol grown to common bean. *NT, no-tillage; CT, conventional tillage; Ch_0, chiseling 2 days before sowing; Ch_1, chiseling 1 year before sowing.

13 consecutive days at 15:00 h, between 35 to 47 days after emergence. A comparison between daily rainfall and average daily air temperature showed that air temperature decreased after rainfall event, indicating the cooling effect of rainfall. Both the average daily air temperature and AW showed obvious monotonically decreasing and increasing trend, although the order was not the same (Fig. 1).

Temporal variability of soil available water under the different tillage systems

Bean plants experienced a wide range of available water throughout the growth cycle under the different tillage systems (Figs. 1, 2). Soon after the emergence of plants (1 DAE), the variation in available water capacity was due to the effect of different tillage systems.

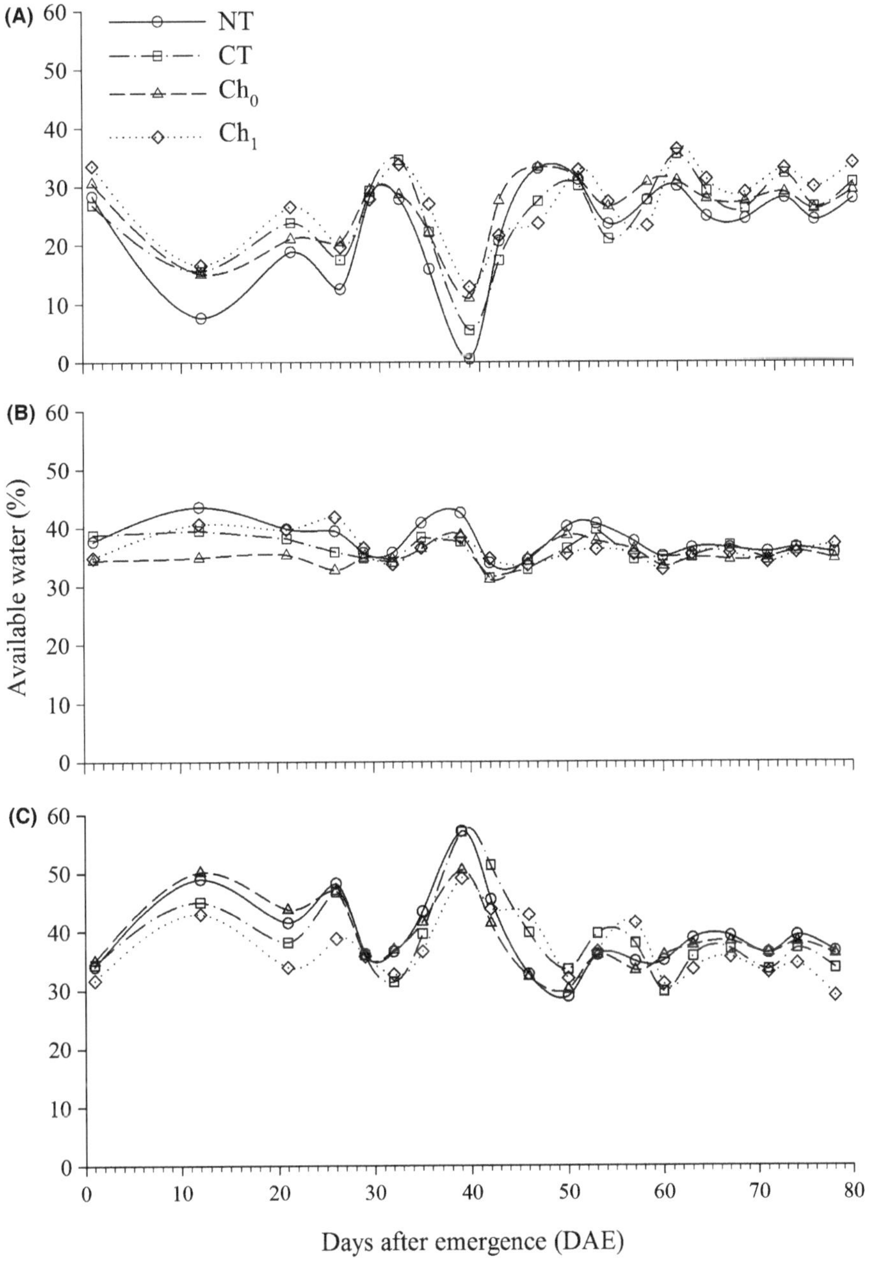

Figure 2. Available water (in percentage) of the total capacity of available water (0–0.30 m soil depth) of the soil layers of 0–0.10, 0.10–0.20, and 0.20–0.30 m of the sandy loam Ultisol grown to common bean crop. *NT, no-tillage; CT, conventional tillage; Ch_0, chiseling 2 days before sowing; Ch_1, chiseling 1 year before sowing.

During the whole growing season, the most consolidated soil condition (NT) generally had the greatest available water capacity, whereas recently chiseled soil (Ch_0) had lower values (Figs. 1, 2). The lowest available water capacity was observed at 57 DAE, during a period of limited rains, in Ch_0 and CT treatments, whereas the highest available water capacity was obtained from NT and Ch_1 systems at 32 DAE after a heavy rainfall (about 82.3 mm). At the end of the flowering stage and early grain formation stage (46–57 DAE), Ch_1 and NT systems had higher available water capacity (Figs. 1, 2). The reason for the low water capacity in Ch_0 and CT systems is attributed to evaporation occurring on the soil surface due to incorporation of residue during tillage operation compared with NT and Ch_1 systems that retain residue on the soil surface and better water retention. Another phenomenon that could cause the lower available water capacity in the tilled plots (Ch_0 and CT) is that the improved soil structure produced by soil mobilization gradually decrease due to natural reconsolidation as a result of alternate wetting and drying cycles (Reichert et al. 2009a). During soil wetting by rainfall, the effective stress in the soil approaches zero, causing the soil matrix to rupture under its own weight, thus decreasing the size and number of water transmission pores. Furthermore, the dynamic forces of water moving through the pores tend to compress the soil matrix thereby reducing these transmission pores (Ahuja et al. 1998). During drainage or redistribution after infiltration, the elevated negative pore-water pressures increases the effective stress on the soil matrix, which further brings the soil particles closer, thus further pore size reduction and limited water movement. In a study on effect of tillage systems and nutrient sources affecting soil cover, temperature, and moisture in a clayey Oxisol under corn by Veiga et al. (2010), differences in moisture in the surface layer of tilled treatments were greater than in untilled treatments, due to differences in the intensity of soil mobilization and residue remaining after soil management. Collares et al. (2006) and Lima et al. (2006) reported that NT had the best soil moisture conditions during the bean crop cycle and soil physical and hydraulic conditions were favorable for plant growth. Fabrizzi et al. (2005) and Schwen et al. (2011) found an increase in soil moisture storage under conservation tillage than tilled soil and attributed it to decrease in evaporation, improved soil infiltration, and enhanced soil protection from raindrop impact.

Lowest percentage of available water was observed in the surface layer (0–0.10 m) for all tillage systems (Fig. 2A). Highest percentage of available water was observed from the 0.20–0.30 cm layer (Fig. 2C), and low amplitude in the percentage of available water during the bean crop cycle was observed in the 0.10–0.20 m soil layer (Fig. 2B). Conventional soil tillage systems (CT and Ch_0), in general, had the highest amplitudes in water storage during the bean crop cycle, resulting mainly from higher variability of water content in the surface layer (Reichert et al. 2009b; Veiga et al. 2010).

Largest decrease in water availability occurred between 33 to 41 days after emergence, a period of 9 days when no rainfall occurred. Lower water availability was observed between 42 to 59 days after emergence, a period when low rainfall volumes occurred (Fig. 1C). Since top soil loses more water (Gubiani et al. 2014) due to combined effect of evaporation and transpiration (Montanari et al. 2010), presence of straw on surface layer in systems without tillage (NT-lowest temperature and amplitude) or minimum tillage (Ch_1-lower temperature than conventional tillage) reduces soil heating (Silva et al. 2006; Reichert et al. 2009b), consequently lowering water loss by evaporation (Andrade et al. 2002) and providing lower temporal variation in water storage. Awe et al. (2015) also reported lower temporal variability of soil water storage in an Alfisol with residue retention. However, even if the structural soil conditions are suitable, sandy soils have lower natural capacity for water storage and availability to plants (Reichert et al. 2009c). Therefore, such soils should be used for crops with greater resistance to the variability of water availability and stress during the growth cycle.

Temporal relationship among soil available water, rainfall, and average daily air temperature

Prior to state-time analysis, the results of autocorrelation of average daily air temperature (Ta) and rainfall (*P*) and of soil available water (AW) for different soil layers under different tillage systems are presented in Table 1 and Figure 3, whereas cross-correlation (CC) functions of AW versus Ta and R are shown in Figures 4 and 5 (only results of the 0–0.10 m surface layer are shown because of the similarity in resulting Figures). Temporal data of

Table 1. Autocorelation lengths, λ, of measured water content of the different soil layers during the growing period of the common bean crop.

Soil layer (m)	System			
	NT	CT	Ch_0	Ch_1
0–0.10	1.818	1.834	1.843	1.825
0.10–0.20	1.713	1.729	1.795	1.742
0.20–0.30	1.750	1.742	1.733	1.773

NT, no-tillage; CT, conventional tillage; Ch_0, chiseling 2 days before sowing; Ch_1, chiseling 1 year before sowing.

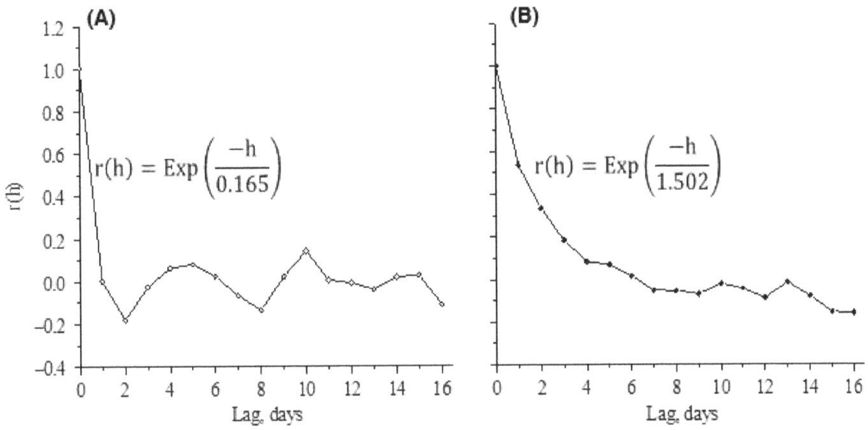

Figure 3. Autocorrelation functions of the (A) daily rainfall and (B) average daily air temperature during the growing period of the common bean crop.

Ta autocorrelated (λ = 1.502) (Fig. 3), AW data of all the soil layers under the four tillage systems also autocorrelated ($\lambda > 1$) (Table 1), whereas average daily R did not autocorrelate ($\lambda < 1$) (Fig. 3). AW did not correlate with R for all soil layers and tillage systems (Fig. 4), whereas there was correlation, although negative, between AW and average daily Ta, for all soil layers and tillage systems (Fig. 5). Timm et al. (2011) also reported autocorrelation ($\lambda < 1$) for rainfall and attributed it to the that rainfall the general nature of rainfall is more of season and rarely or not consistently repeat its magnitude during the growing season or over years. The slight differences in autocorrelation and cross-correlation functions of the soil available water (AW) in Table 1 and Figure 3 showed that the temporal variability of AW measurements is not uniform for the different soil depths and management practices during the crop growth cycle. This behavior is attributed to the fact that AW is rarely in equilibrium

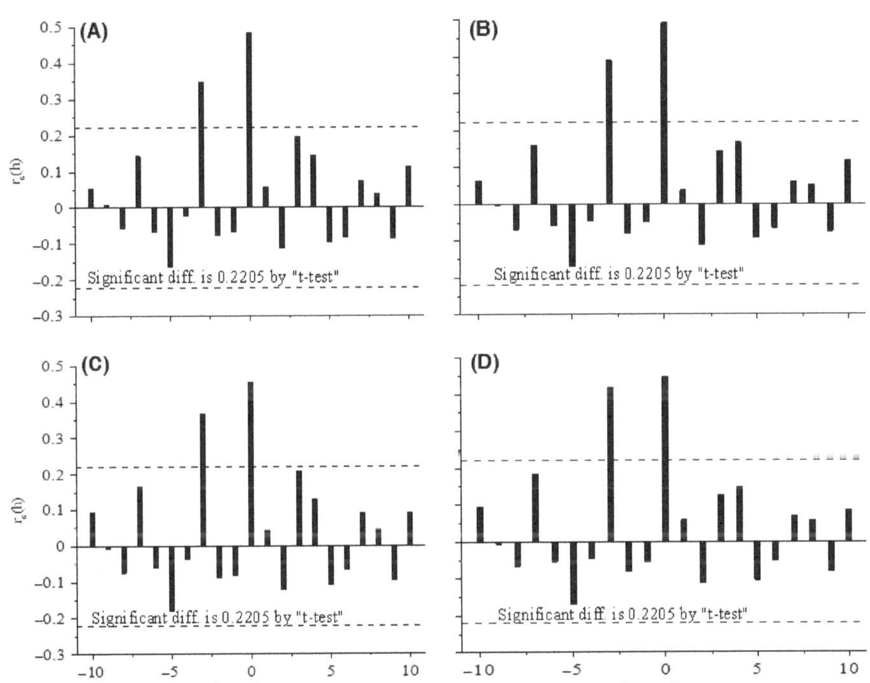

Figure 4. Cross-correlation function between daily rainfall and soil water content of the 0–0.10 m soil surface layer of (A) NT, (B) CT, (C) Ch_0, and (D) Ch_1 systems during the growing period of the common bean crop. *NT, no-tillage; CT, conventional tillage; Ch_0, chiseling 2 days before sowing; Ch_1, chiseling 1 year before sowing.

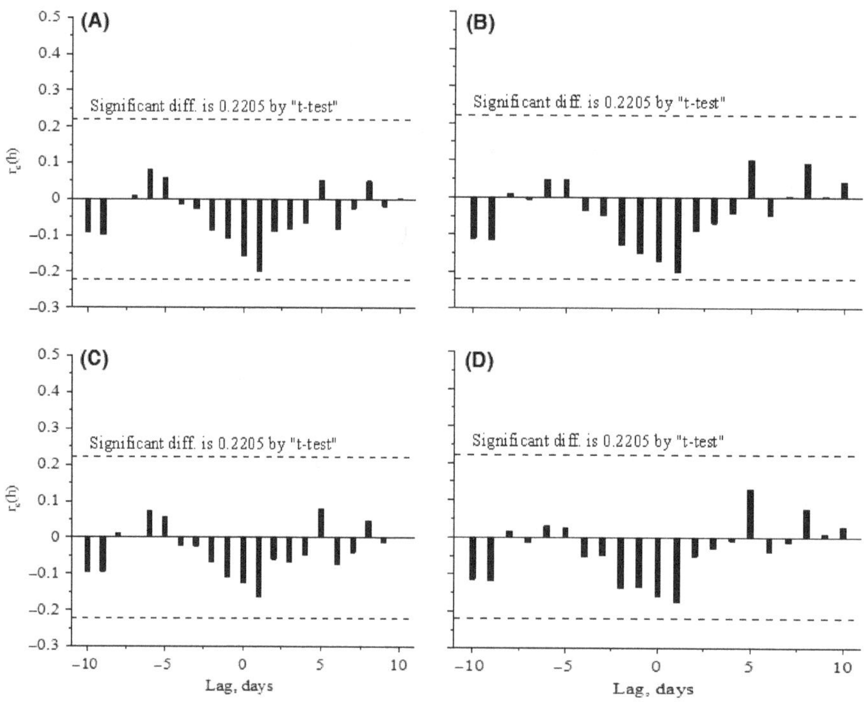

Figure 5. Cross-correlation function between mean daily air temperature and soil water content of the 0–0.10 m surface layer of (A) NT, (B) CT, (C) Ch_0, and (D) Ch_1 systems during the growing period of the common bean crop. *NT, no-tillage; CT, conventional tillage; Ch_0, chiseling 2 days before sowing; Ch_1, chiseling 1 year before sowing.

in the soil profile, as it is constantly being distributed and redistributed due to natural processes including rainfall, evaporation, and infiltration at the soil surface (Timm et al. 2006); extraction by plant roots and deep percolation as well as alterations and disruptions of soil structural formation occasioned by soil management practices (Reichert et al. 2009a). Furthermore, the autocorrelograms and cross-correlograms of AW and related properties, respectively, are not expected to be unique within a field because, according to Nielsen and Wendroth (2003), there is spatial and temporal variability in soil properties due to soil type, slope aspect, vegetal cover, and unstable climatic conditions, respectively.

Because of the observed monotonic decrease and increase trend in daily rainfall and soil available water (Fig. 1), as well as the correlation between AW and Ta (Fig. 5), there is the opportunity of evaluating how both R and Ta would contribute to the estimation of AW using applied statistical time series analysis (ASTSA). Thus, Figures 6, 7 and 8 show the temporal patterns of the state-time analysis of scaled AW for 0–0.10, 0.10–0.20, and 0.20–0.30 m layers of the bean field under four tillage systems. The continuous lines is the scaled estimated AW, the marked points are scaled observed AW, and shaded region represents the 95% confidence limits which is plus or minus two standard deviation.

State equations in Figures 6A–D showed that the contribution of previous soil available water (AW_{t-1}), rainfall (P_{t-1}), and air temperature (Ta_{t-1}) to the estimation of present soil available water capacity (AW_t) of 0–0.10 m surface layer were between 50.9 and 56.6%; 26.1 and 28.3%; 14.2 and 19.2%, respectively, for different tillage systems. For 0.10–0.20 m layer, the contribution of AW_{t-1} increased, between 54.9 and 65.1%, that of P_{t-1} did not change (not more than 28%), and the contribution from Ta_{t-1} slightly decreased, between 11.2 and 15.6% (Fig. 7). For 0.20–0.30 m layer, the contribution of AW_{t-1} to the estimation of AW_t was highest compared with those of upper layers, with values ranging between 63.9 and 67.4%. On the other hand, the contribution from both the P_{t-1} (between 22.9 and 24.5%) and Ta_{t-1} (between 9.2 and 12.1%) decreased. As expected, the contribution from the previous soil available water status to estimation of present soil available water capacity was highest, whereas that from previous rainfall events was higher than that from previous average daily temperature. A comparison of the contributions of the variables to the estimation of AW_{t-1} among different tillage systems did not show any discernible trend.

Considering the temporal associations among AW, R, and Ta for different tillage systems, between 94.9 and 96.3% of the variance of AW was explained by the use of autoregressive state-time analysis for 0–0.10 m surface layer; between 95 and 97.2% of the variance was explained for 0.10–0.20 m soil layer; and a slightly higher variance (between 96.7 and 97.4%) was explained for 0.20–0.30 m soil layer (Figs. 6, 7 and 8). The ω_t term in each equation represents the error term, which is randomly distributed and a mean of zero (data not shown). The higher transition coefficient (known as relative contribution) of AW_{t-1} compared to both P_{t-1} and Ta_{t-1} is an indication that present value of soil available water depends more on previous measurements of itself than previous observations of daily rainfall and average daily air temperature and vice-versa, if either of the rainfall or average daily air temperature is the subject of research.

The state-time analysis is a step ahead of classical statistic because it gives better insight of the relationship among climatic variables that influences AW. The state-time analysis also revealed that, since the observation of a variable is an estimate of its true value, the analysis automatically considers that each measurement possesses an explicit observation as well as incorporation of model error (denoted as ω_t) (Nielsen and Wendroth 2003). The results of this study agreed with findings of similar study by Timm et al. (2011) and Awe et al. (2015) on time series analysis of soil water storage and atmospheric variables.

In this study, the increase in the relative contribution of AW_{t-1} to the estimation of the present values of AW_t with soil depth (Figs. 6, 7, and 8) is attributed to higher and much more stable soil water content in subsurface layers (Fig. 2). This is unexpected because of the higher variability of soil water status recorded in the surface layer as it is the layer prone to evaporation than the deeper layers. According to Timm et al. (2006), the temporal changes in soil water status at the soil surface were as a consequence of the in situ alternate wetting and drying cycles. Rainfall events cause the recharge of soil moisture, thus the higher contribution from rainfall amount with respect to the average daily air temperature. Timm et al. (2011) stated that changes in soil water storage are generally related directly to the amounts of rainfall that infiltrate the soil surface during relatively short-term periods.

As observed in previous studies (e.g., Timm et al. 2011; Awe et al. 2015), the errors (ω_t) were randomly distributed, with mean close to zero (data not shown). Although this study has also proven the applicability of state-time analysis for estimating AW from itself and related variables, however, other soil properties and processes affecting or related to soil water status such as evapotranspiration, soil temperature, etc. should be incorporated in future

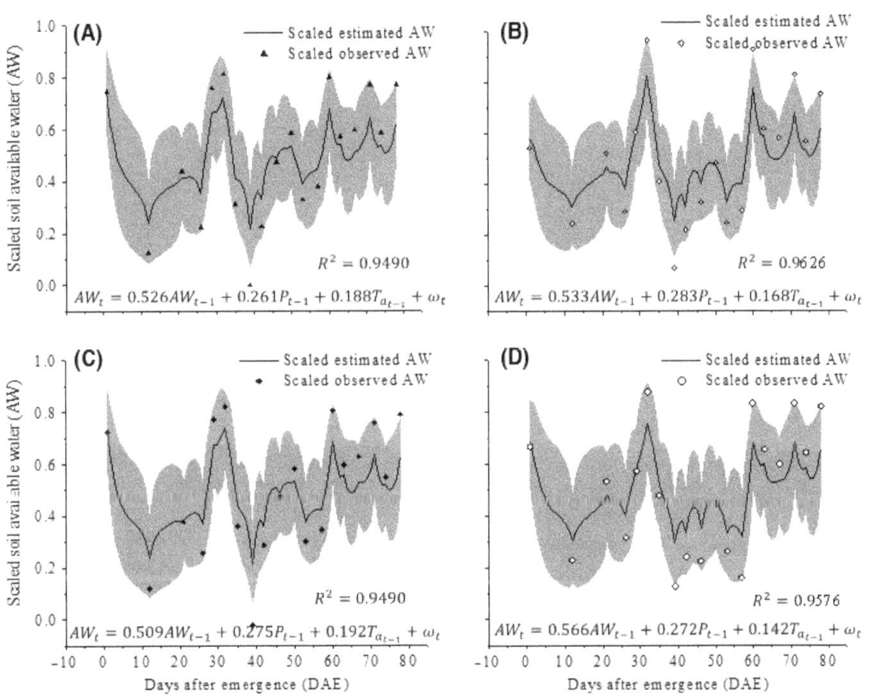

Figure 6. State-time analysis of scaled soil available water (AW) of the 0–0.10 m soil surface layer of the bean crop field under (A) NT, (B) CT, (C) Ch_0 and (D) Ch_1 systems during the growing period of the common bean crop. *NT, no-tillage; CT, conventional tillage; Ch_0, chiseling 2 days before sowing; Ch_1, chiseling 1 year before sowing. **AW_t, scaled present soil available water, mm; AW_{t-1}, scaled previous soil available water, mm; P_{t-1}, scaled previous rainfall, mm; Ta_{t-1}, scaled previous average daily air temperature; ω_t, error term. The shaded region represents ± standard deviation at 95% significant level. R^2, coefficient of determination.

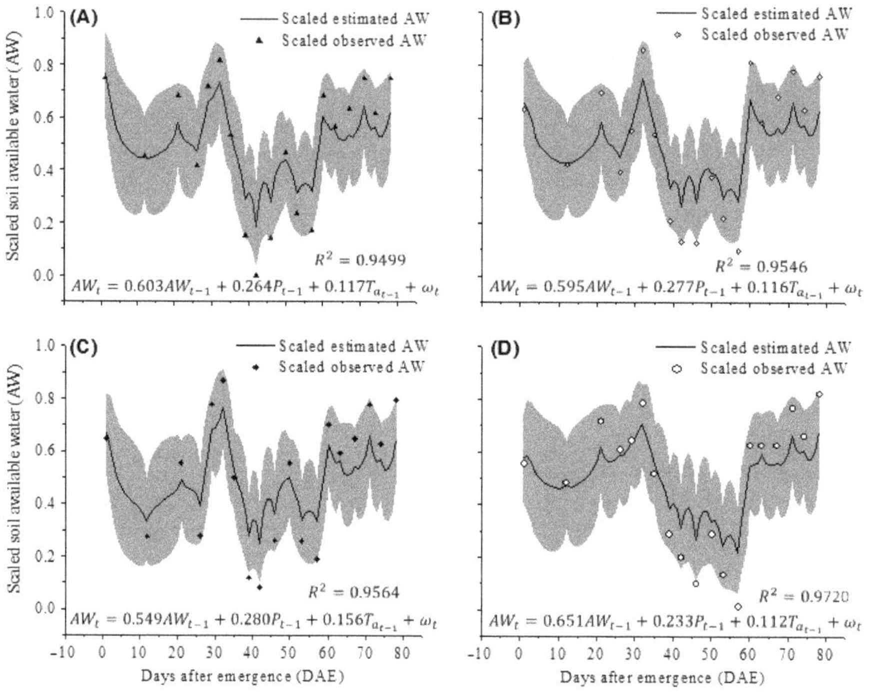

Figure 7. State-time analysis of scaled soil available water (AW) of the 0.10–0.20 m soil layer of the bean crop field under (A) NT, (B) CT, (C) Ch_0, and (D) Ch_1 systems during the growing period of the common bean crop. *NT, no-tillage; CT, conventional tillage; Ch_0, chiseling 2 days before sowing; Ch_1, chiseling 1 year before sowing. **AW_t, scaled present soil available water, mm; AW_{t-1}, scaled previous soil available water, mm; P_{t-1}, scaled previous rainfall, mm; Ta_{t-1}, scaled previous average daily air temperature; ω_t, error term. The shaded region represents ± standard deviation at 95% significant level. R^2, coefficient of determination.

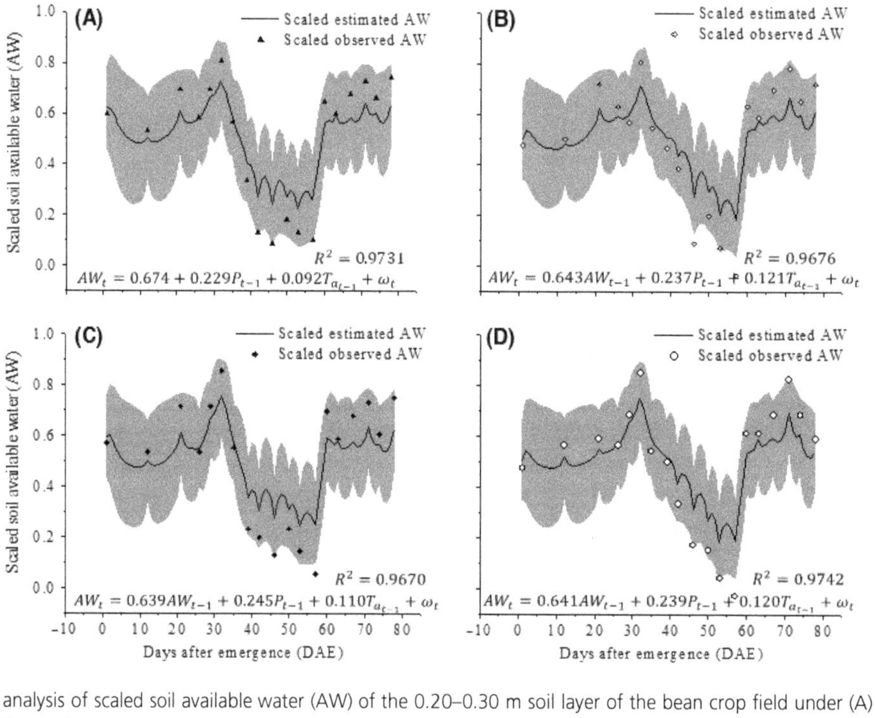

Figure 8. State-time analysis of scaled soil available water (AW) of the 0.20–0.30 m soil layer of the bean crop field under (A) NT, (B) CT, (C) Ch_0, and (D) Ch_1 systems during the growing period of the common bean crop. *NT, no-tillage; CT, conventional tillage; Ch_0, chiseling 2 days before sowing; Ch_1, chiseling 1 year before sowing. **AW_t, scaled present soil available water, mm; AW_{t-1}, scaled previous soil available water, mm; P_{t-1}, scaled previous rainfall, mm; Ta_{t-1}, scaled previous average daily air temperature; ω_t, error term. The shaded region represents ± standard deviation at 95% significant level. R^2, coefficient of determination.

studies for better understanding of the complex soil-plant-water relations. Also studies similar to this should combine with spatial (state-space) analysis so that "time-specific" and "site-specific" decisions would be possible for sustainable management.

Soil physico-hydraulic properties and available water capacity

Based on the classification proposed by Horn and Fleige (2003), the percentage of available water (AWC) with respect to total porosity was low at 18 and 69 DAE for all tillage systems in the 0–0.05 m layer. For 0.10–0.15 m layer, available water was very low for the CT system and low for other tillage systems at 18 and 69 DAE (Table 2). Considering the effect of management systems, the AWC of the 0–0.05 m surface layer was high in CT and Ch_0 treatments compared with NT and Ch_1. For the 0.10–0.15 m layer, the Ch_0 and Ch_1 treatments had the highest AWC (about 9%), whereas the CT had the lowest value (5.9%). Except for CT treatment, the AWC increased with time in both soil layers. Also, unavailable water was low at 18 and 69 DAE for all tillage systems in both 0–0.05 and 0.10–0.15 m layers, and the average values of the unavailable water was at part (about 7%) for the different soil management systems (Table 2). Soil voids ratio was low to medium in the 0-0.05 m layer at 18 and 69 DAE, whereas for the 0.10–0.15 m layer, the voids ratio was low for all the tillage systems at 18 and 69 DAE (Table 3).

The air capacity of the soil when subjected to different water tensions are presented in Table 3. As expected, air capacity increased with increasing water tension. For the 0–0.05 m surface layer, the Ch_1 treatment had the highest air capacity values at all water tensions, whereas Ch_0 had the lowest values shortly after emergence (18 DAE). For the 0.10–0.15 cm subsurface layer, the CT treatment had the highest air capacity values, with Ch_0 treatment also having the lowest values for all water tensions. For this same layer, air capacity reduced at 69 DAE for Ch_0 (medium) and Ch_1 (high). Soil air capacity at field capacity (FC = moisture at 10 kPa) for the 0–0.05 m surface layer was very high for Ch_1 and high for other tillage systems at 18 DAE. For 0.10–0.15 m layer, air capacity at FC at 18 DAE was medium for Ch_0 system and high for the other systems; while at 69 DAE, it was medium for CT system and high for other systems (Table 3). Generally, there was a decrease in air capacity at 69 DAE for the different tillage methods (Table 3). Higher air capacity values were observed at permanent wilting point (PWP = moisture at 1500 kPa) in 0–0.05 m layer for Ch_1 system (39.0%) at 18 DAE and CT system (32.5%) at 69 DAE. For 0.10–0.15 m layer, high air capacity values were observed for CT system (27.4%) at 18 DAE and Ch_1 system (27.4%) at 69 DAE (Table 3).

Soil water available (AW) content under field conditions is influenced by soil structure and texture and the temporal variation is determined by climatic conditions, crop physiology, soil surface condition, and management practices (Alletto and Coquet 2009). The higher AWC in the surface layer of both CT and Ch_0 treatments was as a result of soil loosening which increased the pore space, that is, higher macropore volume which promotes water transmission and retention. Contrary to our results, Alam et al. (2014) reported that AWC was found to be significantly higher in no-tillage than other tillage practices due to the evidence of organic matter buildup and other favorable physical characteristics. Fernandez-Ugalde et al.

Table 2. Soil available and unavailable water as a percentage of saturation water content or total porosity measured on different days after emergence (DAE) of common bean crop from two soil layers of the sandy loam Ultisol under different tillage systems.

System	0–0.05 m				0.10–0.15 m			
	18 DAE		69 DAE		18 DAE		69 DAE	
Available water capacity (%)								
NT	10.4	L[1]	11.3	L	8.2	L	9.3	L
CT	11.1	L	12.9	L	5.9	VL	5.5	VL
Ch_0	10.8	L	12.5	L	9.0	L	9.8	L
Ch_1	7.9	L	13.3	L	9.2	L	12.6	L
Unavailable water (%)								
NT	6.4	L	7.2	L	7.3	L	7.6	L
CT	6.4	L	6.8	L	7.0	L	7.7	L
Ch_0	7.0	L	7.4	L	7.5	L	7.5	L
Ch_1	6.1	L	6.8	L	7.2	L	7.4	L

NT, no-tillage; CT, conventional tillage; Ch_0, chiseling 2 days before sowing; Ch_1, chiseling 1 year before sowing.
[1]Classification proposed by Horn and Fleige (2003): L, low; VL, very low. For available water capacity, VL = <6% and L = 6–14%; and for unavailable water, L = 4–8%.

Table 3. Voids ratio and air capacity at water tensions of 1, 6, 10, 33, and 100 kPa for two soil layers of the sandy loam Ultisol for two distinct days after emergence (DAE) of the common bean crop.

System	0–0.05 m				0.10–0.15 m			
	18 DAE		69 DAE		18 DAE		69 DAE	
Void ratio								
NT	0.688	M[1]	0.576	L	0.487	L	0.484	L
CT	0.613	M	0.650	M	0.526	L	0.437	L
Ch_0	0.572	L	0.528	L	0.456	L	0.493	L
Ch_1	0.844	M	0.646	M	0.515	L	0.539	L
Air capacity (1 kPa)								
NT	3.5	L	4.9	M	3.5	L	4.8	M
CT	2.9	L	5.2	M	3.7	L	3.4	L
Ch_0	3.0	L	4.6	M	3.1	L	4.9	M
Ch_1	5.4	M	5.1	M	3.4	L	5.3	M
Air capacity (6 kPa)								
NT	14.2	H	8.5	M	9.7	M	7.9	M
CT	11.0	M	11.4	M	9.8	M	5.7	M
Ch_0	9.1	M	8.1	M	7.3	M	8.1	M
Ch_1	16.7	H	12.1	H	8.5	M	8.3	M
Air capacity (10 kPa)								
NT	17.3	H	12.4	H	13.4	H	12.7	H
CT	15.0	H	15.6	H	18.8	H	6.9	M
Ch_0	13.2	H	10.4	M	11.3	M	12.5	H
Ch_1	24.2	VH	15.6	H	12.9	H	12.4	H
Air capacity (33 kPa)								
NT	23.8	H	18.0	H	17.2	H	15.7	H
CT	20.4	VH	19.6	H	21.5	VH	17.3	H
Ch_0	18.0	H	14.7	H	14.8	H	15.6	H
Ch_1	31.1	VH	19.1	H	17.4	H	15.1	H
Air capacity (100 kPa)								
NT	29.5	VH	24.9	VH	23.3	VH	21.0	VH
CT	25.0	VH	24.0	VH	24.7	VH	20.5	VH
Ch_0	21.5	VH	18.1	H	19.7	H	20.3	VH
Ch_1	35.5	VH	23.5	VH	22.1	VH	20.2	VH

NT, no-tillage; CT, conventional tillage; Ch_0, chiseling 2 days before sowing; Ch_1, chiseling 1 year before sowing.
[1]Classification proposed by Horn and Fleige (2003); L, low; M, medium; H, high; VH, very high. For void ratio, L = 0.43–0.61 and M = 0.61–0.85; and for air capacity, L = 2–4%, M = 4–12%, H = 12–20%, and VH = >20%.

(2009) stated that soils under no-tillage practices possess greater water storage capacity than tilled soils. The low AWC observed for CT treatment in the 0.10–0.15 cm layer is attributed to negative effect of soil disturbance on pore system. Soil disturbance increases the amount of pores, but it disconnects them (Lamandé et al. 2003), which jeopardizes their efficiency to the flow of water and air. These results agree with the findings of Dalmago et al. (2009) who found CT treatment had higher AWC than NT from 15 cm down to 75 cm deeper layers.

The increase in AWC with time is attributed to the natural consolidation process as the volume of macropores decrease with subsequent increase in micropores. Bamberg et al. (2011) in their study on temporal changes in soil physical and hydraulic properties in strawberry fields in Brazil also found increase in AWC with time. The authors attributed this to decreased Ma and increased Mi, indicating that the soils changed their structure to a more packed configuration.

The unavailable water corresponds to soil water content at permanent wilting point (1500 kPa), which is influenced primarily by soil texture and mineralogy and controlled by adsorption and matrix forces. The insignificant difference in unavailable water among the different tillage systems is attributed to the fact at this potential the effect of soil texture and mineral composition is more on water retention than soil structure, that is, unaffected by management. Reynolds et al. (2002) did not find difference in unavailable water between no-tillage and conventional tillage for the various soil types evaluated. According to these authors, the absence of significant difference between management systems on unavailable water is that it is determined by clay content which is unaffected by management systems. On the contrary, Dalmago et al. (2009)

found elevated unavailable water in conventional tillage compared with no-tillage system but that the difference was significant only in deeper layers.

The void ratio expresses the relationship between the void volume and solid volume of a given soil, thus the higher the void ratio, the more porous and the less dense the soil is. The Ch_0 and NT treatments with high void ratio in the 0–0.05 m surface layer is an indication of porous soil structure, which is explained by the fact that they had the low soil bulk density. For the Ch_1 and CT treatments, the low void ratio is attributed to the disruption of soil matrix by soil disturbance with distort pore volume and continuity. Tavares Filho and Tessier (2009) also found lower void ratio from CT treatment than NT system. According to Ribeiro et al. (2007), in the soil under NT, the soil particulates are still arranged in aggregates, which permit the existence of voids, whereas in the case of soil under CT, the soil particulates are arranged in close contact, which results in a low void ratio and high BD. The lower void ratio observed in the 0.10–0.15 m subsurface layer is attributed to the compact nature of the soil, or in other words, the soil particles are closely packed, characterized by higher bulk density and low biological activity, which may be caused by the existence of tillage pan as well as the effect of the weight of overlying layer. Also, the general decrease in the void ratio with time is as a result of natural reconsolidation process (Reichert et al. 2009b), even in the absence of external stress application such as the NT treatment, as a result of alternate drying and wetting cycles.

Air capacity is one of the important parameters used to assess changes in soil functions resulting from management practices (Tang et al. 2011), such as soil air permeability that characterizes the ability to conduct gas, by mass flow, in response to pressure gradients (Stepniewski et al. 1994), where its behavior is a function of the size and continuity of pores (Tuli et al. 2005). At low water tension, air capacity values were low, indicating low air-filled pores. In a study on clay texture Oxisol cultivated under conventional tillage and no-tillage systems for 23 years, Rodrigues et al. (2011) observed that air capacity may be zero if all the pores are filled with water and/or are not interconnected. The increase in air capacity, as the soil is subjected to higher water tensions, is attributed to increased air-filled pores. Møldrup et al. (2001) in eight different soils with different textures and Schönning et al. (2002) in two sandy loam soils cultivated under different tillage and crop rotation systems for 50 years also reported increase in air permeability and attributed the increase to reduced tortuosity and increase in the number of pores that act in the flow.

Bean growth parameters and yield

Plant emergence was highest in Ch_1 at 6 DAE (Fig. 9) followed by NT compared with other tillage systems. Possibly, this soil condition of Ch_1 and NT systems had better soil-seed contact, but without much difference in soil water content (Fig. 1C). Lima et al. (2010) also recorded a negative effect on the emergence of bean due to soil mobilization, whereas previously chiseled and conventional tillage provided less suitable conditions for crop emergence.

Crop leaf area (Fig. 10A) and crop height (Fig. 10B) were reduced by soil mobilization (CT and Ch_0), which probably occurred due to increased air temperature and the amplitude of soil temperature as a result of tillage (Silva et al. 2006) but mainly by limited availability of water (Gubiani et al. 2014). However, soil disturbance by conventional tillage was associated with greater plant height and leaf area index of bean according to the findings of Lima et al. (2010). Moreover, Gubiani et al. (2014) observed an increase in leaf area index from soil chiseling compared with compacted no-tillage system with irrigation. Therefore, the positive results obtained from tillage systems can be masked by water stress and weather conditions.

Common bean grain yield did not differ statistically among tillage systems, although tilled soils (857 kg ha^{-1} for CT and 859 kg ha^{-1} for Ch_0) had higher yield in relation to consolidated soils (723 kg ha^{-1} for NT and 664 kg ha^{-1} for Ch_1). These yields were similar to the average yield obtained in the state of Rio Grande do Sul in the 2002/2003 cropping season (854 kg ha^{-1}) (CONAB 2014).

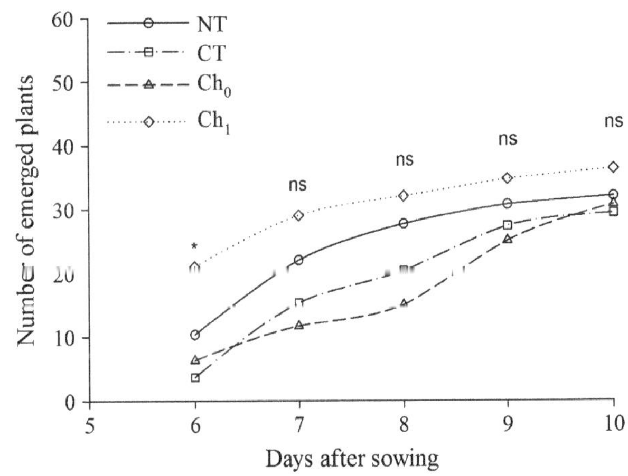

Figure 9. Common bean emergence in the sandy loam Ultisol under different tillage systems. *NT, no-tillage; CT, conventional tillage; Ch_0, chiseling 2 days before sowing; Ch_1, chiseling 1 year before sowing. **Significant and ns: not significant by the Least Significant Difference (LSD) test at 5% level of probability.

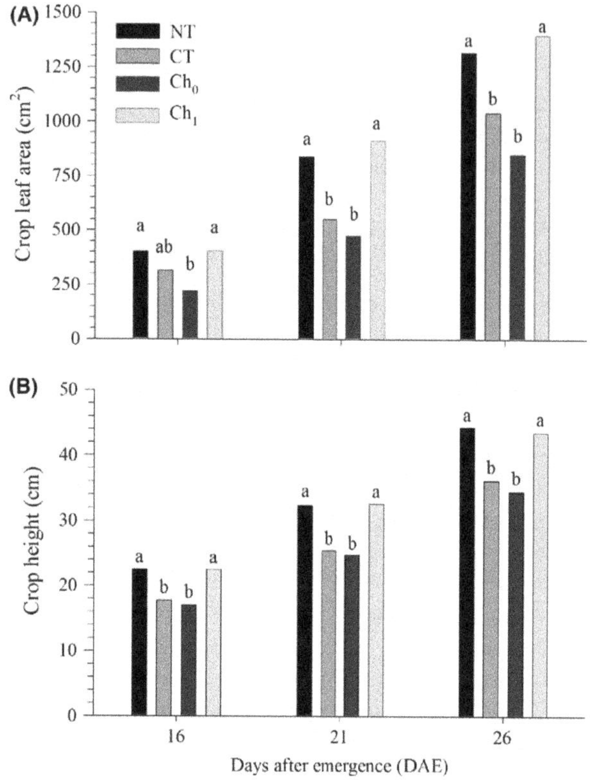

Figure 10. Temporal distribution of leaf area (A) and plant height (B) of common bean cultivated in the sandy loam Ultisol under different tillage systems. *NT, no-tillage; CT, conventional tillage; Ch_0, chiseling 2 days before sowing; Ch_1, chiseling 1 year before sowing. **Means followed by the same letter do not differ by the Least Significant Difference (LSD) test at 5% level of probability.

Common bean growth and development suffered adverse environmental conditions, especially during the flowering stage (38 DAE) and early grain formation (50 DAE), due to limited rainfall and high air temperatures (Fig. 1). For this period, in most days (64 days) during the growing period, average daily air temperature was greater than optimum temperature of 24°C for common beans (Balardin et al. 2000). At this stage of crop development, soil moisture content was close to the permanent wilting point (1500 kPa), mainly in the surface layer of the soil (Reichert et al. 2009b).

Critical air temperature (30°C) (Hoffmann Júnior et al. 2007) occurred at 15:00 h in 40 days, 13 of them were consecutive days between 35 to 47 days after emergence (reproductive period) (Fig. 1B). Temperatures above 30°C during the day provide negative effects on growth and development of common bean (Massignam et al. 1998), because high temperature influences the abscission of flowers and pods, seeds per pod, and low seed mass (Hoffmann Júnior et al. 2007). However, bean is more affected by high temperature in the pre-flowering (Shonnard and Gepts 1994), flowering (Balardin et al. 2000), and pod filling (Shonnard and Gepts 1994) stages, and abscission rate of reproductive organs can reach 50 to 70% of blooming flowers (Hoffmann Júnior et al. 2007). Negative effects on common bean growth and yield may be more severe with larger time where temperature remains above 30°C during the day, maximum temperature, and number of consecutive days with critical temperature (Barbano et al. 2001). High temperature associated with low availability water is the main factor reducing the productive potential of bean (Lima et al. 2006; Abrantes et al. 2011). Maximum temperatures above 28°C associated with water deficit at flowering period may cause yield reduction in common beans for some cultivars (Massignam et al. 1998). Thus, in our study, low soil available water and high air temperature during the flowering phase possibly affected abscission rate of reproductive organs (Fig. 1).

In study by Montanari et al. (2010) in an Oxisol, grain yield was 26.6% lower than the crop yield potential due to soil compaction and lower soil moisture in surface layer (0–0.30 m), despite the adoption of high technological level. Gubiani et al. (2014) found that supplemental irrigation significantly increased grain yield, as high as 2853 kg ha^{-1}, under compacted no-tillage system, with soil bulk density very close to critical (1.72 g cm^{-3}) and despite reduced crop growth. The authors affirmed that the losses in plant growth due to soil compaction can be offset by grain yield if there is enough availability of water, especially during the reproductive phase. Torres et al. (2013) observed water deficit or excess could reduce bean yield, while irrigation with application depth equivalent to 100% evapotranspiration gave the highest productivity.

Under water stress conditions, common bean yield is a function of bulk density and water status of the soil surface layer (Montanari et al. 2013). If water is readily available during the reproductive phase, losses on plant growth in compacted soil can be offset in grain yield (Gubiani et al. 2014), because after full fruiting, drought has little influence on production (Guimarães 1988).

To reduce the effects of uncontrollable environmental factors by soil tillage systems, such as air temperature and rainfall, soil management practices to retain crop residues as mulch on soil surface and eventual chiseling to disrupt soil layers compacted are feasible alternatives. Moreover, irrigation may be performed to reduce the combined effect of high temperature and water deficit on crop growth and development.

Conclusions

The effects of temporal variability of soil water retention and availability, and soil structural condition on growth and yield of common bean were studied under different soil management in a year of low rainfall in a subtropical climate.

The effects of temporal variability of soil water retention and availability, and soil structural condition on growth and yield of common bean were studied under different soil management in a year of low rainfall in a subtropical climate.

Soil tillage significantly affected soil water retention and availability, soil structural quality, and productivity of common bean. During periods of drought, undisturbed soil (NT and Ch_1) retained more water than tilled plots (CT and Ch_0). During grain filling stage, plant available water and void ratio were low for all tillage systems due to low recharge as a result of limited rainfall, with negative effect on crop yield.

Temporal distribution of soil available water correlated with only air temperature. Time series analysis showed its applicability for estimating available water from itself and related variables; however, other soil properties and processes affecting or related to soil water status, for example, evapotranspiration and soil temperature, should be incorporated in future studies for better understanding of the complex soil-plant-water relations.

Conservation agriculture, such as the adoption of no-tillage and old-chiseling improved soil water status. However, when weather conditions impose strong limitations (limited water) to crop growth and development, crop management should incorporate alternatives including irrigation to reduce the combined effect of high temperature and water deficit resulting from the effects of abiotic factors, such as air temperature and rainfall.

Conflict of Interest

None declared.

References

Abrantes, F. A., M. E. Sá, L. C. D. Souza, M. P. Silva, H. M. Simidu, M. Andreotti, et al. 2011. Uso do regulador de crescimento em cultivares de feijão de inverno. Pesq. Agropec. Trop. 42:148–154. (Abstract in English).

Abreu, S. L., J. M. Reichert, and D. J. Reinert. 2004. Escarificação mecânica e biológica para a redução da compactação em Argissolo franco-arenoso sob plantio direto. Rev. Bras. Cienc. Solo 28:519–531. (Abstract in English).

Ahuja, L. R., F. Fiedler, G. H. Dunn, J. G. Benjamin, and A. Garrison. 1998. Changes in soil water retention curves due to tillage and natural reconsolidation. Soil Sci. Soc. Am. J. 62:1228–1233.

Alam, M. K., M. M. Islam, N. Salahin, and M. Hasanuzzaman. 2014. Effect of tillage practices on soil properties and crop productivity in wheat-mungbean-rice cropping system under subtropical climatic conditions. Sci. World J. doi:10.1155/2014/437283. 2014: 15p.

Alletto, L., and Y. Coquet. 2009. Temporal and spatial variability of soil bulk density and near-saturated hydraulic conductivity under two contrasted tillage management systems. Geoderma 152:85–94.

Álvares, C. A., J. L. Stape, P. C. Sentelhas, J. L. M. Gonçalves, and G. Sparovek. 2013. Köppen's climate classification map for Brazil. Meteorol. Zeitschrift. 22:711–728.

Andrade, R. S., J. A. A. Moreira, L. F. Stone, and J. A. Carvalho. 2002. Consumo relativo de água do feijoeiro no plantio direto em função da porcentagem de cobertura morta do solo. Rev. Bras. Eng. Agric. Amb. 6:35–38.

Awe, G. O., J. M. Reichert, L. C. Timm, and O. O. Wendroth. 2015. Temporal processes of soil water status in a sugarcane field under residue management. Plant Soil 387:395–411.

Balardin, R. S., E. C. C. Costa, N. D. Ribeiro, L. M. C. Dutra, and I. F. D. Costa. 2000. Feijão: Recomendações técnicas para cultivo no Rio Grande do Sul. Santa Maria, Comissão Estadual de Pesquisa do Feijão – CEPEF, 80 p.

Bamberg, A. L., W. M. Cornelis, L. C. Timm, D. Gabriels, E. A. Pauletto, and L. F. S. Pinto. 2011. Temporal changes of soil physical and hydraulic properties in strawberry fields. Soil Use Manag. 27:385–394.

Barbano, M. T., O. Brunini, E. B. Wutke, J. L. Castro, P. B. Gallo, R. A. D. Kanthack, et al. 2001. Comparação entre valores observados e estimados de duração dos diferentes subperíodos de desenvolvimento da cultura do feijoeiro. Rev. Bras. Agromet. 9:103–110.

Blake, G. R., and K. H. Hartge. 1986. Bulk density in A. Klute (ed.) Methods of soil analysis. Part 1 - Physical and mineralogical methods. 2nd ed. American Society of Agronomy, Madison, WI. p. 363–375.

BRASIL. Ministério da Agricultura, Pecuária e Abastecimento. 2008. Instrução Normativa nº 2/2008. Adota no zoneamento agrícola de risco climático do Ministério da Agricultura, Pecuária e Abastecimento, as especificações para solo que descreve. MAA, Brasília. Diário Oficial da União de 10/10/2008, Seção 1. p.71.

Camara, K. R., and V. A. Klein. 2005. Propriedades físico hídricas do solo sob plantio direto escarificado e rendimento da soja. Cienc. Rural 35:813–819. (Abstract in English).

Collares, G. L., D. J. Reinert, J. M. Reichert, and D. R. Kaiser. 2006. Qualidade física do solo na produtividade da cultura do feijoeiro num Argissolo. Pesq. Agropec. Bras. 41:1663–1674.

Companhia Nacional de Abastecimento – CONAB. 2014. Levantamento de safra: Séries históricas. Available at: http://www.conab.gov.br/conteudos.php?a=1252&t=&Pagina_objcmsconteudos=2#A_objcmsconteudos (accessed 7 March 2015).

Dalmago, G. A., H. Bergamaschi, J. I. Bergonci, C. A. M. B. Kruger, F. Comiran, and B. M. M. Heckler. 2009. Retenção e disponibilidade de água às plantas, em solo sob plantio direto e preparo convencional. Rev. Bras. Eng. Agric. Amb. 13:855–864. (Abstract in English).

Dourado-Neto, D., L. C. Timm, J. C. M. Oliveira, K. Reichardt, O. O. S. Bacchi, T. T. Tominaga, et al. 1999. State-space approach for the analysis of soil water content and temperature in a sugarcane crop. Sci. Agric. 56:1215–1221.

Empresa Brasileira de Pesquisa Agropecuária – EMBRAPA. 2003. Cultivo do feijoeiro comum: Características da cultura. Embrapa Arroz e Feijão, Sistemas de produção, 2, Versão eletrônica janeiro 2003. Available at http://sistemasdeproducao.cnptia.embrapa.br/FontesHTML/Feijao/CultivodoFeijoeiro/index.htm. (accessed 7 March 2015).

Fabrizzi, K. P., F. O. Garcıa, J. L. Costa, and L. I. Picone. 2005. Soil water dynamics, physical properties and corn and wheat responses to minimum and no-tillage systems in the southern Pampas of Argentina. Soil Till. Res. 81(1):57–69.

FAO, Food and Agricultural Organisation of the United Nations Statistics. 2015. FAOSTAT database. Available at http://faostat3.fao.org. (accessed 7 March 2015).

Faria, L. C., P. G. S. Melo, H. S. Pereira, A. Wendland, S. F. Borges, I. A. Pereira Filho, et al. 2014. Genetic progress during 22 years of black bean improvement. Euphytica 199:261–272.

Fernandez-Ugalde, O., I. Virto, P. Bescansa, M. J. Imaz, A. Enrique, and D. L. Karlen. 2009. No-tillage improvement of soil physical quality in calcareous, degradation-prone, semiarid soils. Soil Till. Res. 106:29–35.

Fontana, D. C., and T. S. Almeida. 2002. Climatologia do número de dias com precipitação pluvial no Estado do Rio Grande do Sul. Rev. Bras. Agromet. 10:135–145.

Freddi, O. S., M. P. Carvalho, J. F. Centurion, and G. F. Barbosa. 2005. Variabilidade espacial da produtividade do feijão em um Latossolo Vermelho distroférrico sob preparo convencional e plantio direto. Acta Sci. Agron. 27:61–67. (Abstract in English).

Gubiani, P. I., J. M. Reichert, C. Campbell, D. J. Reinert, and N. S. Gelain. 2013. Assessing errors and accuracy in dew-point potentiometer and pressure plate extractor measurements. Soil Sci. Soc. Am. J. 76:19–24.

Gubiani, P. I., J. M. Reichert, and D. J. Reinert. 2014. Interação entre disponibilidade de água e compactação do solo no crescimento e na produção de feijoeiro. Rev. Bras. Cienc. Solo 38:765–773. (Abstract in English).

Guimarães, C. M.. 1988. Fatores fisiológicos do estresse hídrico. Pp. 157–174 *in* M. J. O. Zimmermann, M. Rocha, and T. Yamada, eds. Cultura do feijoeiro: Fatores que afetam a produtividade. Associação Brasileira para Pesquisa da Potassa e do Fosfato, Piracicaba, Brasil.

Hoffmann Júnior, L., N. D. Ribeiro, S. S. Rosa, E. Jost, N. L. Poersch, and S. L. P. Medeiros. 2007. Resposta de cultivares de feijão à alta temperatura do ar no período reprodutivo. Cienc. Rural 37:1543–1548. (Abstract in English).

Horn, R., and H. Fleige. 2003. A method for assessing the impact of load on mechanical stability and on physical properties of soils. Soil Till. Res. 73:89–99.

Hui, S., O. Wendroth, M. B. Parlange, and D. R. Nielsen. 1998. Soil variability-Infiltration relationships of agroecosystems. J. Balkan. Ecol. 1:21–40.

Jia, Y., M. Shao, and X. Jia. 2013. Spatial pattern of soil moisture and its temporal stability within profiles on a loess slope in northwestern China. J. Hydrol. 495:150–161.

Kaiser, D. R., D. J. Reinert, J. M. Reichert, G. L. Collares, and M. Kunz. 2009. Intervalo hídrico ótimo no perfil explorado pelas raízes de feijoeiro em um Latossolo sob diferentes níveis de compactação. Rev. Bras. Cienc. Solo 33:845–855. (Abstract in English).

Klein, V. A., J. M. Reichert, and D. J. Reinert. 2006. Água disponível em um Latossolo Vermelho argiloso e murcha fisiológica de culturas. Rev. Bras. Eng. Agríc. Amb., 10:646–650. (Abstract in English).

Klute, A.. 1986. Water retention: laboratory methods. Pp. 635–662 *in* C. A. Black. Methods of Soil Analysis. I. Physical and mineralogical methods. American Society of Agronomy, Soil Science Society of America, Madison.

Lamandé, M., V. Hallaire, P. Curmi, G. Peres, and D. Cluzeau. 2003. Changes in pore morphology, infiltration and earthworm community in a loamy soil under different agricultural management. Catena 54:637–649.

Lima, C. L. R., D. J. Reinert, J. M. Reichert, L. E. A. S. Suzuki, and P. I. Gubiani. 2006. Qualidade físico-hídrica e rendimento de soja (*Glycine max* L.) e feijão (*Phaseolus vulgaris* L.) de um Argissolo Vermelho distrófico sob diferentes sistemas de manejo. Cienc. Rural 36:1172–1178. (Abstract in English).

Lima, C. L. R., D. J. Reinert, J. M. Reichert, and L. E. A. S. Suzuki. 2010. Produtividade de culturas e resistência à penetração de ArgissoloVermelho sob diferentes manejos. Pesq. Agropec. Bras. 45:89–98. (Abstract in English).

Massignam, A. M., H. J. Vieira, S. Hemp, and R. D. Flesch. 1998. Ecofisiologia do feijoeiro. II – Redução do rendimento pela ocorrência de altas temperaturas no florescimento. Rev. Bras. Agromet. 6:41–45. (Abstract in English).

Møldrup, P., T. Olesen, T. Komatsu, P. Schjønning, and D. E. Rolston. 2001. Tortuosity, diffusivity, and permeability in the soil liquid and gaseous phases. Soil Sci. Soc. Am. J. 65:613–623.

Montanari, R., M. P. Carvalho, M. Andreotti, F. C. Dalchiavon, L. H. Lovera, and M. A. O. Honorato. 2010. Aspectos da produtividade do feijão correlacionados com atributos físicos do solo sob elevado nível tecnológico de manejo. Rev. Bras. Cienc. Solo 34:1811–1822. (Abstract in English).

Montanari, R., D. M. P. Pellin, C. A. Silva Júnior, M. Passos e Carvalho, M. T. Gioia, F. C. Dalchiavon, et al. 2013. Correlação entre produção de feijão e atributos físicos de um Latossolo em Mato Grosso do Sul. Rev. Ceres 60:772–784. (Abstract in English).

Moreno, J. A. 1961. Clima do Rio Grande do Sul. Secretaria da Agricultura, Diretoria de Terras e Colonizações, Seção de Geografia, Porto Alegre 46 p.

Nicoloso, R. S., T. J. C. Amado, S. Schneider, M. E. Lanzanova, V. C. Girardello, and J. Bragagnolo. 2008. Eficiência da escarificação mecânica e biológica na melhoria dos atributos físicos de um Latossolo muito argiloso e no incremento do rendimento de soja. Rev. Bras. Cienc. Solo 32:1723–1734. (Abstract in English).

Nielsen, D. R., and O. Wendroth. 2003. Spatial and temporal statistics: sampling field soils and their vegetation. Catena Verlag, Reiskirchen 398 pp.

Portela, J. C., P. L. Libardi, and Q. J. Van Lier. 2001. Retenção de água em solo sob diferentes usos no ecossistema tabuleiros costeiros. Rev. Bras. Eng. Agríc. Amb. 5:49–54. (Abstract in English).

Reichert, J. M., L. E. A. S. Suzuki, D. J. Reinert, R. Horn, and I. Håkansson. 2009a. Reference bulk density and critical degree-of-compactness for no-till crop production in subtropical highly weathered soils. Soil Till. Res. 102:242–254.

Reichert, J. M., D. R. Kaiser, D. J. Reinert, and U. F. B. Riquelme. 2009b. Variação temporal de propriedades físicas do solo e crescimento radicular de feijoeiro em quatro sistemas de manejo. Pesq. Agropec. Bras. 44:310–319. (Abstract in English).

Reichert, J. M., J. A. Albuquerque, D. R. Kaiser, D. J. Reinert, F. L. Urach, and R. Carlesso. 2009c. Estimation of water retention and availability in soils of Rio Grande do Sul. Rev. Bras. Cienc. Solo 33:1547–1560.

Reichert, J. M., C. M. P. Bervald, M. F. Rodrigues, O. R. Kato, and D. J. Reinert. 2014. Mechanized land preparation in eastern Amazon in fire-free forest-based fallow systems as alternatives to slash and burn practices: hydraulic and mechanical soil properties. Agric. Ecosyst Environ. 19:47–60.

Reinert, D. J., and J. M. Reichert. 2006. Coluna de areia para medir a retenção de água no solo – protótipos e teste. Cienc. Rural 36:1931–1935. (Abstract in English).

Reynolds, W. D., B. T. Bowman, C. F. Drury, C. S. Tan, and X. Lu. 2002. Indicators of good soil physical quality and storage parameters. Geoderma 110:131–146.

Ribeiro, K. D., S. M. Menezes, M. G. B. F. Mesquita, and F. M. T. Sampaio. 2007. Propriedades físicas do solo, influenciadas pela distribuição de poros, de seis classes de solos da região de Lavras-MG. Cienc. Agrotec. 31:1167–1175. (Abstract in English).

Rodrigues, S., A. P. Silva, N. F. B. Giarola, and J. A. Rosa. 2011. Permeabilidade ao ar em Latossolo Vermelho sob diferentes sistemas de manejo. Rev. Bras. Cienc. Solo. 35:105–114.

Rosa, D. P., J. M. Reichert, A. Sattler, D. J. Reinert, M. I. Mentges, and D. A. Vieira. 2008. Relações entre solo e haste sulcadora de semeadora em Latossolo escarificado em diferentes épocas. Pesq. Agropec. Bras. 43:395–400. (Abstract in English).

Santos, H. G., P. K. T. Jacomine, L. H. C. Anjos, V. A. Oliveira, J. F. Lumbreras, M. R. Coelho, et al. 2013. Sistema Brasileiro de Classificação de Solos. 2 ed. rev. ampl. Brasília, DF, Embrapa. 353 p.

SAS Institute. 1991. Statistical Analysis System Institute – SAS/STAT Procedure guide for personal computers. Version 5, SAS Inst., Cary, NC.

Schjønning, P., L. J. Munkholm, P. Moldrup, and O. H. Jacobsen. 2012. Modeling soil pore characteristics from measurements of air exchange change: the long term effects of fertilization and crop rotation. Eur. J. Soil Sci. 53:331–339.

Schwen, A., G. Bodner, P. Scholl, G. D. Buchan, and W. Loiskandl. 2011. Temporal dynamics of soil hydraulic properties and the water-conducting porosity under different tillage. Soil Till. Res. 113:89–98.

Shonnard, G. C., and P. Gepts. 1994. Genetic of heat tolerance during reproductive development in common bean. Crop Sci. 34:1168–1175.

Shumway, R. H. 1988. Applied statistical time series analysis. Prentice Hall, Englewood Cliffs.

Silva, V. R., D. J. Reinert, and J. M. Reichert. 2004. Variabilidade espacial da resistência à penetração em plantio direto. Cienc. Rural 34:399–406. (Abstract in English).

Silva, V. R., J. M. Reichert, and D. J. Reinert. 2006. Variação na temperatura do solo em três sistemas de manejo na cultura do feijão. Rev. Bras. Cienc. Solo 30:391–399. (Abstract in English).

SOIL SURVEY STAFF – USDA. 1999. Soil Taxononmy – a basic system of soil classification for making and interpreting soil survey. 2nd edn. USDA, Washington 871 p.

Stepniewski, W., J. Glinski, and B. C. Ball. 1994. Effects of compaction on soil aeration properties. Pp. 167–189 *in* B. D. Soane and C. van Ouwekerk, eds. Soil Compaction in Crop Production. Elsevier, the Netherlands.

Sur, C., Y. Jung, and M. Choi. 2013. Temporal stability and variability of field scale soil moisture on mountainous hillslopes in Northeast Asia. Geoderma 207–208:234–243.

Tang, A. M., Y. Cui, G. Richard, and P. Défossez. 2011. A study on the air permeability as affected by compression of three French soils. Geoderma 162:171–181.

Tavares Filho, J., and D. Tessier. 2009. Characterization of soil structure and porosity under long-term conventional tillage and no-tillage systems. Rev. Bras. Cienc. Solo 33:1837–1844.

Timm, L. C., L. F. Pires, R. Roveratti, R. C. J. Arthur, K. Reichardt, J. C. M. Oliveira, et al. 2006. Field spatial and temporal patterns of soil water content and bulk density changes. Sci. Agric. 63:55–64.

Timm, L. C., D. Dourado-Neto, O. O. S. Bacchi, W. Hu, R. P. Bortolotto, A. L. Silva, et al. 2011. Temporal variability of soil water storage evaluated for a coffee field. Soil Res 49:77–86.

Torres, J. L. R., M. J. Santana, A. Pizolato Neto, M. G. Pereira, and D. M. S. Vieira. 2013. Produtividade de feijão sobre lâminas de irrigação e coberturas de solo. Bioscienc. J. 29:833–841.

Tuli, A., J. W. Hopmans, D. E. Rolston, and P. Moldrup. 2005. Comparison of air and water permeability between disturbed and undisturbed soils. Soil Sci. Soc. Am. J. 69:1361–1371.

Veiga, M., D. J. Reinert, J. M. Reichert, and D. R. Kaiser. 2008. Short and long-term effects of tillage systems and nutrient sources on soil physical properties of a Southern Brazilian Hapludox. Rev. Bras. Cienc. Solo 32:1437–1446.

Veiga, M., D. J. Reinert, and J. M. Reichert. 2010. Tillage systems and nutrient sources affecting soil cover, temperature and moisture in a clayey Oxisol under corn. Rev. Bras. Cienc. Solo 34:2011–2020.

8

Fruiting efficiency: an alternative trait to further rise wheat yield

Gustavo A. Slafer[1,2], Mónica Elia[1], Roxana Savin[1], Guillermo A. García[3,4], Ignacio I. Terrile[5], Ariel Ferrante[6], Daniel J. Miralles[3,4] & Fernanda G. González[4,5]

[1]Department of Crop and Forest Sciences and AGROTECNIO (Center for Research in Agrotechnology), University of Lleida, Av. Rovira Roure 191, 25198 Lleida, Spain
[2]ICREA, Catalonian Institution for Research and Advanced Studies, Barcelona, Spain
[3]IFEVA and Department of Plant Production, Faculty of Agronomy, University of Buenos Aires, Av. San Martin 4453, C1417DSE Buenos Aires, Argentina
[4]CONICET (National Scientific and Technical Research Council), Buenos Aires, Argentina
[5]EEA Pergamino, INTA (National Institute for Agricultural Technology)Ruta 32 km 4.5, B2700WAA Pergamino, Buenos Aires, Argentina
[6]Queensland Alliance for Agriculture and Food Innovation (QAAFI), The University of Queensland, P.O. Box 102, Toowoomba, Queensland 4350, Australia

Keywords
Floret development, grain number, partitioning, spike dry weight, *Triticum aestivum*, *Triticum durum*, yield component.

Correspondence
Gustavo A. Slafer, Department of Crop and Forest Sciences and AGROTECNIO (Center for Research in Agrotechnology), University of Lleida, Av. Rovira Roure 191, 25198 Lleida,

E-mail: slafer@pvcf.udl.cat

Funding Information
The work on fruiting efficiency has been funded by grants from the "National Plan" of Research in Spain, grants AGL2009-11964 and AGL2012-35300 at the Crop Physiology Lab of the University of Lleida; National Agency of Scientific and Technical Promotion (PICT-2008-1039) and INTA (PNCER-1336) at CONICET and INTA Lab, and PICT RAICES 1368 and UBACyT G-076 competitive grants at the University of Buenos Aires

Abstract

Further improvements in wheat yields are critical, for which increases in grain number would be required. In the recent past, higher grain number was achieved through increased growth of the juvenile spikes before anthesis, due to the reduction in stem growth. As current cultivars have already an optimum height, alternatives must be identified for further increasing grain number. One of them is increasing fruiting efficiency (grains set per unit of spike dry weight at anthesis). Fruiting efficiency is the final outcome of the fate of floret development and differences in this trait within modern cultivars would be related to higher survival of floret primordia. Then there are two alternative physiological pathways to improve fruiting efficiency by allowing a normal development of most vulnerable floret primordia: an increased allocation of assimilates for the developing florets before anthesis, or reduced demand of the florets for maintaining their normal development. Both alternatives may be possible, and it might be critical to recognize which of them is the actual cause of differences in fruiting efficiency. When considering this trait in breeding we must be aware of potential trade-offs and therefore it must be avoided that increases in fruiting efficiency be constitutively related to decreases in either spike dry weight at anthesis or grain weight. In this review we described fruiting efficiency and its physiological bases, analyzing genetic variation and considering potential drawbacks that must be taken into account to avoid increases in fruiting efficiency being compensated by other traits.

Introduction

Overall context

Wheat is critical for food security, as it is the crop most widely grown and is the primary source of protein for the world population (Chand 2009; Braun et al. 2010). Genetic gains in wheat yield have been outstanding during the second half of the 20th century (Calderini et al. 1999; Foulkes and Reynolds 2015; and references quoted therein). This success has been instrumental for, roughly contributing 50% (Slafer and Andrade 1991) to the increases in production, more than matching the increased demand produced by an unprecedented growth in world population (more than doubling in just half-century the population built up in the previous 2000 centuries, since the beginning of our species). Thus genetic gains were critical for the improved levels of food security in the last half century.

The population is still growing fast and, in addition, there is an on-going change in diet habits, both factors determining that the demand will further increase over the next decades. This time, the satisfaction of the increased demand must mainly come from yield improvement as the amount of land for agriculture will not increase, and may even decline (Albajes et al. 2013). It may be required that yield of wheat (and other major crops) be increased by at least 50% in the next few decades (Reynolds et al. 2009), which will depend on improving yield potential (Fischer and Edmeades 2010; Hall and Richards 2012). As crop management should be environmentally more sustainable in the future (Godfray 2011; Tilman et al. 2011), the contribution from genetic gains should be similar to that of the few decades following the Green Revolution. But, in the last 2 decades wheat yield grew less than the demand and in several regions showed signs of stagnation (Calderini and Slafer 1998; Brisson et al. 2010; Ray et al. 2012). Likewise, genetic gains reported for different countries for the last decades also seem to have been increasing less than required (Shearman et al. 2005; Acreche et al. 2008; Sadras and Lawson 2011; Lopes et al. 2012).

How was yield improved in the past?

It seems reasonable to speculate that a better understanding of crop yield physiology would help to increase the current rates of genetic gains (Slafer 2003; Araus et al. 2008; Reynolds et al. 2012). Based on the model proposed long time ago by Fischer (1984), and well developed in a recent review (Fischer 2011), the diagram in Figure 1 shows that:

- Wheat yield is most commonly sink-limited during grain filling; which means that the amount of available assimilates (current photosynthesis plus remobilization of preanthesis reserves) are in excess to the demand of the growing grains (Borras et al. 2004; González et al. 2014), not only under high-yielding but also across a wide range of conditions (Pedro et al. 2011; Serrago et al. 2013). Thus, genotypic differences in yield are frequently strongly related to those in grain number (Garcia et al. 2013; Slafer et al. 2014), as it is more plastic than grain weight (Sadras and Slafer 2012). But for a particular number of grains there is variation in grain size which does also affect yield though not as primarily (in most conditions) as grain number (Slafer et al. 2014). In fact, genetic gains in wheat yield have been more related to improvements in the number than in the size of the grains (e.g., Canevara et al. 1994; Calderini et al. 1995; Sayre et al. 1997; Shearman et al. 2005; Acreche et al. 2008).
- Grain number, in turn, is strongly related to spike dry weight at anthesis; which is quite reasonable as: (1) wheat is a cleistogamous plant and therefore most fertile florets set grains and consequently grain number is related to the number of fertile florets; (2) final number of fertile florets depends on the developmental process of floret generation/degeneration; (3) this developmental process occurs in the growing spike before anthesis; and (4) is related to the availability of resources (González et al. 2011a; Ferrante et al. 2013b; Dreccer et al. 2014). Therefore, the final number of grains is source-limited during preanthesis and depends mechanistically on the growth of the juvenile spikes in which floret primordia are developing (Slafer et al. 2005) and, for particular levels of spike dry weight, on the efficiency with which the resources are used to set grains or fruiting efficiency[1] (González et al. 2011b; Ferrante et al. 2012; García et al. 2014). Therefore, breeding through the Green Revolution improved spike growth (and consequently spike dry weight at anthesis) bringing about a reduced rate of floret mortality increasing the number of fertile florets (e.g., Miralles et al. 1998).
- Spike growth during the preflowering period is the result of crop growth during that period and the proportion of that growth partitioned to the spikes. Virtually all papers analyzing these aspects in breeding during the second half of the 20th century agreed in that there were not trends to systematically and consistently modify growth of the crop whilst there was a consistent trend to increase partitioning of biomass to the reproductive organs (e.g., Siddique et al. 1989b; Slafer and Andrade 1993).

When considering how wheat breeding has successfully achieved large gains in the second half of the 20th century, it seems quite straightforward. The tremendous success of the Green Revolution was achieved through relatively simple (seen retrospectively) interventions in yield physiology (Fig. 1, open arrows on left panels and panels on the right). Although exceptional cases could eventually be found, the vast majority of studies of physiological attributes explaining genetic gains in wheat yield coincides in that breeding during the second half of the last century:

- Did not systematically affect biomass production of the crop (Austin et al. 1980, 1989; Perry and D`Antuono 1989;

Siddique et al. 1989a; Slafer and Andrade 1991; Sayre et al. 1997; Calderini et al. 1999), but increased the partitioning toward the developing juvenile spike due to genetic restrictions to stem growth (Siddique et al. 1989a,b; Slafer and Andrade 1993), so that during the stem elongation phase, the stem requirements for extension were reduced and partitioning to the spikes favoured,

- Consequently, even with no trends to improve crop growth, spike dry weight at anthesis consistently increased whilst stem length was reduced from traditional to semi-dwarf plant statures; and the increased growth of the juvenile spikes brought about improvements in floret development increasing spike fertility (Miralles et al. 1998),
- As spike fertility increased, the number of grains set by the crop in modern cultivars improved compared to their predecessors and the increased postanthesis sink strength resulted in concomitant increases in yield (Sayre et al. 1997; Abbate et al. 1998; Calderini et al. 1999).

Consequently, when comparing cultivars released at different times from before to after the Green Revolution, or when comparing isogenic lines for dwarfism (Rht lines), it can be seen that modern cultivars (or semi-dwarfing lines) out-yield the older ones (or the tall lines) due to having more grains associated with higher spike dry weight at anthesis (e.g., Fischer and Stockman 1980; Brooking and Kirby 1981; Siddique et al. 1989b; Slafer and Andrade 1993; Miralles et al. 1998; Acreche et al. 2008).

The problem that fruiting efficiency could help to solve

Past breeding during the Green Revolution and the few decades following it was extremely successful. In the process plant type was modified optimizing its height and consequently current high-yielding cultivars possess a height that is within the optimum range to maximize yield (c. 0.7–1.0 m: Richards 1992; Miralles and Slafer 1995a, 1997; Flintham et al. 1997). This means that the main trait genetically manipulated to obtain remarkable genetic gains in yield cannot be further deployed: taller plants would have penalties due to poor partitioning, and higher risk of lodging; shorter plants would result in lower crop growth as radiation use efficiency would be reduced due to poor radiation distribution within the canopy (Miralles and Slafer 1995a). For future genetic gains to recover the pace that characterized breeding in the recent

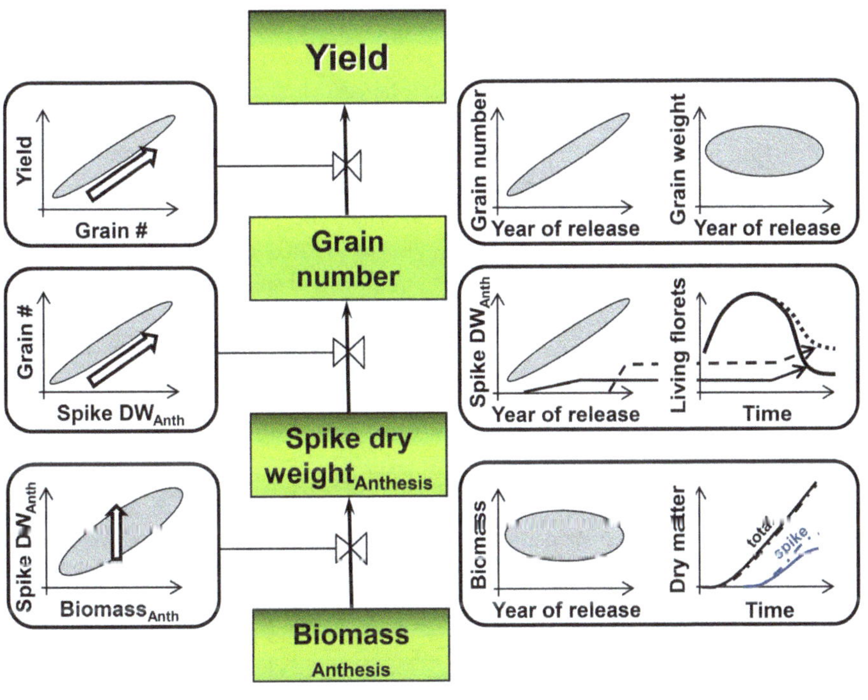

Figure 1. Simplified conceptual model of the physiological components of yield in wheat (Fischer 1984, 2011; Slafer et al. 2005). On the left the most common relationships considering wide ranges in yield produced by genotypic or environmental factors are shown. The open arrows point out the main pathways used by breeding to produce genetic gains in yield during the second half of the 20th century. Panels on the right exemplify relationships most commonly found when the source of variation has been cultivars released at different years during the 20th century (when the abscissa is "time" it refers to the time from the onset of stem elongation to anthesis, and in these cases the solid and dashed lines stand for cultivars released before or after the Green Revolution, respectively). For more detail see Calderini et al. (1999) and Foulkes and Reynolds (2015).

past, alternative traits putatively related with grain number and yield, must be identified and exploited.

One avenue would be identifying sources of variation to increase crop photosynthesis, so that higher biomass at anthesis would result in more spike dry weight at anthesis bringing about more grains set and higher yields (Parry et al. 2011; Reynolds et al. 2012). Another alternative, not mutually exclusive with the previous one, could be to lengthen the duration of the spike growth period (or more broadly the duration of the stem elongation phase) so that for a constant rate of growth and level of partitioning, the longer the phase the higher the spike dry weight at anthesis (Slafer et al. 2001; Miralles and Slafer 2007; González et al. 2011b). Scaling up one step in Figure 1, there is another alternative which is to increase fruiting efficiency; that is the number of grains produced per unit of spike dry weight at anthesis. In other words, the efficiency with which the resources allocated to the spikes are used to produce a certain number of grains (Abbate et al. 1998, 2013; González et al. 2011b; Ferrante et al. 2012; Garcia et al. 2014).

Objective

The aim of this paper is to review the state-of-the-art in wheat fruiting efficiency. For that purpose we (1) described the trait and its physiological bases, (2) revised to what degree fruiting efficiency has been deployed in past breeding, (3) reviewed the literature to quantify the degree of variation available for the trait, particularly within elite material including modern cultivars, and finally (4) considered potential drawbacks that must be taken into account to avoid increases in fruiting efficiency not resulting in yield gains by being compensated by other traits.

Pathways to Increase Fruiting Efficiency

As indicated above, fruiting efficiency is estimated as the ratio between grain number (determined after grain set, normally at maturity) and spike dry weight at anthesis. It reflects the overall efficiency with which a certain amount of resources, allocated to the juvenile spike growing before anthesis, are used to set grains. Grains are the product of pollinated fertile florets, which in turn are the result of floret primordia development; which occurs within the spike (within the spikelets of the spike) precisely during the period when the juvenile spike grows before anthesis (Kirby 1988; González et al. 2011a; Slafer et al. 2015). It has been many times shown that floret primordia development is rather responsive to the availability of assimilates.

Studying the dynamics of floret development for contrasting availability of resources can illustrate the responsiveness of this developmental process. For instance, one of the treatments imposed by Ferrante et al. (2013a) consisted in detillering plants during stem elongation. Then, the fate of floret primordia in the main-shoot spikes of plants tillering freely (and then growing under competition with other spikes) was compared with the ones being detillered permanently (and then growing in the absence of the competition from other spikes). Detillering the plants did not affect the developmental patterns of the most proximal florets (which do develop to produce fertile florets in virtually any condition), but more vulnerable florets in less proximal positions within the spikelets (those that develop to produce fertile florets or die in the process, depending on the conditions; for example, florets 3, 4, and 5 from the rachis) developed more in the spikes of the detillered plants than in the spikes of the plants with free tillering (Fig. 2). This increased survival of the distal florets was associated with increased assimilate availability to the growth of the main stem spike in the detillered plants (Ferrante et al. 2013a).

This is the main reason why it has been widely documented that the number of grains is drastically sensitive to spike growth immediately before anthesis, and therefore it decreases sharply in response to shading during the period when spike growth takes place (Fischer1985; Savin and Slafer 1991; Slafer et al. 1994; Abbate et al. 1997; Demotes-Mainard et al. 1999; Demotes-Mainard and Jeuffroy 2004), as well as to changes in duration of the spike growth phase as affected by manipulation of day-length in sensitive cultivars (Miralles et al. 2000; González et al. 2003, 2005b; Fischer 2007; Serrago et al. 2008). These responses are mostly related to the positive relationship between grain number and spike dry weight at anthesis, that also explains for instance grain number (and yield) responsiveness to nitrogen fertilization (Fischer 1993; Abbate et al. 1995; Demotes-Mainard et al. 1999; Demotes-Mainard and Jeuffroy 2001; Prystupa et al. 2004; Ferrante et al. 2010, 2012, 2013a) or to general environmental changes affecting yield (Marti and Slafer 2014), as well as to improvements in biomass partitioning to the juvenile spikes (as discussed above, explaining the main physiological reason for the Green Revolution). More recently, it has been also shown the mortality of floret primordia (the main component explaining the final number of fertile florets and grains) seems to be clearly dependent upon the availability of assimilates (González et al. 2011a; Ferrante et al. 2013b).

Fruiting efficiency is the final outcome of floret developmental rates (determining the number of floret primordia that reach the stage of fertile florets at anthesis) and the proportion of grain set per fertile floret (the opposite to grain abortion), per unit of spike weight

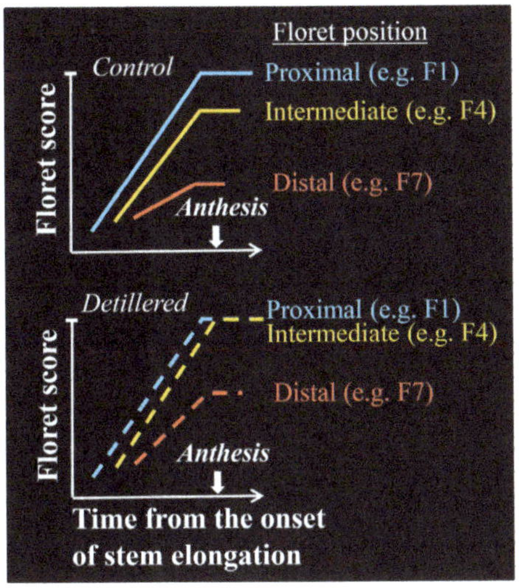

Figure 2. Schematic diagram highlighting changes in rate of development of florets at different positions within the spikelets (from the rachis, being F1 the most proximal) for main-shoot spikes of control or detillered plants (the spikes of the latter growing with more availability of resources). See details in Ferrante et al. (2013a).

(i.e., more efficient genotypes would show a higher survival of floret primordia, and/or a reduced level of grain abortion). As modern wheat cultivars show low grain abortion (Siddique et al. 1989a,b; González et al. 2003, 2005a), differences in fruiting efficiency in modern cultivars maybe likely related to higher survival of floret primordia.

Therefore, there are two alternative physiological pathways (Fig. 3) to improve fruiting efficiency, by allowing a normal development of most vulnerable floret primordia (those in more distal positions of the spikelets), through maintaining cell division in these florets and avoiding the initiation of autophagy[2] (Ghiglione et al. 2008). These are:

- An increased allocation of assimilates for the florets developing during spike growth before anthesis (for a particular level of spike growth). This would result in less investments in the structural pieces of the spike (rachis, glumes) in favour of an increased allocation to the growth of the florets, or
- A reduced demand of the florets for maintaining their normal development. If florets constitutively demand less assimilates for normal development, the proximal florets would leave more resources available for more distal florets which would maintain their growth and development normally for longer. Then, the intermediate florets would increase their likelihood to become fertile florets at anthesis (and to produce a grain). The most likely consequence of this would be that fertile florets at anthesis would be smaller in the genotype with higher fruiting efficiency.

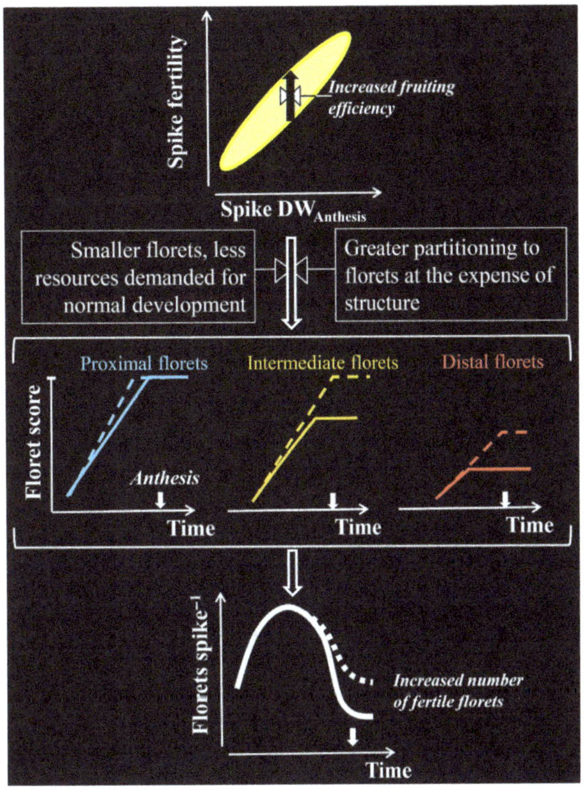

Figure 3. Scheme presenting the two alternative physiological pathways that could bring about increases in fruiting efficiency (dashed lines), either by reducing the demands of individual florets or by increasing the spike dry matter partitioning towards florets. Proximal florets would be fertile in any case, but intermediate florets would increase their likelihood to become fertile florets and even distal florets which will not be fertile florets in any case would reach a higher level of development if fruiting efficiency is increased.

As in some studies differences in fruiting efficiency seemed associated with dry matter partitioning within the spike at anthesis (Slafer and Andrade 1993; Abbate et al. 1997), whilst in others such relationship was not identified (Abbate et al. 1998; Fischer 2007), both alternatives may be possible. It might be critical to recognize which of them is the actual cause of differences in fruiting efficiency as the potential trade-offs with grain weight are quite different (see later discussion on *Potential drawbacks*).

One inconvenience of fruiting efficiency is that it is a trait requiring the sampling and processing of samples at two different stages: spike dry weight must be determined at anthesis[3] and grain number when it has been fixed, at least 2–3 weeks later, most normally at maturity. To make this simpler, sometimes the dry weight of the chaff at maturity (structural parts of the spike after removing the grains) is used to provide an estimate of the dry weight of the spikes at anthesis (e.g., Abbate et al. 2013; Marti and Slafer 2014). For this to be accurate there should be no growth in the nongrain parts of the spikes during grain filling. Although to the best of our knowledge reasons have not been understood, there are clear indications that chaff weight at maturity is consistently higher than spike dry weight at anthesis (or than the nongrain spike dry weight a week after anthesis). Fischer (2011) estimated this difference to be neither minor nor constant (chaff weight can be 20–50% greater), based on relatively old studies (Wall 1979; Fischer and Stockman 1980; Stockman et al. 1983) in which the focus was not genotypic variation in fruiting efficiency. Not many studies analyzed (or at least reported) genetic variation in spike dry weight at anthesis and chaff weight at maturity within high-yielding modern cultivars. The two of which we are aware of showed data that broadly confirmed that the estimated difference reported by Fischer (2011) was not an exception. In the papers by González et al. (2011b) and Abbate et al. (2013) chaff weight at maturity was, without a single exception, greater than spike dry weight at anthesis (averaging across genotypes and experiments a difference of 32 and 43%, respectively), with a substantial variation due to genotypes within each of the experiments; and there was a consistent trend to increase the magnitude of the underestimation with increases in fruiting efficiency as the slope was substantially lower than 1 (Fig. 4). Expanding further the analysis, when the variation was not restricted to modern cultivars but to a much larger population of double haploid lines (Garcia et al. 2014), the general pattern that using chaff at maturity does consistently underestimate fruiting efficiency (and the higher the efficiency the greater the magnitude of the overestimation) is further emphasized (Fig. 4). Thus, even though there was a significant positive relationship between both estimates, it seems clear that using chaff at maturity could provide only a rough proxy to fruiting efficiency.

In some cases it has been preferred to determine spike dry weight few days after anthesis to consider the period of grain set (during the lag-phase of grain filling; Loss et al., 1989) as well. Although this is rather sensible, it has two major inconveniences. Firstly a huge amount of extra work is required to remove the tiny grains that might have already started to grow (otherwise the spike dry weight would be strongly overestimated and the overestimation would vary between genotypes (and environments) depending on the number of grains set and the potential size of the grains. Secondly, the estimated spike dry weight would include any eventual growth occurring during that extra week, which is naturally positive, but would exclude the weight contributed by the fertile florets to the spike dry weight at anthesis, which would be a drawback. All in all, we believe that it would be convenient to standardize the determination of fruiting efficiency using spike dry weight at anthesis.

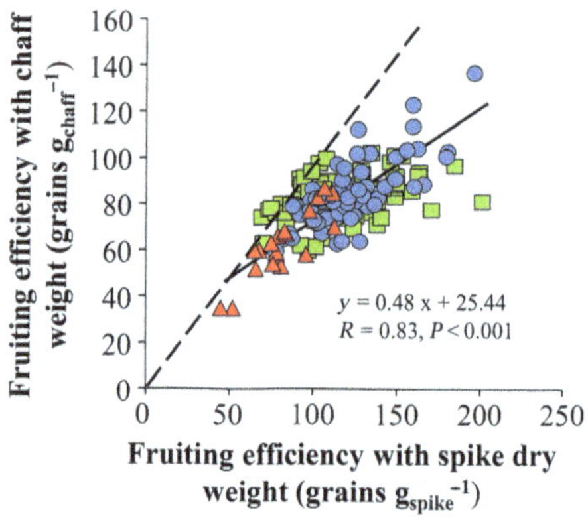

Figure 4. Relationship between fruiting efficiency estimated with the chaff weight measured at maturity and estimated with spike dry weight at anthesis. Data adapted from Abbate et al. (2013, red triangles) and González et al. (2011b, blue circles) comparing modern cultivars. Solid line stands for the regression of the pooled data of these two sources, dashed line is the 1:1 ratio ($y = x$). In addition, we added data adapted from Garcia et al. (2014) who analyzed variation among a mapping population with a much larger degree of variation than modern cultivars that broadly fall within the same cloud of data-points but with an even smaller slope; data from the study in which spike dry weight was measured at anthesis and values at chaff were unpublished).

Contributions of Fruiting Efficiency to Wheat Breeding

Was fruiting efficiency improved during past breeding?

Due to the dominating effect of reducing stem growth (semi-dwarfism), favoring the growth of the juvenile spike bringing about improvements in spike fertility and yield (see above), there have been only limited attempts to quantify the impact of past breeding on fruiting efficiency. In this section we revisited the very few cases in which this trait was analyzed in experiments comparing side-by-side cultivars released at different times (covering at least 20 years of breeding) under field conditions (Table 1).

Within the few cases available, there were no consistent trends: no increases in fruiting efficiency were observed at all in Argentina, there seemed to have been a clear positive trend in Spain; and the relationship was not significant but a positive trend might be inferred in the UK (Fig. 5). Grain number was increased in all the three countries analyzed, in Argentina the improvement in grain number was exclusively due to parallel increases in spike growth before anthesis (Slafer and Andrade 1993; González et al. 2003), while in Spain it was associated with spike growth before anthesis as well but also with fruiting efficiency of the cultivars (Acreche et al. 2008). In the UK (Shearman et al. 2005) increases in grain number were also consequence of increased

Table 1. Description of studies in which breeding impact on fruiting efficiency was estimated (or the data are available for the estimation). All experiments carried out under high-yielding conditions.

Location	Cultivar	Year of release	Grain number[1] (10^{-3} m^{-2})	Fruiting efficiency[1] (grains g_{spike}^{-1})	Reference
Argentina, Buenos Aires	Klein Favorito	1920	12.7/20.8	122/130	Slafer and Andrade (1993)
	Eureka Ferrocarril Sur	1939	15.1/29.2	89/134	
	Buck Pucará	1980	22.8/37.4	118/114	
Argentina, Buenos Aires	Eureka Ferrocarril Sur	1939	13.6	128	González et al. (2003)
	Buck Manantial	1964	16.3	126	
	ProINTA Puntal	1994	19.5	139	
United Kingdom; Leicestershire	Maris Huntsman	1972	15.1	98	Shearman et al. (2005)
	Avalon	1980	19.1	91	
	Norman	1981	19.6	89	
	Galahad	1983	20.4	73	
	Riband	1989	19.1	102	
	Haven	1990	19.7	84	
	Brigadier	1993	23.91	128	
	Rialto	1995	19.5	92	
Spain, Lleida	Aragon 03	1940	9.7/16.1	68/79	Acreche et al. (2008)
	Pane 247	1950	11.8/18.1	74/93	
	Estrella	1963	17.7/21.2	116/108	
	Siete Cerros	1966	21.8/23.6	122/93	
	Anza	1971	22.4	105	
	Marius	1976	26.2	93	
	Soisson	1988	28.2	126	
	Isengrain	1998	26.1	100	

[1]When two figures were included they represent the values of two different growing seasons.

spike dry weight, although again there was a nonsignificant positive trend between grain number and fruiting efficiency for the cultivars released during the last three decades of the 20th century. In a much shorter period of analysis, Abbate et al. (1998) reported for cultivars released from 1984 to 1994 in Argentina that improvement of grain number was mostly associated with differences in fruiting efficiency.

Potential usefulness of fruiting efficiency in future breeding

In the past, fruiting efficiency was not even considered as a possibility because huge gains could be obtained relatively simply by selecting for reduced height (increasing partitioning to the spikes before anthesis and gaining in sink strength and yield, see above). In the future, it may be more relevant as a trait to consider, at least for identifying prospective parents for a cross designed to increase further yield potential. Selecting for fruiting efficiency would be much more difficult, unless trustworthy molecular markers are identified or high throughput tools could be developed. Abbate et al. (2013) have shown that selecting few individual spikes at maturity from a crop may produce sensible results reducing enormously the amount of work required for a more refined determination of fruiting efficiency. However, even measuring fruiting efficiency in few spikes would be a difficult task in realistic

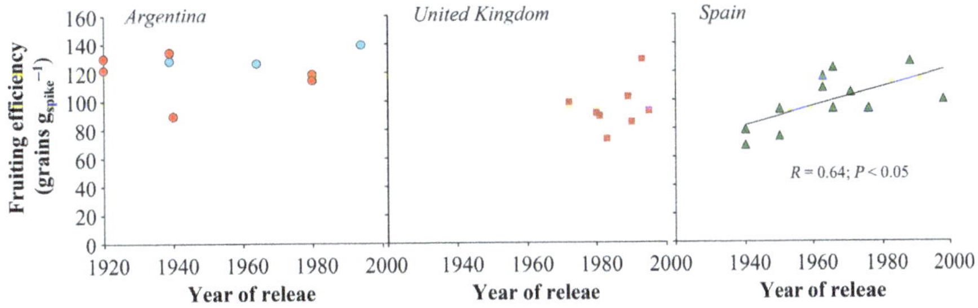

Figure 5. Relationship between fruiting efficiency and the year of release of the cultivars for studies conducted in Argentina (left panel; Slafer and Andrade 1993; red circles; González et al. 2003, blue circles), UK (middle panel, Shearman et al. 2005), and Spain (right panel, Acreche et al. 2008).

breeding programs selecting thousands of plots per season.

In any case, there is an indication that if it were possible to identify proper markers (or to develop a high throughput screening method) it could be effective to select for fruiting efficiency. Pedro et al. (2012) have conducted an experimental selection procedure (selecting divergently for higher or lower fruiting efficiency) on segregating mutants (Fig. 6). These mutants were a TILLING population of durum wheat generated by Martin Parry's group at Rothamsted Research (RRes), UK (Parry et al. 2009). As determination of fruiting efficiency is destructive it would be impracticable in early generations when segregation is high and, therefore, the authors used the number of grains per unit stem length as a close proxy (see Pedro et al. 2012). The process started with a selection made on the M2 generation at RRes under glasshouse conditions and then continued for three further generations (M3, M4, and M5) under field conditions at the University of Lleida (UdL), although the M5 generation was grown and selected at RRes under glasshouse conditions as well. Finally the selected lines, already largely stabilized M6 populations, were grown in field plots at normal plant density at the UdL. Selection was systematically applied through successive generations (M2 to M5).

There was a positive response to selection through the entire process. At any single selecting generation the number of grains of the selected offspring evidenced response to the divergent selection, being the difference between the offspring selected for high or low fruiting efficiency enlarged throughout the generations (Fig. 6). In addition, these results were consistent, disregarding whether the performance of the offspring were analyzed under greenhouse or field conditions (Fig. 6, grain number in M5 plants). Finally in the M6 generation it was clear not only that lines selected for high fruiting efficiency produced more grains and higher yield than the sister lines selected for low fruiting efficiency, but also that the best lines from the population selected for high fruiting efficiency out-yielded (because they produced a higher number of grains) the controls in the field plot experiment (Fig. 6). Thus, selecting for a proxy of fruiting efficiency produced individual plants with enhanced reproductive output which did translate into the performance at the crop level of organization.

Genetic Variation in Fruiting Efficiency Within elite Germplasm

It was shown (above) that when considering the physiological bases for yield gains in past breeding, fruiting efficiency was not consistently relevant: although there was genetic variation between cultivars released at different times, fruiting efficiency was not regularly improved across studies (Fig. 5) during a period when improving partitioning of dry matter to the growing spikes was dominant (Calderini et al. 1999; Foulkes and Reynolds 2015). More relevant for future breeding is to count with genetic variation for this trait among modern, high-yielding genotypes (which are the elite material in most breeding programmes from which breeders attempt further pyramiding genes contributing to yield).

Genetic variation among modern cultivars or advanced lines

Again the number of studies reporting on fruiting efficiency for elite genotypes is rather scarce but the few published papers that we found showed a wide range of variability (Table 2). Differences among studies are expected as the actual values may depend on particular methodological details (e.g., how and when spike dry weight was measured). But what is really relevant is the consistently large degree of genotypic variation found within particular studies. It is clear that fruiting efficiency varies largely and, whenever different cultivars are compared a wide range of variation can be evidenced (Table 2).

In a paper that was just published, Mirabella et al. (2015) reported new results confirming wide variation in fruiting efficiency among modern cultivars and revealing that this variation was consistently larger than GxE interaction.

As genotypes within particular studies do differ in many other traits that may affect spike dry weight at anthesis, and then grain number determination, it would not be necessarily expected a strong correlation between grain number and fruiting efficiency in every single case. However, in the studies analyzed comparing modern cultivars or high-yielding lines there was in general a positive relationship: considering only the cases in which there were more than 10 genotypes compared (to have at least 8 degrees of freedom for the regression analysis) the relationship between grain number and fruiting efficiency was significantly positive in the studies published by Abbate et al. (2013); González et al. (2011b); Garcia et al. (2014) in the experiment carried out in Argentina; and Foulkes et al. (2011). The unique exception was the experiment conducted in Mexico by Garcia et al. (2014) in which a nonsignificant positive relationship was found.

Thus, in the cases analyzed not only was there substantial genetic variation among modern cultivars but also, unlike what occurs when comparing modern versus old cultivars, the differences among them in grain number tended to be due to their differences in fruiting efficiency; emphasizing the potential value of the trait for future breeding.

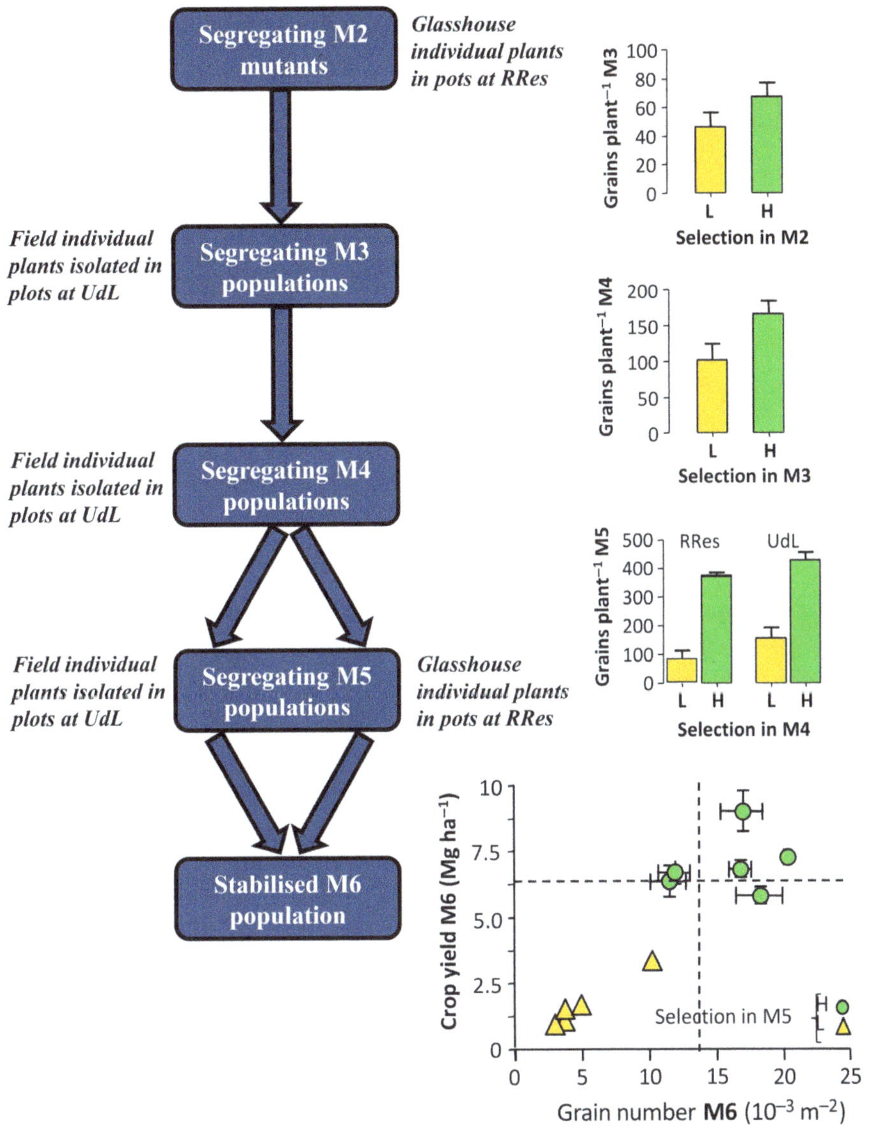

Figure 6. Diagram showing the selection process throughout the different seasons and conditions. At each generations plants were selected divergently for low (L) and high (H) fruiting efficiency and at each following generation there was a response of the number of grains per plant to the selection process (right bar plots). When lines selected were in the M6 generation (largely stabilized), a field plot experiment including the selected lines together with four commercial controls (dashed lines) was carried out. See details in Pedro et al. (2012).

Relevance of fruiting efficiency in a population designed to increase grain number and yield

Beyond the degree of variation available in elite germplasm, further variation can be expected if the trait shows transgressive segregation (observation of extreme individual phenotypes which exceed parental phenotypic values in either a positive or negative direction; Rieseberg et al. 1999). Transgressive phenotypes can be produced when alleles at multiple loci present in parental lines (some of them reducing and others increasing the parental phenotypic value) recombine in the segregating progeny (Bell and Travis 2005).

Phenotyping of wheat elite population, that is, population obtained by the cross of well-adapted and high-yielding parental lines, have reported positive transgressive segregation in traits related to phenology and tillering dynamics (Borràs Gelonch et al. 2012), in drought-adaptive traits (Olivares-Villegas et al. 2007), or in reserves and yield numerical components (Rebetzke et al. 2008; Rattey et al. 2009). Regarding fruiting efficiency, it has been recently phenotyped a doubled-haploid population derived from a cross between two high-yielding wheat cultivars with

Table 2. Genotypic variation (ranges) reported in the literature for grain number and fruiting efficiency in experiments in which different genotypes were compared side by side. Information is presented on whether fruiting efficiency was estimated with spike dry weight (including when and how) or chaff weight at maturity. Plant material type, number of genotypes, years of release (when it was mentioned) of these genotypes, country of evaluation, level of input in the experiment were also included.

Measurement used to estimate fruiting efficiency	Plant material	Number of genotypes	Year of release	Country	Input level	Grain number (10^{-3} m^{-2})	Fruiting efficiency (grains g^{-1})	Reference
Chaff at maturity	SW – CC	16	Not indicated	Argentina	High	11.2–24.6	35–87	Abbate et al. (2013)
	SW – CC	39	2000–2008	Argentina	High	15–29.7	59–137	González et al. (2011b)
Spike dry weight (SDW) at anthesis	SW – DH, CC	4	Not indicated	Chile	High	20.1–38.8	86–136	Bustos et al. (2013)
	DW – CC	4	1987–2002	Spain	High	27–30.9	80–115	Ferrante et al. (2012)
	DW – CC	4	1987–2003	Spain	Medium	18.5–21.3	64–90	Ferrante et al. (2012)
	SW – DH, CC	87	Not indicated	Argentina	High	11.5–28	69–201	Garcia et al. (2014)
	SW – CC	39	2000–2008	Argentina	High	15–29.7	78–196	González et al. (2011b)
	DW – CC, RIL	6	Not indicated	Spain	High	12.6–16.8	64–84	Pedro et al. (2011)
SDW at 1 week after anthesis	RIL	2	Not indicated	Australia	High	18.7–24.1	162–201	Dreccer et al. (2009)
	SW – CC, DH, RIL, EL	30	Not indicated	Mexico	High	13.3–22.4	40–76	Foulkes et al. (2013)
	SW – DH	8	Not indicated	Mexico	High	16.4–26	44–75	Gaju et al. (2014)
	SW – DH, CC	107	Not indicated	Mexico	High	9.9–20	42–91	Garcia et al. (2014)
SDW at 1 week after anthesis (without grains)	SW – CC, EL	6	Not indicated	Argentina	High	13.3–21.3	61–106	Abbate et al. (1998)
	SW – CC	16	Not indicated	Argentina	High	11.2–24.6	45–113	Abbate et al. (2013)
	SW – CC	4	1984–1994	Argentina	High	11.5–22.5	45–126	Lazaro and Abbate (2012)

SW, spring wheat; DW, durum wheat; CC, commercial cultivars; EL, experimental line; DH, doubled-haploid line; RIL, recombinant inbred line.

similar phenology but consistently differing in yield components: Bacanora (possessing high grain number) and Weebil (possessing heavy grains), under two different evaluation environments (Garcia et al. 2014). It was reported that 20–34% (depending on the experiment) of the DH lines exhibited higher values of fruiting efficiency than Bacanora (the parent characterized by having high fruiting efficiency; Garcia et al. 2014). Furthermore, focusing on top DH lines, simulating the selection carried out in a breeding program aimed to improve grain number, transgressive segregation was also evident (i.e., lines with grain number increased by 25% over the parental mean had also higher fruiting efficiency, consistently across experiments; Garcia et al. 2014). This suggests that crossing parents from the elite germplasm of high-yielding lines of a breeding programme may be a valuable strategy to further increase grain number through the expression of transgressive variation in fruiting efficiency.In a paper that just became available, Martino et al. (2015) also found that fruiting efficiency exhibited transgressive segregation and would be reasonably heritable.

Potential Drawbacks

As with any other potentially valuable trait, when considering the usefulness of fruiting efficiency to achieve yield gains through increasing grain number it must be considered whether it may cause important trade-offs with other yield determining components. The two most obvious trade-offs that must be analyzed are (1) that increases in fruiting efficiency be not constitutively related to decreases in spike dry weight at anthesis (the other major determinant of grain number), resulting fruiting efficiency irrelevant to bring about actual increases in grain number; and (2) that increases in grain number produced by higher fruiting efficiency is not constitutively related to decreases in grain weight (the other major yield component), resulting fruiting efficiency irrelevant to bring about actual increases in yield.

Does high fruiting efficiency require low spike dry weights?

The conceptual model proposed by Fischer (1984) includes the idea that fruiting efficiency and spike dry weight at anthesis are, at least partially, independent. The strong relationship between grain number and spike dry weight at anthesis when the variations were imposed either by genetic factors like introgressing semi-dwarfing genes (e.g., Fischer and Stockman 1980; Brooking and Kirby 1981; Miralles et al. 1998), management factors like nitrogen fertilization (e.g., Fischer 1993; Prystupa et al. 2004; Ferrante 2012), or artificial manipulations of the growing conditions like extending the photoperiod or shading the crops during stem elongation (e.g., Stockman et al. 1983; Savin and Slafer 1991; González et al. 2003, 2005a; Serrago et al. 2008) seem to support the idea that there would be no major trade-offs between these two components of grain number. However, studies directly analyzing trade-offs between fruiting efficiency and spike dry weight at anthesis are, once again, rather scarce.

From the few direct evidences reported in the literature, there are no evidences for a consistent conclusion. There are studies reporting negative relationship between these traits, supporting the idea of a likely trade-off (Dreccer et al. 2009; Ferrante et al. 2012; Lázaro and Abbate 2012); while in other cases there were no significant relationships, supporting the idea of independence (González et al. 2011b; Bustos et al. 2013; Garcia et al. 2014). Thus, there are no evidences for postulating a constitutive, and therefore inexorable, trade-off between fruiting efficiency and spike dry weight at anthesis and even in the cases in which such trade-off emerged, the cause is uncertain and the interpretation not possible at this stage of knowledge on the trait (Foulkes and Reynolds 2015).

Therefore, it is reinforced the idea that it may be possible to select for fruiting efficiency not necessarily expecting any trade-off in spike dry weight at anthesis. For instance, in the work by Garcia et al. (2014) it was shown that high yield was associated with grain number due to higher fruiting efficiency, not compensated by reductions in spike dry weight, when the top yielding lines (which represent those that would have been selected in a breeding program) were analyzed. Similarly, Pedro et al. (2012) showed that selecting divergently for (a proxy to) fruiting efficiency, after five generations there were two clear populations of either lower yielding with fewer grains or higher yielding with more grains than commercial controls used in the study.

Does high fruiting efficiency constitutively reduce the size of the grains?

Increasing grain number via fruiting efficiency would result in low yield advantages if average grain weight is highly reduced due to a trade-off between fruiting efficiency and average grain weight. It seems likely to expect a negative relationship between the average weight of the grains and fruiting efficiency (Ferrante et al. 2012) although it is not necessarily so always (González et al. 2014). However, a negative relationship does not always represent a true compensation; this would depend upon the nature of the causes behind the negative relationship. For this analysis we will assume that the basis for the negative relationship would hardly be the increased competition for resources to maximize grain size as the availability of assimilates during the effective period of grain filling is most frequently enough to fill the grains (see above in section How was yield improved in the past?).

In this context, there are two alternative hypothetical causes for the negative relationship between average grain weight and fruiting efficiency, one representing a clear trade-off while in the other the trade-off is only apparent (Fig. 7). The mechanism behind the true trade-off would be that the cause of the increased fruiting efficiency would be a reduced demand of individual florets to develop normally and consequently the final size of the fertile florets would be smaller (i.e., the same amount of resources may sustain the normal growth or more, though smaller, florets). As final grain weight would be related, albeit not necessarily linearly, to the size of the florets (Calderini et al. 2001; Hasan et al. 2011) increasing fruiting efficiency would concomitantly reduce the size of all grains and consequently the average grain weight (Fig. 7, left side). Should this hypothesis reflect the reality, increasing fruiting efficiency would increase number of grains which would be of smaller potential size and consequently would not necessarily produce any yield gain. But the trade-off could be only apparent if the reason behind the increased fruiting efficiency would be independent of the size of the florets (and then potential size of the grains); which may require an improved partitioning of the resources within the juvenile spike toward the florets at the expense of the more-structural parts. Should this hypothesis be true, increasing fruiting efficiency would not affect the potential size of individual grains, but as it would reduce the mortality of more distal florets and these florets have lower size potential than the proximal ones, it would increase the proportion of florets of smaller size potential (Miralles and Slafer 1995b; Acreche and Slafer 2006) and consequently the average size would be reduced but yield gains would be certain (Fig. 7, right side). In this case, fruiting efficiency could be used as a trustworthy selection criterion to further increase yield, if molecular markers were identified or high throughput tools were developed for making this possible in practice.

As far as we are aware there is only one paper reporting the relationship between potential grain weight (obtained after 50% trimming of the spike at the onset of grain filling) and fruiting efficiency for cultivars differing in this last trait (González et al. 2014), in which it was not found a general negative relationship between potential grain weight and fruiting efficiency. Furthermore, in a rather comprehensive study comparing two cultivars contrasting in fruiting efficiency and average grain weight under 8 different environmental conditions (Ferrante et al. 2015), it was evident that the cultivar with consistently higher fruiting efficiency and lower average grain weight had grains at particular positions within the spike of very similar weight to those of the cultivar with consistently lower fruiting efficiency and higher average grain weight. It was clear in that case that the negative relationship would not be constitutive and there

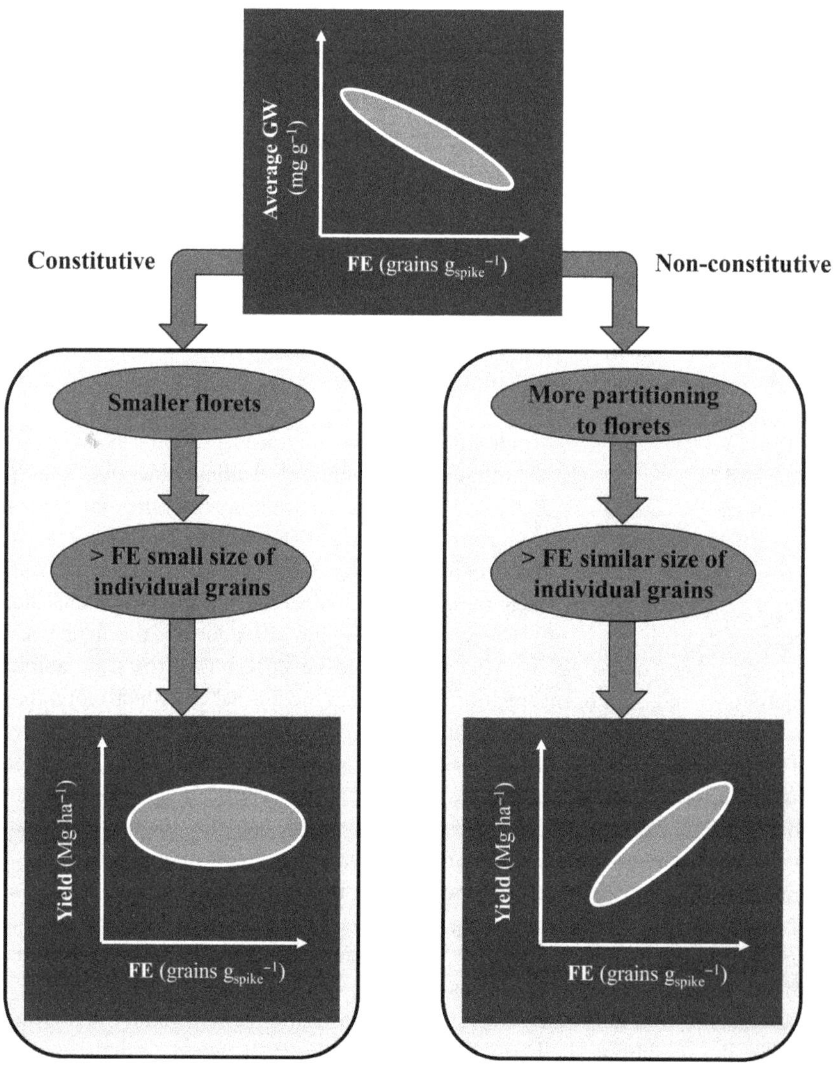

Figure 7. Schematic representation of the two alternative hypotheses to explain the negative relationship between average grain weight and fruiting efficiency. Left: a constitutive reduction in floret size bringing about a trade-off between weight of each of the grains (not just the average weight of all grains) and fruiting efficiency resulting in no yield gain from increased fruiting efficiency. Right: a nonconstitutive alternative hypothesis in which the size of individual grains are not affected but the proportion of grains of smaller size potential is increased and then increasing fruiting efficiency does produce yield gain, even when the average grain weight is reduced. Adapted from Ferrante et al. (2015)

would be an apparent, not real, trade-off produced by the increased proportion of distal grains in the cultivar with high fruiting efficiency (Ferrante et al. 2015).

Concluding Remarks

Alternative ways to further increase yield by breeding are urgently required to maintain current levels of food security. Fruiting efficiency is among the traits which could be exploited to achieve the needed gains in wheat yield. It has not been exploited consistently in past breeding and shows a wealth of genetic variation, particularly within the elite germplasm for breeding programmes. An empiric divergent selection exercise evidenced that it is likely to obtain yield gains through selecting for (a proxy to) fruiting efficiency. With the current understanding of the physiology of fruiting efficiency we can only characterize elite germplasm for breeders to be able to choose prospective parents when deciding their strategic crosses (in which case the characterization must consider the nature of the high fruiting efficiency to discard trade-offs). However, to be able to select for this trait in realistic breeding programmes molecular markers must be identified or high throughput phenotyping systems must be developed.

Finally, we focused this paper on wheat as it has been in this crop where this trait has been treated more explicitly. However, the concept is equally applicable to other cereals (and likely to any grain crop), as fruiting efficiency is an integral component to understand grain number determination through the general physiological model that establishing that grain number is related to growth during the "critical period for grain number determination" (e.g., for barley: Miralles et al. 2000; Prystupa et al. 2004; Arisnabarreta and Miralles 2010; maize: Otegui and Bonhomme 1998; Vega et al. 2001; Rattalino Edreira and Otegui 2012; and soybean: Jiang and Egli 1995; Kantolic and Slafer 2001, 2005; Egli 2010; Kantolic et al. 2013). Briefly, in all grain crops, due to evolutionary and breeding reasons, grain number is far more relevant than grain weight in determining yield (Sadras 2007) and grain number is far more dependent on floret survival than on floret initiation processes (Sadras and Slafer 2012). Therefore, in all grain crops grain number is related to growth in that critical period and genotypic differences in the efficiency of transforming that growth into grain number would likely be expected. Thus, what we exemplified in this paper for wheat is simply a general process linking reproductive biology and agronomic performance of crops that could likely be manipulated to further increase yield. However, comparisons among different species should be avoided as in this case differences in fruiting efficiency may be tightly related to constitutive differences in grain size among crops as shown by Gambín and Borrás (2010) for a comprehensive comparison among widely different species; and by Marti and Slafer (2014) for a constitutive difference between durum and bread wheat.

Acknowledgments

The work on fruiting efficiency has been funded by grants from the "National Plan" of Research in Spain, grants AGL2009-11964 and AGL2012-35300 at the Crop Physiology Lab of the University of Lleida; National Agency of Scientific and Technical Promotion (PICT-2008-1039) and INTA (PNCER-1336) at CONICET and INTA Lab, and PICT RAICES 1368 and UBACyT G-076 competitive grants at the University of Buenos Aires.

Conflict of Interest

None declared.

Notes

1 This trait has been also termed "spike fertility index" (e.g. Fischer 2011; Foulkes et al. 2011; and references therein) but the term index in crop physiology usually refers to ratios of variables with similar units (like harvest index or leaf area index), while the term efficiency is used to reflect the translation of certain amount of resources into variables of agronomic interest (e.g. water or nitrogen use efficiency). Thus, the term fruiting efficiency has been increasingly employed to describe the efficiency with which resources allocated to the spikes translate to the generation of a certain number of grains (e.g. González et al. 2011a; Pedro et al. 2011; Ferrante et al. 2012; Reynolds et al. 2012; Bustos et al. 2013; González et al. 2014; Marti and Slafer 2014; Garcia et al. 2014; Foulkes and Reynolds 2015; Sadras and Calderini, 2015).

2 a massive decrease in the expression of genes involved in cell proliferation, a decrease in soluble carbohydrate levels, and an increase in the expression of genes involved in programmed cell death; Ghiglione et al. 2008).

3 In some cases it has been preferred to determine spike dry weight few days after anthesis to consider the period of grain set (during the lag-phase of grain filling; Loss et al., 1989) as well. This is rather sensible but it has two major inconveniences. Firstly a huge amount of extra work is required to remove the tiny grains that might have already started to grow (otherwise the spike dry weight would be strongly overestimated and the overestimation would vary between genotypes (and environments) depending on the number of grains set and the potential size of the grains. Secondly, the estimated spike dry weight would include any eventual growth occurring

during that extra week, which is naturally positive, but would exclude the weight contributed by the fertile florets -as the small grains would have been removed- to the spike dry weight at anthesis, which would be a drawback. All in all, we believe that it would be convenient to standardize the determination of fruiting efficiency using spike dry weight at anthesis.

References

Abbate, P. E., F. H. Andrade, and J. P. Culot. 1995. The effects of radiation and nitrogen on number of grains in wheat. J. Agric. Sci. 124:351–360.

Abbate, P. E., F. H. Andrade, J. P. Culot, and P. S. Bindraban. 1997. Grain yield in wheat: effects of radiation during spike growth period. Field Crop Res. 54:245–257.

Abbate, P. E., F. H. Andrade, L. Lázaro, J. H. Bariffi, H. G. Berardocco, V. H. Inza, et al. 1998. Grain yield increase in recent argentine wheat cultivars. Crop Sci. 38:1203–1209.

Abbate, P. E., A. C. Pontaroli, L. Lazaro, and F. Gutheim. 2013. A method of screening for spike fertility in wheat. J. Agric. Sci. 151:322–330.

Acreche, M. M., G. Briceno-Felix, J. A. M. Sanchez, and G. A. Slafer. 2008. Physiological bases of genetic gains in Mediterranean bread wheat yield in Spain. Eur. J. Agron. 28:162–170.

Acreche, M. M., and G. A. Slafer. 2006. Grain weight response to increases in number of grains in wheat in a Mediterranean area. Field Crop Res. 98:52–59.

Albajes, R., C. Cantero-Martinez, T. Capell, P. Christou, A. Farre, J. Galceran, et al. 2013. Building bridges: an integrated strategy for sustainable food production throughout the value chain. Mol. Breeding 32:743–770.

Araus, J. L., G. A. Slafer, C. Royo, and M. Dolores Serret. 2008. Breeding for yield potential and stress adaptation in cereals. Crit. Rev. Plant Sci. 27:377–412.

Arisnabarreta, S., and D. J. Miralles. 2010. Nitrogen and radiation effects during the active spike-growth phase on floret development and biomass partitioning in 2- and 6-rowed barley isolines. Crop Past. Sci. 61:578–587.

Austin, R. B., J. Bingham, R. D. Blackwell, L. T. Evans, M. A. Ford, C. L. Morgan, et al. 1980. Genetic improvements in winter-wheat yields since 1900 and associated physiological-changes. J. Agric. Sci. 94:675–689.

Austin, R. B., M. A. Ford, and C. L. Morgan. 1989. Genetic-improvement in the yield of winter-wheat - a further evaluation. J. Agric. Sci. 112:295–301.

Bell, M. A., and M. P. Travis. 2005. Hybridization, transgressive segregation, genetic covariation, and adaptive radiation. Trends Ecol. Evol. 20:358–361.

Borrás, L., G. A. Slafer, and M. E. Otegui. 2004. Seed dry weight response to source-sink manipulations in wheat, maize and soybean: a quantitative reappraisal. Field Crop Res. 86:131–146.

Borràs-Gelonch, G., G. J. Rebetzke, R. A. Richards, and I. Romagosa. 2012. Genetic control of duration of pre-anthesis phases in wheat (Triticum aestivum L.) and relationships to leaf appearance, tillering, and dry matter accumulation. J. Exp. Bot. 63:69–89.

Braun, H.-J., G. Atlin, and T. Payne. 2010. Multi-location testing as a tool to identify plant response to global climate change. Pp. 115–138 in M. P. Reynolds, ed. Climate change and crop production. CABI, Wallingford, UK.

Brisson, N., P. Gate, D. Gouache, G. Charmet, F.-X. Oury, and F. Huard. 2010. Why are wheat yields stagnating in Europe? A comprehensive data analysis for France. Field Crop Res. 119:201–212.

Brooking, I. R., and E. J. M. Kirby. 1981. Interrelationships between stem and ear development in winter-wheat – the effects of a norin-10 dwarfing gene, gai-Rht2. J. Agric. Sci. 97:373–381.

Bustos, D. V., A. K. Hasan, M. P. Reynolds, and D. F. Calderini. 2013. Combining high grain number and weight through a DH-population to improve grain yield potential of wheat in high-yielding environments. Field Crop Res. 145:106–115.

Calderini, D. F., M. F. Dreccer, and G. A. Slafer. 1995. Genetic improvement in wheat yield and associated traits. A re-examination of previous results and the latest trends. Plant Breeding 114:108–112.

Calderini, D. F., M. P. Reynolds, and G. A. Slafer. 1999. Genetic gains in wheat yield and associated physiological changes during the twentieth century. Pp. 351–377 in E. H. Satorre and G. A. Slafer, eds. Wheat: ecology and physiology of yield determination. Food Products Press, New York.

Calderini, D. F., R. Savin, L. G. Abeledo, M. P. Reynolds, and G. A. Slafer. 2001. The importance of the period immediately preceding anthesis for grain weight determination in wheat. Euphytica 119:199–204.

Calderini, D. F., and G. A. Slafer. 1998. Changes in yield and yield stability in wheat during the 20th century. Field Crop Res. 57:335–347.

Canevara, M., M. Romani, M. Corbellini, M. Perenzin, and B. Borghi. 1994. Evolutionary trends in morphological, physiological, agronomical and qualitative traits of Triticum aestivum L. cultivars bred in Italy since 1900. Eur. J. Agron. 3:175–185.

Chand, R. 2009. Challenges to ensuring food security through wheat. CAB Rev. Persp. Agricult. Vet. Sci. Nutr. Nat. Res. 4:1–13.

Demotes-Mainard, S., and M. H. Jeuffroy. 2001. Partitioning of dry matter and nitrogen to the spike throughout the spike growth period in wheat crops subjected to nitrogen deficiency. Field Crop Res. 70:153–165.

Demotes-Mainard, S., and M. H. Jeuffroy. 2004. Effects of nitrogen and radiation on dry matter and nitrogen accumulation in the spike of winter wheat. Field Crop Res. 87:221–233.

Demotes-Mainard, S., M. H. Jeuffroy, and S. Robin. 1999. Spike dry matter and nitrogen accumulation before anthesis in wheat as affected by nitrogen fertilizer: relationship to kernels per spike. Field Crop Res. 64:249–259.

Dreccer, M. F., A. F. van Herwaarden, and S. C. Chapman. 2009. Grain number and grain weight in wheat lines contrasting for stem water soluble carbohydrate concentration. Field Crop Res. 112:43–54.

Dreccer, M. F., K. B. Wockner, J. A. Palta, C. L. McIntyre, M. G. Borgognone, M. Bourgault, et al. 2014. More fertile florets and grains per spike can be achieved at higher temperature in wheat lines with high spike biomass and sugar content at booting. Funct. Plant Biol. 41:482–495.

Egli, D. B. 2010. Soybean reproductive sink size and short-term reductions in photosynthesis during flowering and pod set. Crop Sci. 50:1971–1977.

Ferrante, A., R. Savin, and G. A. Slafer. 2010. Floret development of durum wheat in response to nitrogen availability. J. Exp. Bot. 61:4351–4359.

Ferrante, A., R. Savin, and G. A. Slafer. 2012. Differences in yield physiology between modern, well adapted durum wheat cultivars grown under contrasting conditions. Field Crop Res. 136:52–64.

Ferrante, A., R. Savin, and G. A. Slafer. 2013a. Floret development and grain setting differences between modern durum wheats under contrasting nitrogen availability. J. Exp. Bot. 64:169–184.

Ferrante, A., R. Savin, and G. A. Slafer. 2013b. Is floret primordia death triggered by floret development in durum wheat? J. Exp. Bot. 64:2859–2869.

Ferrante, A., R. Savin, and G. A. Slafer. 2015. Relationship between fruiting efficiency and grain weight in durum wheat. Field Crop Res. 177:109–116.

Fischer, R. A. 1984. Wheat. In: Smith, W. H., J., B. S. (Eds.), Symposium on potential productivity of field crops under different environments IRRI, Los Baños, Philippines pp.129–154 in 'Proc. Symp. on potential productivity of field crops under different environments, Sep. 1980'. IRRI, Los Baños.

Fischer, R. A. 1985. Number of kernels in wheat crops and the influence of solar-radiation and temperature. J. Agric. Sci. 105:447–461.

Fischer, R. A. 2007. Understanding the physiological basis of yield potential in wheat. J. Agric. Sci. 145:99–113.

Fischer, R. A. 2011. Wheat physiology: a review of recent developments. Crop Past. Sci. 62:95–114.

Fischer, R. A., and G. O. Edmeades. 2010. Breeding and cereal yield progress. Crop Sci. 50:S85–S98.

Fischer, R. A., 1993. Irrigated spring wheat and timing and amount of nitrogen fertilizer. II. physiology of grain yield response Field Crop Res. 33:57–80.

Fischer, R. A., and Y. M. Stockman. 1980. Kernel number per spike in wheat (Triticumaestivum) responses to preanthesis shading. Aust. J. Plant Physiol. 7:169–180.

Flintham, J. E., A. Borner, A. J. Worland, and M. D. Gale. 1997. Optimizing wheat grain yield: effects of Rht (gibberellin-insensitive) dwarfing genes. J. Agric. Sci. 128:11–25.

Foulkes, M. J., Rivera, C., Trujillo, E., Sylvester-Bradley, R., Slafer, G. A., Reynolds, M. P., 2013. Optimizing harvest index through increasing partitioning to spike growth and maximizing grain number. In: Reynolds, M., Braun, H. (Eds.), 3rd International Workshop of the Wheat Yield Consortium. CIMMYT, Cd. Obregón, Mexico, pp. 26–31.

Foulkes, M. J., and M. P. Reynolds. 2015. Breeding challenge: improving yield potential. Pp. 397–421 in V. O. Sadras and D. F. Calderini, eds. Crop physiology applications for genetic improvement and agronomy, 2nd edn. Elsevier, Amsterdam, The Netherlands.

Foulkes, M. J., G. A. Slafer, W. J. Davies, P. M. Berry, R. Sylvester-Bradley, P. Martre, et al. 2011. Raising yield potential of wheat. III. Optimizing partitioning to grain while maintaining lodging resistance. J. Exp. Bot. 62:469–486.

Gaju, O., M. P. Reynolds, D. L. Sparkes, S. Mayes, G. Ribas-Vargas, J. Crossa, et al. 2014. Relationships between physiological traits, grain number and yield potential in a wheat DH population of large spike phenotype. Field Crop Res. 164:126–135.

Gambín, B. L., and L. Borrás. 2010. Resource distribution and the trade-off between seed number and seed weight: a comparison across crop species. Anna. Appl. Biol. 156:91–102.

García, G. A., A. K. Hasan, L. E. Puhl, M. P. Reynolds, D. F. Calderini, and D. J. Miralles. 2013. Grain yield potential strategies in an elite wheat double-haploid population grown in contrasting environments. Crop Sci. 53:2577–2587.

Garcia, G. A., R. A. Serrago, F. G. González, G. A. Slafer, M. P. Reynolds, and D. J. Miralles. 2014. Wheat grain number: identification of favourable physiological traits in an elite doubled-haploid population. Field Crop Res. 168:126–134.

Ghiglione, H. O., F. G. González, R. Serrago, S. B. Maldonado, C. Chilcott, J. A. Cura, et al. 2008. Autophagy regulated by day length determines the number of fertile florets in wheat. Plant J. 55:1010–1024.

Godfray, H. C. J. 2011. Food for thought. Proc. Natl Acad. Sci. U.S.A. 108:19845–19846.

González, F. G., M. L. Aldabe, I. I. Terrile, and D. P. Rondanini. 2014. Grain weight response to different postflowering source: sink ratios in modern high-yielding Argentinean wheats differing in spike fruiting efficiency. Crop Sci. 54:297–309.

González, F. G., D. J. Miralles, and G. A. Slafer. 2011a. Wheat floret survival as related to pre-anthesis spike growth. J. Exp. Bot. 62:4889–4901.

González, F. G., G. A. Slafer, and D. J. Miralles. 2003. Grain and floret number in response to photoperiod during stem elongation in fully and slightly vernalized wheats. Field Crop Res. 81:17–27.

González, F. G., G. A. Slafer, and D. J. Miralles. 2005a. Floret development and survival in wheat plants exposed to contrasting photoperiod and radiation environments during stem elongation. Funct. Plant Biol. 32:189–197.

González, F. G., G. A. Slafer, and D. J. Miralles. 2005b. Pre-anthesis development and number of fertile florets in wheat as affected by photoperiod sensitivity genes Ppd-D1 and Ppd-B1. Euphytica 146:253–269.

González, F. G., I. I. Terrile, and M. O. Falcon. 2011b. Spike fertility and duration of stem elongation as promising traits to improve potential grain number (and yield): variation in modern Argentinean wheats. Crop Sci. 51:1693–1702.

Hall, A. J., and R. A. Richards. 2012. Prognosis for genetic improvement of yield potential and water-limited yield of major grain crops. Field Crop Res. 143:18–33.

Hasan, A. K., J. Herrera, C. Lizana, and D. F. Calderini. 2011. Carpel weight, grain length and stabilized grain water content are physiological drivers of grain weight determination of wheat. Field Crop Res. 123:241–247.

Jiang, H., and D. B. Egli. 1995. Soybean seed number and crop growth rate during flowering. Agron. J. 87:264–267.

Kantolic, A. G., and G. A. Slafer. 2001. Photoperiod sensitivity after flowering and seed number determination in indeterminate soybean cultivars. Field Crop Res. 72:109–118.

Kantolic, A. G., and G. A. Slafer. 2005. Reproductive development and yield components in indeterminate soybean as affected by post-flowering photoperiod. Field Crop Res. 93:212–222.

Kantolic, A. G., G. E. Peralta, and G. A. Slafer. 2013. Seed number responses to extended photoperiod and shading during reproductive stages in indeterminate soybean. Eur. J. Agron. 51:91–100.

Kirby, E. J. M. 1988. Analysis of leaf, stem and ear growth in wheat from terminal spikelet stage to anthesis. Field Crop Res. 18:127–140.

Lazaro, L., and P. E. Abbate. 2012. Cultivar effects on relationship between grain number and photothermal quotient or spike dry weight in wheat. J. Agric. Sci. 150:442–459.

Loss, S. P., Kirby, E. J. M., Siddique, K. H. M., Perry, M. W., 1989. Grain growth and development of old and modern Australian wheats. Field Crops Res. 21, 131–146.

Lopes, M. S., M. P. Reynolds, Y. Manes, R. P. Singh, J. Crossa, and H. J. Braun. 2012. Genetic yield gains and changes in associated traits of CIMMYT spring bread wheat in a "historic" set representing 30 years of breeding. Crop Sci. 52:1123–1131.

Marti, J., and G. A. Slafer. 2014. Bread and durum wheat yields under a wide range of environmental conditions. Field Crop Res. 156:258–271.

Martino, D. L., P. E. Abbate, M. E. Cendoya, F. Gutheim, N. E. Mirabella, and A. C. Pontaroli. 2015. Wheat spike fertility: inheritance and relationship with spike yield components in early generations. Plant Breeding, in press.

Mirabella, N. E., P. E. Abbate, I. A. Ramirez, and A. C. Pontaroli. 2015. Genetic variation for wheat spike fertility in cultivars and early breeding materials. J. Agric. Sci., in press.

Miralles, D. J., S. D. Katz, A. Colloca, and G. A. Slafer. 1998. Floret development in near isogenic wheat lines differing in plant height. Field Crop Res. 59:21–30.

Miralles, D. J., R. A. Richards, and G. A. Slafer. 2000. Duration of the stem elongation period influences the number of fertile florets in wheat and barley. Aust. J. Plant Physiol. 27:931–940.

Miralles, D. J., and G. A. Slafer. 1995a. Yield, biomass and yield components in dwarf, semi-dwarf and tall isogenic lines of spring wheat under recommended and late sowing dates. Plant Breed. 114:392–396.

Miralles, D. J., and G. A. Slafer. 1995b. Individual grain weight responses to genetic reduction in culm length in wheat as affected by source-sink manipulations. Field Crop Res. 43:55–66.

Miralles, D. J., and G. A. Slafer. 1997. Radiation interception and radiation use efficiency of near-isogenic wheat lines with different height. Euphytica 97:201–208.

Miralles, D. J., and G. A. Slafer. 2007. Sink limitations to yield in wheat: how could it be reduced? J. Agric. Sci. 145:139–149.

Olivares-Villegas, J. J., M. P. Reynolds, and G. K. McDonald. 2007. Drought-adaptive attributes in the Seri/Babax hexaploid wheat population. Funct. Plant Biol. 34:189–203.

Otegui, M. E., and R. Bonhomme. 1998. Grain yield components in maize. Field Crop Res. 56:247–256.

Parry, M. A. J., P. J. Madgwick, C. Bayon, K. Tearall, A. Hernandez-Lopez, M. Baudo, et al. 2009. Mutation discovery for crop improvement. J. Exp. Bot. 60:2817–2825.

Parry, M. A. J., M. Reynolds, M. E. Salvucci, C. Raines, P. J. Andralojc, X.-G. Zhu, et al. 2011. Raising yield potential of wheat. II. Increasing photosynthetic capacity and efficiency. J. Exp. Bot. 62:453–467.

Pedro, A., R. Savin, D. Z. Habash, and G. A. Slafer. 2011. Physiological attributes associated with yield and stability in selected lines of a durum wheat population. Euphytica 180:195–208.

Pedro, A., R. Savin, M. A. J. Parry, and G. A. Slafer. 2012. Selection for high grain number per unit stem length through four generations from mutants in a durum wheat population to increase yields of individual plants and crops. Field Crop Res. 129:59–70.

Perry, M. W., and M. F. D`Antuono. 1989. Yield improvement and associated characteristics of some Australian spring wheat cultivars introduced between 1860 and 1982. Aust. J. Agric. Res. 40:457–472.

Prystupa, P., R. Savin, and G. A. Slafer. 2004. Grain number and its relationship with dry matter, N and P in the spikes at heading in response to N x P fertilization in barley. Field Crop Res. 90:245–254.

Rattalino Edreira, J. I., and M. E. Otegui. 2012. Heat stress in temperate and tropical maize hybrids: Differences in crop growth, biomass partitioning and reserves use. Field Crop Res. 130:87–98.

Rattey, A., R. Shorter, S. Chapman, F. Dreccer, and A. van Herwaarden. 2009. Variation for and relationships among biomass and grain yield component traits conferring improved yield and grain weight in an elite wheat population grown in variable yield environments. Crop Past. Sci. 60:717–729.

Ray, D. K., N. Ramankutty, N. D. Mueller, P. C. West, and J. A. Foley. 2012. Recent patterns of crop yield growth and stagnation. Nat. Commun. 3:1–7.

Rebetzke, G. J., A. F. van Herwaarden, C. Jenkins, M. Weiss, D. Lewis, S. Ruuska, et al. 2008. Quantitative trait loci for water-soluble carbohydrates and associations with agronomic traits in wheat. Aust. J. Agric. Res. 59:891–905.

Reynolds, M., M. J. Foulkes, G. A. Slafer, P. Berry, M. A. J. Parry, J. W. Snape, et al. 2009. Raising yield potential in wheat. J. Exp. Bot. 60:1899–1918.

Reynolds, M., J. Foulkes, R. Furbank, S. Griffiths, J. King, E. Murchie, et al. 2012. Achieving yield gains in wheat. Plant Cell Environ. 35:1799–1823.

Richards, R. A. 1992. The effect of dwarfing genes in spring wheat in dry environments. I. Agronomic characteristics. Aust. J. Agric. Res. 43:517–527.

Rieseberg, L. H., M. A. Archer, and R. K. Wayne. 1999. Transgressive segregation, adaptation and speciation. Heredity 83:363–372.

Sadras, V. O. 2007. Evolutionary aspects of the trade off between seed size and number in crops. Field Crop Res. 100:125–138.

Sadras, V. O., and C. Lawson. 2011. Genetic gain in yield and associated changes in phenotype, trait plasticity and competitive ability of South Australian wheat varieties released between 1958 and 2007. Crop Past. Sci. 62:533–549.

Sadras, V. O., and G. A. Slafer. 2012. Environmental modulation of yield components in cereals: Heritabilities reveal a hierarchy of phenotypic plasticities. Field Crop Res. 127:215–224.

Sadras, V. O., Calderini, D. F., 2015. Crop physiology: applications for breeding and agronomy. In: Sadras, V. O., Calderini, D. F. (Eds.), Crop Physiology Applications for Genetic Improvement and Agronomy. Second Edition. Elsevier, Amsterdam, The Netherlands, pp. 1–14.

Savin, R., and G. A. Slafer. 1991. Shading effects on the yield of an Argentinean wheat cultivar. J. Agric. Sci. 116:1–7.

Sayre, K. D., S. Rajaram, and R. A. Fischer. 1997. Yield potential progress in short bread wheat in Northwest Mexico. Crop Sci. 37:36–42.

Serrago, R. A., I. Alzueta, R. Savin, and G. A. Slafer. 2013. Understanding grain yield responses to source-sink ratios during grain filling in wheat and barley under contrasting environments. Field Crop Res. 150:42–51.

Serrago, R. A., D. J. Miralles, and G. A. Slafer. 2008. Floret fertility in wheat as affected by photoperiod during stem elongation and removal of spikelets at booting. Eur. J. Agron. 28:301–308.

Shearman, V. J., R. Sylvester-Bradley, R. K. Scott, and M. J. Foulkes. 2005. Physiological processes associated with wheat yield progress in the UK. Crop Sci. 45:175–185.

Siddique, K. H. M., E. J. M. Kirby, and M. W. Perry. 1989a. Ear stem ratio in old and modern wheat-varieties – relationship with improvement in number of grains per ear and yield. Field Crop Res. 21:59–78.

Siddique, K. H. M., R. K. Belford, M. W. Perry, and D. Tennant. 1989b. Growth, development and light interception of old and modern wheat cultivars in a Mediterranean-type environment. Aust. J. Agric. Res. 40:473–487.

Slafer, G. A. 2003. Genetic basis of yield as viewed from a crop physiologist's perspective. Ann. Appl. Biol. 142:117–128.

Slafer, G. A., L. G. Abeledo, D. J. Miralles, F. G. González, and E. M. Whitechurch. 2001. Photoperiod sensitivity during stem elongation as an avenue to raise potential yield in wheat. Euphytica 119:191–197.

Slafer, G. A., and F. H. Andrade. 1991. Changes in physiological attributes of the dry-matter economy of bread wheat (triticum-aestivum) through genetic-improvement of grain-yield potential at different regions of the world: a review. Euphytica 58:37–49.

Slafer, G. A., and F. H. Andrade. 1993. Physiological attributes related to the generation of grain-yield in bread wheat cultivars released at different eras. Field Crop Res. 31:351–367.

Slafer, G. A., J. L. Araus, C. Royo, and L. F. G. Del Moral. 2005. Promising eco-physiological traits for genetic improvement of cereal yields in Mediterranean environments. Ann. Appl. Biol. 146:61–70.

Slafer, G. A., D. F. Calderini, D. J. Miralles, and M. F. Dreccer. 1994. Preanthesis shading effects on the number of grains of 3 bread wheat cultivars of different potential number of grains. Field Crop Res. 36:31–39.

Slafer, G. A., A. G. Kantolic, M. L. Appendino, G. Tranquilli, D. J. Miralles, and R. Savin. 2015. Genetic and environmental effects on crop development determining adaptation and yield. Pp. 285–319 in V. O. Sadras and D. F. Calderini, eds. Crop physiology applications for genetic improvement and agronomy, 2nd edn.. Elsevier, Amsterdam, The Netherlands.

Slafer, G. A., R. Savin, and V. O. Sadras. 2014. Coarse and fine regulation of wheat yield components in response to genotype and environment. Field Crop Res. 157:71–83.

Stockman, Y. M., R. A. Fischer, and E. G. Brittain. 1983. Assimilate supply and floret development within the spike of wheat (Triticum aestivum L). Aust. J. Plant Physiol. 10:585–594.

Tilman, D., C. Balzer, J. Hill, and B. L. Befort. 2011. Global food demand and the sustainable intensification of agriculture. Proc. Natl Acad. Sci. U.S.A. 108:20260–20264.

Vega, C. R., F. Andrade, and V. Sadras. 2001. Reproductive partitioning and seed set efficiency in soybean, sunflower and maize. Field Crop Res. 72:163–175.

Wall, P. C. 1979. An analysis of factors limiting grain numbers and yield of spring wheat in a low-latitude environment. Ph.D. thesis, University of Reading, UK.

9

Genetically modified crops, regulatory delays and international trade

Stuart J. Smyth

Department of Agricultural and Resource Economics, University of Saskatchewan, 51 Campus Drive, Saskatoon, Saskatchewan S7N 5A8, Canada

Keywords

Adoption impacts, biotechnology, Cartagena Protocol on Biosafety, environmental nongovernmental organizations, food security, risk.

Correspondence

Stuart J. Smyth, Department of Agricultural and Resource Economics, University of Saskatchewan, 51 Campus Drive, Saskatoon, Saskatchewan S7N 5A8, Canada.

E-mail: stuart.smyth@usask.ca

Funding Information

No funding information provided.

Abstract

Genetically modified (GM) crops have been produced in the initial adopting countries for 20 years. Over this period of time, hundreds of articles and reports have been published by academic journals, government regulatory agencies, and national science organizations on the safety aspects of biotechnology and GM crops. In addition to this, there is a growing body of quantified peer reviewed literature on the economic and environmental benefits following the adoption of GM crops in both developed and developing countries. Some estimates place the economic benefits in the billions of dollars a year range. In spite of the documentation of these economic and environmental benefits, GM crops face a challenging future. Environmental nongovernmental organizations (eNGOs) are relentless in their campaigns of misinformation about the dangers and hazards of GM crops. While eNGOs are unable to quantify their claims and accusations, their political and policy influences continue, particularly in Europe and numerous developing nations. The result of this is regulatory delays for the approval of new GM crops and frequent international commodity trade failures, where shipments have been rejected due to the low-level presence of a GM crop. Taken in combination, the regulatory and trade challenges facing GM crops are having a detrimental impact on improving food security. This article quantifies the benefits of GM crops, highlights the regulatory costs of delayed approval, and provides insights into the spillover effects from GM crop trade.

Introduction

Genetically modified (GM) crops have been internationally produced for 20 years and in spite of the substantial quantification of the economic, environmental, and health benefits from the adoption of GM crops, opposition to GM crops remains firmly rooted within many organizations and countries. While GM crops have undergone risk assessments in over 30 different countries without the detection of any increased risk to humans, livestock, or the environment, myths, rumors, and outright lies continue to be perpetuated by environmental nongovernmental organizations (eNGOs) about the damages, dangers, and threats posed by GM crops. The eNGO movement has interfered with biosafety regulatory approval frameworks in many developing countries to ensure that no GM crops are ever approved. When trace amounts of GM crops are detected in non-GM shipments, borders are slammed shut, causing trade conflicts lasting several years.

Given the 20 years of GM crop adoption, numerous studies have reported on the substantial benefits, yet regulatory and trade barriers abound. GM crops have proven to be safe and yet the eNGO community spends billions lobbying and fighting against their use (Byrne 2015). Why is this the case? What is responsible for the continued opposition? The answer is politics. Science-based regulation of innovation in numerous jurisdictions has been replaced by socioeconomic regulation, a nefarious scheme advanced by eNGOs through the liberal interpretation of the Cartagena Protocol on Biosafety (CPB). The CPB has been manipulated to such an extent by eNGOs, that it is only now revealing the hidden objective of the CPB, which is to put a halt to all forms of agricultural research and development in developing countries. Developing countries are expected to reach food security through small-scale, largely organic crop production (UNCTAD 2013).

The question of how to improve global food security when opposition to one of the proven beneficial technologies is so seemingly entrenched is indeed a modern day dilemma. This article summarizes the literature on the global economic, environmental, and health benefits of GM crops. This is followed by a discussion of the regulatory challenges that have been precipitated following the advent of the CPB. Documentation of the trade barriers that exist regarding GM crops rounds out the evidence compiled on the mistreatment of GM crops. The remainder of the article discusses the crucial steps that will be required to truly advance global food security in the face of such organized opposition.

Global Benefits of GM Crops

While GM crops have undergone risk assessments in over 30 countries, at present, they are only produced in 28 countries (James 2015). James identifies that 180 million hectares of GM crop production occurred in 2015, involving over 17 million farmers. Two thirds of the countries adopting GM crops are developing countries. While some GM crop varieties have limited production, such as papaya, eggplant, alfalfa, and sugar beets, the benefits of GM crops focus on the observed effects of the four dominant GM crops: canola, corn, cotton, and soybeans. The benefits from GM crops can be differentiated into three main categories, economic, environmental, and health.

Economic benefits of GM crops

With the adoption of GM crops rapidly expanding over the first two decades of availability, the benefits from the technology must be significant, otherwise farmers would have begun to return to previous crop options. One team of authors has annually examined the economic impacts of GM crops and their most recent data on the 2014 cropping year estimates that the global economic benefits of GM crops exceeded US$17 billion (Brookes and Barfoot 2016). Cumulatively, over the period from 1996 to 2014, the authors estimate that the economic benefits of GM crops have reached US$150 billion. The production increases from GM crops have a substantial contribution to improving global food security as Brookes and Barfoot estimate that over the 19-year period of GM crop production, an additional 158 million tons of soybeans and 322 million tons of corn have been produced that would not have resulted without the technology.

Studies on the economic benefits from GM crops first began to appear about the turn of the millennium, with many of these based on the results from experimental field trial data that were then extrapolated to the crop production area, using an estimated adoption curve. These were rudimentary studies that provided an initial sense of what degree of benefits might be observed, particularly in developing countries. By the time the technology had been adopted for a decade, studies of greater rigor began to appear, based on farmer surveys.

Several articles undertook a detailed assessment of the numerous studies that began to regularly appear following the first decade of GM crop adoption. The first of these was conducted by Carpenter (2010), which examined 168 studies on GM crop yields, finding that 124 of the studies reported yield increases, 32 no change, and only 13 reporting lower yields. Finger et al. (2011) examined 203 peer reviewed studies, concluding that yield increases are evident with GM crop adoption, but noted that these increases were due to reduced insect and weed population pressures and not due to actual genomic yield increases. Areal et al. (2013) examined 97 studies comparing yield increases between GM and non-GM, finding GM crops outperformed conventional crop in both developed and developing countries. The most recent study was conducted by Klümper and Qaim (2014), who undertook a meta-analysis of 147 studies on the impacts of GM crops, finding that chemical pesticide use decreased by 37%, crop yields increased by 22%, and farmer profits increased by 68%.

While these overarching studies provide an international sense as to the distribution of benefits observed from GM crops, it is useful to draw on some specific examples to further illustrate the level of benefits that have been observed. Subramanian and Qaim (2010) examined Bt cotton adoption in India, finding that it raised vulnerable household incomes (those defined as living on <$2/day) by 134%. Hutchison et al. (2010) led a study on the economic benefits of GM corn adoption in the United States, finding that GM corn created $6.8 billion in extra value, with 60% going to nonadopters due to lower insect pressures. Based on a farm survey of nearly 600 canola producers, Gusta et al. (2011) found that GM canola resulted in annual benefits of $350–$400 million in Western Canada. Yorobe and Smale (2012) identified that GM corn adoption in the Philippines increased household income from $400/year to $600/year, a 50% increase. Vitale et al. (2014) found Bt cotton adoption in Burkina Faso resulted in a profit of $150/ha versus $70/ha for conventional cotton. Alston et al. (2014) estimate the cumulative global benefits from GM soybeans to be $46 billion over the 15-year period from 1996 to 2010.

Environmental benefits of GM crops

The first studies examining the environmental benefits of GM crops were undertaken following the adoption of Bt cotton in China. Pray et al. (2001) surveyed Chinese farmers in 1999, finding that the adoption of Bt cotton allowed farmers to spray less frequently, in some instances dropping from 30 per season to 3, but more commonly from 12 to 3–4. Huang et al. (2010) update the Chinese Bt cotton story observing that across the entire sample region insecticide applications dropped from 14 kg/ha to 4 kg/ha.[1] The spillover of environmental benefits from the lengthy adoption of Bt cotton began to be observed as the authors reported finding that in some non-Bt cotton fields the amount of insecticide used dropped from in excess of 40 kg/ha to >10 kg/ha. Similar environmental benefits from the adoption of Bt cotton in India were quantified by Subramanian and Qaim (2010), where they found Bt cotton reduced pesticide use by 41%.

A detailed study of the environmental changes following a decade of GM canola production in Western Canada identified substantial benefits. Prior to the commercialization of GM canola, the chemicals available to be applied to canola for weed control were limited, with the majority requiring soil application. The commercialization of herbicide-tolerant varieties of GM canola allowed producers to begin to use foliar chemical applications. The result of this was that the environmental impact of the chemicals applied to canola dropped by 53% when compared to the chemicals that were previously used on canola (Smyth et al. 2011a). When comparing chemical applications between GM canola production and the alternate scenario where GM canola had not been commercialized, the authors found the total volume of chemicals applied to canola dropped by 1.3 million kg/year. The rapid adoption of GM canola provided improved weed control options, which reduced the use of tillage as a form of weed control, resulting in one million tons of carbon being either sequestered by the soil or no longer released from implement passes (Smyth et al. 2011b). Brookes and Barfoot (2016) show CO_2 emission reductions equal to removing 10 million cars for 1 year.

Recent research shows the environmental costs of not adopting GM crops. Australia approved GM canola in 2003. However, in 2004 a moratorium was implemented across Australia against growing GM canola. It was not until 2008 when the central canola producing states of New South Wales and Victoria lifted the moratorium, followed by Western Australia in 2010. An examination of the environmental costs of the Australian moratoriums was undertaken that estimated what the adoption level for GM canola could have been after one decade of production, from 2004 to 2014, had the moratorium not been implemented (Biden 2016). The delayed adoption of GM canola production in Australia cumulatively resulted in the application of an additional 6.5 million kg of chemical active ingredient. The application of these additional chemicals were done through an additional 7 million field passes, requiring 8.7 million liters of diesel. The environmental impact of the additional chemicals applied was 14% higher than would have been the case if GM canola had not been subjected to an adoption moratorium. Finally, an estimated 24 million additional kilograms of greenhouse gases were released due to the nonadoption of GM canola.

Health benefits of GM crops

Some of the most significant benefits from GM crops, yet least widely recognized, are the health and lifestyle benefits. Identified benefits in this area range from a reduction in pesticide poisonings to less time spent hand weeding fields.

One continuous criticism advanced by critics of biotechnology and GM crops, such as Vandana Shiva, is that GM cotton is responsible for thousands of small landholder farmers committing suicide every year. Shiva and her ilk intentionally ignore the facts about Indian farmer suicide and continue to perpetuate lies about the unfortunate situation of farmer suicides. Research on this subject was conducted by Gruère and Sengupta (2011) and documented a one-third reduction in the suicide rate following the release of Bt cotton among Indian farmers based on extrapolating the pre-Bt cotton commercialization suicide rate.

Research from South Africa following the adoption of GM corn documents the first lifestyle benefits of the technology. Gouse (2013) surveyed small landholders finding that GM corn adoption had resulted in 10–12 fewer days of female hand weeding per season. This research identified that the women farmers spent this time predominantly in one of two areas, hauling more water for their vegetable gardens or spending time with their children.

While the economic and environmental benefits of GM crops are important, they definitely pale in comparison to the health benefits of reduced pesticide poisoning. Vitale et al. (2014) surveyed Bt cotton farmers in Burkina Faso over the first 3 years of adoption. While the economic benefits discussed above are significant, the health benefits far outweigh the economic benefits. Their farm survey results identified that an estimated 30,000 fewer cases of pesticide poisoning per year were occurring among Bt cotton farmers.

While the health benefits in Burkina Faso are important, they are dwarfed by the health benefits identified by Kouser and Qaim (2011) regarding pesticide poisoning reductions following Bt cotton adoption in India. Their research estimated that the number of pesticide poisonings was reduced by between 2.4 million and 9 million cases a year. They estimated that the financial savings for the Indian Ministry of Health ranged between US$14 and 51 million.

Every developed and developing country that has adopted one of the available GM crops has experienced at least one of, if not all of increased yield, reduced chemical use, and/or fewer cases of pesticide poisoning. With hundreds of studies now quantifying the various benefits, where is the holdup? Why is adoption in developing countries not happening at a faster rate than it is? What or who is responsible? The article now turns its attention to a discussion of the regulatory barriers that have been systematically structured to frustrate innovations in agriculture.

Regulatory Challenges Facing GM Crops

Domestic regulatory systems are now part of corporate investment strategies (Smyth et al. 2014). In January 2012, BASF announced that it was moving its agricultural research division from Europe to the United States due to the timeliness of regulatory decisions in Europe (BASF, 2012). It is even possible to view the Bayer acquisition of Monsanto as being a mechanism for Bayer to move underutilized, if not stranded, capital invested in plant variety development in Europe to the United States, given the European aversion to agriculture innovations in plant technologies.

Regulatory systems in some jurisdictions are being driven by politics, not science. Politicization of risk has occurred in the European Union (EU) as they have fully incorporated a rigid version of the precautionary principle into their regulatory framework for GM products (Smyth et al. 2015). The EU's rigid application is an inappropriate interpretation of the precautionary principle. As was established by the World Trade Organization's (WTO) ruling in the 2006 case by Argentina, Canada, and the USA against the EU's moratorium on GM crops, the precautionary principle requires some form of evidence to be invoked.

This has reduced the EU regulatory system for GM crop approvals to utter gridlock with only a single variety being approved for commercial production in the past decade. Even decisions regarding approval for import take longer in the EU than in the United States. Smart et al. (2016) found that approvals in the United States required an average of 686 days compared to 995 days in the EU. Currently, >10% of global industry research and development (R&D) on agriculture is spent in the EU versus one-third 20 years ago (Little 2015). While it would not be possible to attribute all of this loss in R&D investment to the EU's present regulatory regime, the regulatory framework has to be considered as having a substantial impact on this investment decline. Pardey et al. (2016) have completed a thorough analysis of agricultural R&D spending, finding that in 2011, global investments totaled US$69 billion. High-income countries accounted for 55% (down from 69% in 1980), while middle-income countries accounted for 43% (up from 29% in 1980).

The lack of coordinated regulatory capacity for GM crops is causing untold delays in the adoption and diffusion of the technology. For example, by October 2015, EU Member States had to inform the European Commission as to whether they would opt out of EU-wide GM crop approvals as has been mandatory since 2003. By this deadline, 19 of 28 MS indicated they would opt out (Smyth et al. 2016). The EU is no longer capable of approving GM crops for production and is struggling to manage to approve GM crops for import. Varieties of GM crops submitted in 2005 for import approval (predominantly for animal feed use) still have not been approved over 11 years later.

While existing and previous technologies to develop new crop varieties has posed a significant quandary for EU regulators, new plant breeding techniques may be an even greater challenge. New plant breeding techniques that provide no identifiable markers that can be used to verify the technology utilized to create the new variety now have their initial varieties undergoing risk assessment and approval. Examples of these technologies are targeted mutagenesis techniques such as oligonucleotide-directed mutagenesis (ODM), zinc finger nuclease (ZFN), meganuclease technique, and transcriptional activator-like effector–nuclease (TALEN), and gene silencing techniques such as clustered regularly interspaced short palindromic repeats (CRISPR). Many of these technologies have been classified as forms of mutagenesis in the United States, with the resulting products not subject to GM crop regulations. Examples such as Innate potatoes developed by Simplot and Arctic apples developed by Okanagan Specialty Fruits document the acceptance of innovative plant breeding technologies in some parts of the world. In an open letter to the European Commission, leading eNGO groups across Europe have attacked these new breeding techniques (NBTs), calling for every single new NBT to be classified as a GMO technology and therefore rejected for use within the EU (Panella et al. 2015).

The easing of regulatory burdens on innovative agricultural biotechnologies does not seem to be on the horizon as recent and ongoing trade negotiations have not satisfactorily addressed the subject. The recent Comprehensive Economic and Trade Agreement (CETA) between Canada and the

EU does have provisions for addressing regulatory issues, but in the era of biotechnology, there is no ability to resolve disputes. Therefore, regulatory issues will simply be discussed and discussed for years with no mechanism for resolution. The CETA clearly states that trade barriers such as the low-level presence of EU-unapproved GM varieties will be referred to the Dialogue on Biotech Market Access Issues. Even though it is clearly acknowledged that there are outstanding trade and regulatory barriers, CETA does nothing to resolve these other than to refer the topics to an ongoing dialog process that lacks any formal definition. The language used in the CETA documents is also present in the language of the Trans-Pacific Partnership and the current negotiations between the US and the EU on the Transatlantic Trade and Investment Partnership (T-TIP).

Perhaps the single biggest regulatory barrier to biotechnology is the inappropriate interpretation and application of the Cartagena Protocol on Biosafety (CPB). The CPB was negotiated at the end of the last millennium as an environmental response to the WTO's unwillingness to open to the Sanitary and Phytosanitary (SPS) Agreement to consumer aversion to biotechnology and GM crops. In December 2000, agreement was reached on the CPB, as a supplement agreement to the Convention on Biological Diversity. The CPB came into effect as an international agreement in 2003, focusing solely on the trans-boundary movement of living modified organisms (LMOs). The EU pushed heavily for the establishment of the CPB, which incorporates the precautionary principle to the fullest extent. The EU had previously lobbied the WTO to have the SPS Agreement revised to reflect consumers' socioeconomic concerns, however, they were unsuccessful in these efforts. Currently, 170 countries have ratified the CPB, while leading GM crop-producing countries such as Argentina, Australia, Canada, and the United States are not parties.

One consequence that was not clearly evident at the time of the CPB's negotiation and coming into effect was the disastrous effect that would be created from the movement away from science-based regulation. With the hidden agenda of undermining agricultural research and development (R&D) in developing countries, the eNGOs that lobbied vociferously for the ratification of the CPB have been successful at ensuring the inclusion of socioeconomic issues that have no data, no methodology, and no ability to contribute meaningfully to biosafety (Ludlow et al. 2014). The eNGO community has advanced nebulous socioeconomic considerations (SECs) such as gender impacts; farmers' rights; cultural, spiritual, and ethical aspects; land tenure; labor and employment; rural–urban migration; impacts on consumer choice; and impacts on market access as issues that require assessment prior to the approval of a GM crop. Virtually all of these have limited or nonexistent data in developing nations, lack credible methodologies that would be required to undertake an assessment, and lack any benchmark capability to know if the resulting impact would be positive or negative. Given all of these attributes, it is clear that SECs are simply a disguised objective to ensure that all agricultural R&D is delayed, if not outright prevented, in most developing nations. The extreme of SEC inclusion is evidenced by the biosafety regulations of Mali, whereby no negative socioeconomic impact from the commercialization of a GM crop is allowable (CBD 2014). In the history of innovative products and technologies, there is not a single one that has resulted in zero adverse effects occurring. The adoption of fire to heat caves resulted in burns; learning how to sail boats resulted in people drowning; and the invention of cell phones resulted in deaths due to texting while driving. Mali's regulatory barrier to GM crops makes it abundantly clear that the CPB is a disguised barrier to all crop R&D in developing countries.

Regulatory barriers of this nature have substantial impacts on R&D investment decisions, as demonstrated by BASF's decision to transfer all of its agricultural R&D assets out of Europe to North America. While private companies have the luxury of being able to relocate assets in an effort to secure a higher return on investment (ROI), public sector investors are not afforded such luxury. In most developing countries, public research centers invest in the development of local crop variety improvements and will frequently partner with international agencies, organizations, and companies in the development of new plant varieties. These centers will have minimal ROI expectations as the royalties from new crop varieties will be a low percentage of seed prices. A 2-year delay in the regulatory approval of a new crop variety with a ROI rate of 20% has been shown to eliminate all positive returns (Smyth et al. 2014). With national research centers relying on ROI revenue to fund new variety R&D, the revenue loss will have devastating effects.

Not only is biotech research jeopardized in developing countries, but all agriculture research is in danger of being ended, especially if the eNGO community is successful in getting the various new breeding techniques to be regulated as GMO technologies. Should the full list of NBTs be classified as GMOs by eNGOs for developing country regulators, investment in agricultural R&D could be expected to virtually cease, given the very low probability that any resulting product could ever be approved through an SEC-based regulatory system.

Trade Impacts From GM Crops

International commodity trade has never taken place at a purity level of 100%. This is particularly impossible as the purity level for certified seed, the seed that the pedigreed seed industry sells to farmers, is sold at a purity level of 99.5%. At this level, there is an allowance of one quarter of a percent for weed seeds and one quarter of a percent for other crop varieties. Given that the grain, pulse, and oilseed production process starts with 0.5% impurities, commodities cannot make it through the entire growing, harvesting, and transportation stages without an increase in the presence of impurities or off-types. While challenging to verify due to the private nature of commodity contracts, conversations with those in the grain trade industry indicate that contracts are typically written with impurity thresholds of 3–5%. The lower the threshold, the higher the price the importer has to pay.

Shortly after the commercialization of GM crops, adoption and import moratoriums began to be enacted. The most significant of these was the GM crop moratorium that the EU implemented from 1998 to 2003. While this was ultimately ruled as a trade barrier in 2006 by the WTO, the damage had already been done. Global commodity trade has been adversely affected by the low-level presence (LLP) of approved GM varieties as well as the adventitious presence (AP) detection of unapproved GM varieties in non-GM markets.

Prior to 2003, regulation and approval of GM crops in Europe was done individually by Member States. When issues of LLP and AP occurred, the resulting fields were either destroyed or harvested and the resulting crop contained (Hobbs et al. 2014). Incidents of this nature happened over several years without international borders being closed to commodity trade. In 2003, the EU removed its GM crop moratorium and established the European Food Safety Authority (EFSA), which coincided with an end to rational approaches to LLP and AP detection. Following 2003, any LLP or AP detection results in an immediate closure of EU borders to commodity trade, costing billions in lost trade opportunities (Smyth et al. 2015). In some instances, this has resulted in shipments of conventional commodities being destroyed due to false-positive GM crop detections (Kalaitzandonakes et al. 2016). Examples exist in American exports of corn and rice to the EU, while Canada's flax market was adversely affected by the closure of the EU border to Canadian flax exports.

In September 2009, GM flax was detected in Canadian flax exports to the EU. The EU immediately closed the border to all flax shipments from Canada. The border closure lasted 4 months, until an industry stewardship testing protocol could be developed, agreed upon, and implemented. This stewardship protocol ensures that every field of Canadian flax is tested for GM flax and this testing protocol continues all the way along the flax handling system, rail, interlake freighters, and transoceanic vessels. The 4-month export loss cost the Canadian flax industry C$30 million, while in the EU the border closure resulted in losses of €40 million to their flax processing industry and over 600 jobs (Ryan and Smyth 2012). The initial agreement regarding the stewardship testing protocol was to test all shipments for a period of 5 years, costing $1 million per year. This testing protocol was to have ended in 2014; however, no incentives to end the testing program exist and it will continue to cost the Canadian flax industry $1 million per year for the foreseeable future. The EU was Canada's leading flax export market, so Canada cannot risk ending testing protocols given the importance of the EU market, while on the other hand, the EU is not going to suddenly drop requirements that Canadian flax no longer be tested, given how political the issue of GM comingling is in Europe. Unless a sunset clause is agreed to upon negotiation of a testing protocol, the testing for LLP of GM crops will continue to cost the commodity trade industry in perpetuity.

American commodity exports have similarly been adversely affected. Herculex I corn was approved in the United States and was submitted to the EU for import approval in 2000. When import approval was not shortly received, the industry undertook export testing to ensure this variety did not reach the EU. The EU's regulatory system managed to approve the import of Herculex I corn, taking 6 years to do so; however, during this period, traces of Herculex R corn were detected in EU shipments. The corn industry's challenge of perfectly ensuring that zero Herculex R corn entered the EU animal feed market caused corn gluten feed exports from the US to the EU to drop by 30–40%. By 2009, Herculex R was approved for import, but the corn market was disrupted for 3 years, all over a variety of corn that was used as an ingredient in livestock feed.

Rice exports from the US to the EU suffered a similar fate. Trace amounts of LL601 rice were detected in EU in 2006. The problem in this instance was that LL601 had not received variety approval in the United States, so its AP in rice shipments was a major concern for the rice industry. Over 1000 lawsuits were launched against Bayer, the developer of LL601, ultimately costing over $2 billion in settlement costs. By 2012, the issue was finally resolved with the last of the court cases being completed, however, the US rice export market to the EU was only 40% of what it had previously been.

The message that these commodity trade problems send to developing countries considering the adoption of GM crops is resoundingly negative. If industrial countries like Canada and the United States struggle to manage identity preservation systems and export GM-free shipments to the EU, how would developing countries be able to exceed the Canadian or American systems, especially given the limited resources and inland testing capability? Developing countries that rely on EU markets for exported products see the GM crop LLP trade backlash and ongoing costs, which negatively impacts developing country decisions to adopt GM crops. In fact, the EU has gone as far as threatening developing countries into ensuring that they do not adopt GM crops. In 2014, the EU's Ambassador to Kenya, Lodewijk Briet was quoted as saying: "Local farmers will find it difficult to export their crops to Europe if they adopt the Genetically Modified (GM) crops" (Gathura 2014).

If threatening developing countries is not despicable enough, the EU also blackmails countries into ensuring that they adopt the innovation-killing interpretation of the CPB if they want to export agricultural products to the EU. The WTO allows countries to apply different tariff rates to specific countries for specific products that are not granted to all other trading partners. Essentially this allows countries to export their products at a lower tariff rate than competing export countries. The EU requires that any country wanting the lower tariff rate to have ratified the CPB: "the CPB aims to ensure a balance between the sustainable use of modern biotechnology and economic interests. The protocol governs the trans-boundary movement of Genetically Modified Organisms (GMOs), and authorizes states to prevent the import of GMOs according to the precautionary principle" (European Commission, 2016: 13).

The CPB is the biggest international barrier to innovative agricultural technologies, GM crops, and NBTs presently in existence. In addition to this, the EU forces developing countries wanting preferred tariff rates to be a member of the CPB and the EU threatens developing countries that if they adopt GM crops, they will lose access to EU markets. Developing countries witness how the EU shuts its borders to Canadian and US exports when the LLP of approved and unapproved GM crops are detected. The EU and its vigorous application of the CPB leaves most developing nations feeling they simply have no option when it comes to domestically deciding whether or not to adopt GM crops and rather that this decision has already been made for them by the EU and particularly by EU-based eNGOs.

If all this was not deemed a significant enough deterrent to developing countries, in the spring of 2016, the European Parliament completely abandoned a previous G7 commitment to improving food security. In 2012, the G7 group of countries launched the New Alliance for Food Security and Nutrition program in an effort to lift 50 million people out of poverty. In June 2016, Members of the European Parliament (MEP) called on developing countries to reject the program, calling it "a mistake." MEPs demanded "that the G7 abandon its commitment to GMOs in this public-private partnership" and this type of farming practice "destroys family farming and reduces biodiversity" (The Guardian, 2016).

The EU and its eNGOs spend billions annually to prevent GM crops from being adopted in developing countries. In 2015, it is estimated that the anti-GM movement spent US$10 billion fighting biotech around the globe (Byrne 2015). In 2011, the big six multinational biotechnology development companies spent US$8.6 billion on agriculture R&D (Hobbs 2014). Clearly, GM crops and their benefits is a political issue for eNGOs and given their investments opposing the technology, it is too financially lucrative for them to ever back away from. With the relentless pressure from the eNGO community to erect regulatory barriers against commercializing GM crops taken together with the EU's rejection of the peer reviewed scientific evidence GM crop benefits, its intimidation of developing countries into signing the CPB, and blackmailing developing countries to not adopt GM crops, global food security is not going to increase given the present state of international regulation and trade for GM crops.

Breaking the Food Security Barriers

Politics, not science, is driving the international agenda pertaining to GM crop approval, adoption, and trade. If science was the determining factor for the debate about the benefits of GM crops, then the opposition to GM crops should have waned considerably over the past decade. Sadly, it has not. If, as Byrne (2015) identifies, the eNGO movement is spending in the region of US$10 billion annually fighting against GM crops, then one would expect that these eNGOs would be generating even larger sums of revenue. The continuation of eNGO opposition to GM crops in the face of mounting economic, environmental, and health benefits can be attributed to two factors. First, it is obviously profitable for the eNGO movement to maintain its opposition to GM crops. This allows them to raise more revenue, hire additional staff, and grow their movements, thus increasing their ability to ensure that developing countries implement socioeconomic-based regulatory systems that must reject GM crop variety approval applications. Second, the eNGO movement has gained substantial political influence with governments regarding biotechnology. Nowhere is this more clearly evident than in Europe, where when given the choice of approving or rejecting GM crops at the domestic level, two thirds of EU members chose to reject the technology. Continued eNGO opposition to GM crops is both profitable and powerful.

The eNGOs have a 20-year head start on those in industry and academia in vociferously opposing GM crops and have seemingly successfully adopted the tobacco industry's strategy whereby they criticize those that engage in research on the benefits of GM crops. The CPB has been operating for nearly 15 years to entrench structural opposition to agriculture R&D and innovation. Agricultural innovations face a decade worth of work to remove the barriers to improving global food security so that there is a chance to feed the world of 2050. Global food security can improve, if and only if, two radical changes are made.

The first change will be to aid and assist developing countries in opting out of the Cartagena Protocol on Biosafety. As long as the CPB is a tool for the EU and its cadre of eNGOs, to dictate and manipulate food security in developing countries, these countries will continue to face starvation and malnourishment. Academics need to be engaged in quantifying the benefits of various types of GM crop adoption (and NBTs) as well as the costs of not adopting GM crops. Australia's GM canola moratorium cost farmers nearly $500 million in lost revenue (Biden 2016). Federal governments and philanthropic organizations need to partner to establish matching funding programs, allowing for the submission of peer reviewed research grant applications, thus ensuring that rigorous methodologies are employed to undertake research on the cost of delayed GM crop adoption. Grants of this nature could fund the research necessary for developing countries to better appreciate the costs that the CPB is inflicting and to develop the research and data required to opt out of the CPB.

The second requirement will be to develop a means of fencing Europe out of global commodity trade. GM crop adopting nations like Canada, USA, and Australia need to seek new regional trade agreements with developing countries regarding agricultural commodity trade. This will especially be the case for products developed via NBTs. All agriculture trade ties between the EU and developing countries need to be severed. The EU threat must be removed so that developing countries can freely decide for themselves whether or not to adopt new technologies such as GM crops. The eNGO community's ability to dictate first world food choices to malnourished developing countries has to end. To successfully achieve this, political lobbying will be required to encourage governments to the benefits of developing country trade agreements.

Mid-century food security is achievable, but it will require some difficult decisions to be made. The political power wielded by European-based eNGOs has thoroughly corrupted these organizations as they have successfully manipulated the CPB into becoming an agreement that encourages the rejection of scientific innovation and one that ultimately supports environmental and biodiversity degradation. Continuing with the situation of EU and eNGO manipulation and corruption will only ensure that the future of food security is thoroughly jeopardized, placing the lives of hundreds of millions of developing world citizens at risk. Humanity dictates that the EU and these nefarious eNGOs be expeditiously removed from the formula for food security.

Conflict of Interest

None declared.

Note

[1]The authors do not state whether these figures are kilograms of active ingredient applied. It is expected that this reduction would be in kilograms of active ingredient applied per hectare, but this is not explicitly stated by the authors.

References

Alston, J. M., N. Kalaitzandonakes, and J. Kruse. 2014. The size and distribution from the adoption of biotech soybean varieties. Chapter 45. Pp. 728–751 *in* S. J. Smyth, P. W. B. Phillips, D. Castle, eds. Handbook on agriculture, biotechnology and development. Edward Elgar Publishing Ltd, Cheltenham, U. K..

Areal, F. J., L. Riesgo, and E. Rodriguez-Cerezo. 2013. Economic and agronomic impact of commercialized GM crops: a meta analysis. J. Agric. Sci. 151:7–33.

BASF. 2012. News release: BASF to concentrate plant biotechnology activities on main markets in North and south America. Available at http://www.basf.com/group/pressrelease/ P-12-109.2012. (accessed 17 September 2012).

Biden, S. 2016. Comparing the Adoption of Genetically Modified Canola in Canada and Australia: The Environmental and Economic Opportunity Costs of Delay. University of Saskatchewan Masters' of Science Thesis.

Brookes, G., and P. Barfoot. 2016. Global income and production impacts of using GM crop Technology. 1996–2014. GM Crops Food 7:38–77.

Byrne, J. 2015. Presentation by President of v-Fluence Interactive at the Agricultural Bioscience International Conference. Melbourne.

Carpenter, J. 2010. Peer-reviewed surveys indicate positive impact of commercialized GM crops. Nat. Biotechnol. 28:319–321.

Convention on Biological Diversity. 2014. Global overview of the information on socioeconomic considerations arising from the impact of living modified organisms on the conservation and sustainable use of biological diversity. UNEP/CBD/BS/AHTEG-SEC/1/2, Montreal, Canada.

European Commission. 2016. The EU Special Incentive Arrangement for Sustainable Development and Good Governance ('GSP+') covering the period 2014 – 2015. Available at http://trade.ec.europa.eu/doclib/docs/2016/January/tradoc_154178.pdf (accessed 26 September 2016).

Finger, R., N. El Benni, T. Kaphengst, C. Evans, S. Herbert, B. Lehmann, et al. 2011. A meta analysis on farm-level costs and benefits of GM crops. Sustainability 3:743–762.

Gathura, G. 2014. Europe to shun Kenya's GM crops, farmers told. The Standard. June 20th. Available at: http://www.standardmedia.co.ke/article/2000125446/europe-to-shun-kenya-s-gm-crops-farmers-told (accessed 26 September 2016).

Gouse, M. 2013. An evaluation of the gender differentiated impact of genetically modified crop adoption: A pilot study in South Africa - GM maize and gender: Evidence from smallholder farmers in KwaZulu-Natal, South Africa. Project report to the Program for Biosafety Systems, International Food Policy Research Institute.

Gruère, G., and D. Sengupta. 2011. Bt cotton and farmer suicides in India: an evidence-based assessment. J. Dev. Stud. 47:316–337.

Gusta, M., S. J. Smyth, K. Belcher, P. W. B. Phillips, and D. Castle. 2011. Economic benefits of genetically-modified herbicide-tolerant canola for producers. AgBioForum 14:1–13.

Hobbs, J. E. 2014. The private sector: MNEs and SMEs. Chapter 4. Pp. 56–70 *in* S. J. Smyth, P. W. B. Phillips, D. Castle, eds. Handbook on agriculture, biotechnology and development. Edward Elgar Publishing Ltd, Cheltenham, U. K..

Hobbs, J. E., W. A. Kerr, and S. J. Smyth. 2014. The perils of zero tolerance: technology management, supply chains and thwarted globalisation. Int. J. Technol. Globalisation 7:203–216.

Huang, J., J. Mi, H. Lin, Z. Wang, R. Chen, R. Hu, et al. 2010. A decade of Bt cotton in Chinese fields: assessing the direct effects and indirect externalities of Bt cotton adoption in China. Sci. China Life Sci. 53:981–991.

Hutchison, W. D., E. C. Burkness, P. D. Mitchell, Moon, R. D., Leslie, T. W., Fleischer, S. J., et al. 2010. Areawide suppression of European corn borer with Bt maize reaps savings to non-Bt maize growers. Science 330:222–225.

James, C. 2015. 20th Anniversary (1996 to 2015) of the Global Commercialization of Biotech Crops and Biotech Crop Highlights in 2015. ISAAA Brief 51. Available at: http://www.isaaa.org/resources/publications/briefs/51/default.asp (Accessed 14 September 2016).

Kalaitzandonakes, N., P. W. B. Phillips, J. Wesseler, and S. J. Smyth, eds. 2016. The coexistence of genetically modified, organic and conventional foods: government policies and market practices. Springer Science+Business Media LLC., New York.

Klümper, W., and M. Qaim. 2014. A meta-analysis of the impacts of genetically modified crops. PLoS ONE 9:1–7

Kouser, S., and M. Qaim. 2011. Impact of Bt cotton on pesticide poisoning in smallholder agriculture: a panel data analysis. Ecol. Econ. 70:2105–2113.

Little, J. 2015. Smart regulation and innovation for EU agriculture. Presentation to the 7th European Innovation Summit. December 7-10.

Ludlow, K., S. J. Smyth, and J. Falck-Zepeda, eds. 2014. Socio-economic considerations in biotechnology regulations. Springer Publishers, New York.

Panella, F., N. Holland, R. Steinbrecher, A. Ferrante, M. Schimpf, H. Wallace, et al. 2015. Open letter to the Commission on new genetic engineering methods. Available at: http://www.genewatch.org/uploads/f03c6d66a9b354535738483c1c3d49e4/New_Breeding_Techniques___Open_Letter_27_Jan_2015.pdf (Accessed 23 September 2016).

Pardey, P. G., C. Chan-Kang, S. P. Dehmer, and J. M. Beddow. 2016. Agricultural R&D is on the move. Nature 537:301–303.

Pray, C., D. Ma, J. Huang, and F. Qiao. 2001. Impact of Bt cotton in China. World Dev. 30:931–948.

Ryan, C. D., and S. J. Smyth. 2012. Economic implications of low-level presence in a zero-tolerance European import market: the case of Canadian Triffid Flax. AgBioForum 15:21–30.

Smart, R. D., M. Blum, and J. Wesseler. 2016. Trends in approval times for genetically engineered crops in the United States and the European Union. J. Agric. Econ. doi:10.1111/1477-9552.12171.

Smyth, S. J., M. Gusta, K. Belcher, P. W. B. Phillips, and D. Castle. 2011a. Changes in herbicide use following the adoption of HR Canola in Western Canada. Weed Technol. 25:492–500.

Smyth, S. J., M. Gusta, K. Belcher, P. W. B. Phillips, and D. Castle. 2011b. Environmental impacts from herbicide tolerant canola production in Western Canada. Agric. Syst. 104:403–410.

Smyth, S. J., J. McDonald, and J. Falck-Zepeda. 2014. Investment, regulation and uncertainty: managing new plant breeding techniques. GM Crops Food 5:1–14.

Smyth, S. J., P. W. B. Phillips, and W. A. Kerr. 2015. Food security and the evaluation of risk. Global Food Secur. 4:16–23.

Smyth, S. J., P. W. B. Phillips, and W. A. Kerr. 2016. EU failing FAO challenge to improve global food security. Trends Biotechnol. 34:521–523.

Subramanian, A., and M. Qaim. 2010. The impact of Bt cotton on poor households in rural India. J. Dev. Stud. 46:295–311.

The Guardian. 2016. European parliament slams G7 food project in Africa. June 8th. Available at: https://www.theguardian.com/global-development/2016/jun/08/european-parliament-slams-g7-food-project-in-africa (Accessed 26 September 2016)

United Nations Conference on Trade and Development. 2013. Trade and environmental review 2013: wake up before it is too late. United Nations, New York.

Vitale, J., G. Vognan, and M. Ouattarra. 2014. Cotton. Chapter 38. Pp. 604–620 *in* S. J. Smyth, P. W. B. Phillips, D. Castle, eds. Handbook on agriculture, biotechnology and development. Edward Elgar Publishing Ltd, Cheltenham, U. K..

Yorobe, J. M., and M. Smale. 2012. Impacts of Bt maize on smallholder income in the Philippines. AgBioForum 15:152–162.

10

Vertical farming increases lettuce yield per unit area compared to conventional horizontal hydroponics

Dionysios Touliatos, Ian C. Dodd & Martin McAinsh

The Lancaster Environment Centre, Lancaster University, Lancaster, UK

Keywords
Land use efficiency, plant factory, urban agriculture, vertical column grower.

Correspondence
Dionysios Touliatos, The Lancaster Environment Centre, Lancaster University, Lancaster, UK.
E-mail: d.touliatos@lancaster.ac.uk

Funding Information
Biotechnology and Biological Sciences Research Council (Grant/Award Number: 'BB/J012718/1')

Abstract

Vertical farming systems (VFS) have been proposed as an engineering solution to increase productivity per unit area of cultivated land by extending crop production into the vertical dimension. To test whether this approach presents a viable alternative to horizontal crop production systems, a VFS (where plants were grown in upright cylindrical columns) was compared against a conventional horizontal hydroponic system (HHS) using lettuce (*Lactuca sativa L.* cv. "Little Gem") as a model crop. Both systems had similar root zone volume and planting density. Half-strength Hoagland's solution was applied to plants grown in perlite in an indoor controlled environment room, with metal halide lamps providing artificial lighting. Light distribution (photosynthetic photon flux density, PPFD) and yield (shoot fresh weight) within each system were assessed. Although PPFD and shoot fresh weight decreased significantly in the VFS from top to base, the VFS produced more crop per unit of growing floor area when compared with the HHS. Our results clearly demonstrate that VFS presents an attractive alternative to horizontal hydroponic growth systems and suggest that further increases in yield could be achieved by incorporating artificial lighting in the VFS.

Introduction

The global population is expected to reach 9 billion by 2050, a significant proportion of which will be urban dwellers, requiring a 70% increase in agricultural productivity (Corvalan et al. 2005; Tilman et al. 2011). Continued rural to urban migration is predicted to drive the expansion of urban landscapes and accelerate the loss of cultivated land surrounding towns and cities (Pandey and Seto 2015). Coupled with land degradation and loss of soil fertility, due to land-use intensification and climate change, agricultural land is increasingly becoming a scarce resource (Foley et al. 2011; Lambin et al. 2013) in addition to being a threat for biodiversity (Chaplin-Kramer et al. 2015). In light of the above, the need for innovation in land-use efficiency for crop production is therefore increasingly important (Lambin and Meyfroidt 2011).

Vertical farming has been proposed as an engineering solution to increase productivity per area by extending plant cultivation into the vertical dimension, thus enhancing land use efficiency for crop production (Eigenbrod and Gruda 2014). The large-scale implementation of vertical farming involves stacking growth rooms, such as glasshouses and controlled environment rooms, on top of each other to construct food-producing high-rise buildings (Despommier 2011). The same concept can be applied at a smaller scale through vertical farming systems (VFS). These growth systems expand crop production into the vertical dimension to produce a higher yield using less floor area (Hochmuth and Hochmuth 2001; Resh 2012). Examples of VFS include the use of vertical columns (Linsley-Noakes et al. 2006), vertically suspended grow bags (Neocleous et al. 2010), conveyor-driven stacked growth systems (Mahdavi et al. 2012), A-frame designs (Hayden 2006), and plant factory approaches (Kato et al. 2010).

Although these studies have quantified crop production, there have been few direct comparisons with horizontal systems of similar cropping density and little information is available on whether vertical column systems present a viable alternative to horizontal crop production systems. In addition, previous yield comparisons of VFS with conventional horizontal systems have confounded other factors with crop orientation. For example, yield increases of 129–200% in VFS and increased profits of 3.6–5.5 US dollar·m^{-2} compared to conventional soil cultivation have been reported (Liu et al. 2004). However, their VFS utilized a soilless growing medium; rendering the comparison essentially invalid. Similarly, significantly higher yields have been reported for strawberry grown in a vertical column VFS compared to conventional grow bags and a multi-tiered VFS (Ramírez-Gómez et al. 2012); although no information was provided regarding the root zone volume of the growth systems.

The aim of this study was to compare a vertical column VFS and a conventional horizontal hydroponic system (HHS) with similar fertigation regimes, root zone volumes, and planting densities to determine whether VFS represents a viable alternative to HHS. Lettuce was used as a model plant as it is widely grown in hydroponics as a rapidly growing leafy vegetable (Safaei et al. 2015) thereby avoiding some of the complexities of changes in crop biomass allocation during the reproductive process (Heller et al. 2014). The study was conducted indoors using only artificial lighting, as this is the dominant approach found in most urban vertical farming projects, especially in plant factory designs (Kang et al. 2014; He et al. 2015) and allows more precise control of environmental conditions (Poorter et al. 2012). Our results show that VFS increased lettuce yield per unit area compared to HHS and suggest that variation in light intensity between cropping systems of different spatial orientation could explain differences in crop yield.

Materials and Methods

Location

The study was conducted in a 3.4 m × 4.15 m walk-in Controlled Environment room (CE room) at the Lancaster Environment Centre (LEC, Lancaster University, UK). Illumination was provided by 12 400 W metal halide lamps (HQI-T 400N; Osram, St Helens, UK) for a 16 hour photoperiod (06.00 h to 22.00 h). Highly reflective plastic film (LBS Horticulture Ltd, Lancashire, UK) was placed on the walls of the room in order to increase the diffusion of light. Room air temperature ranged between 16 and 18 °C and relative humidity ranged from 60% to 80%. Room temperature and humidity were recorded by an Ektron II C sensor (HortiMaX B.V., Pijnacker, the Netherlands), which was hanging from the ceiling in the middle of the CE room, at 1.83 m above the ground. The CE room accommodated 2 VFS and 2 HHS, with one of each arranged on each side of the room. Preliminary measurements of photosynthetic photon flux density (PPFD) (before and after installing the VFS) revealed no shading effect of the VFS on the PPFD in the HHS (data not shown).

The vertical farming system

Plants were grown vertically in upright cylindrical columns comprised of individual modular units stacked on top of each other to reduce the system footprint. Each modular unit consisted of two stackable elements: a growing container (10.5 cm high and 7.5 cm radius) and a spacing collar (20 cm high and 7.5 cm radius). Five growing containers and six spacing collars were sterilized in TriGene disinfectant (MediChem International Ltd., Sevenoaks, UK) prior to filling each container with 130 g ± 0.5 g medium grade perlite (LBS Horticulture Ltd, Lancashire, UK). Highly porous perlite was the substrate of choice in order to minimize the risk of root-zone hypoxia and the resultant accumulation of ethylene within the airspace of the vertical column. The perlite was held in place by horticultural frost fleece (LBS Horticulture Ltd, Lancashire, UK) that was placed at the bottom of each growing container. The perlite in each container was levelled and seedlings were placed on the perlite at 90° to the horizontal. Each VFS contained 20 lettuce plants in total. Distance between the top of the VFSs and the light was 80 cm.

The horizontal hydroponic system

The HHS comprised of five cylindrical PVC pipes (45.5 cm high and 3.6 cm radius), which were sterilized in TriGene disinfectant (MediChem International Ltd), filled with 130 g ±0.5 g of perlite (LBS Horticulture Ltd, Lancashire, UK) and placed in parallel at 20 cm apart center to center. Horticultural frost fleece (LBS Horticulture Ltd, Lancashire, UK) was placed in the outlet of each pipe to hold the perlite in place. Each pipe held four plants placed in 4.4 cm square holes, in rows. In order to prevent the growth of algae black nylon fabric was used to cover the outlet channel of the system. Each HHS contained 20 lettuce plants in total. PVC pipes were mounted on commercial growth benches (90 cm from the ground and 130 cm from the lights); which was equivalent in height to VFS's layer 2 (Fig. 1).

Fertigation

Each growth system was supplied with recirculated half-strength Hoagland's solution (Hoagland and Arnon 1950) from a 18 L Titan PC4R Tank (Kingspan Environmental Ltd, Armagh, UK), through a 1.27 cm double-walled PVC hose (LBS Horticulture Ltd). The composition of the nutrient solution was 0.5 mmol·L^{-1} NH_4NO_3, 1.75 mmol·L^{-1} $Ca(NO_3)_2·4H_2O$, 2.01 mmol·L^{-1} KNO_3, 1.01 mmol·L^{-1} KH_2PO_4, 0.5 mmol·L^{-1} $MgSO_4·7H_2O$, 1.57 μmol·L^{-1}

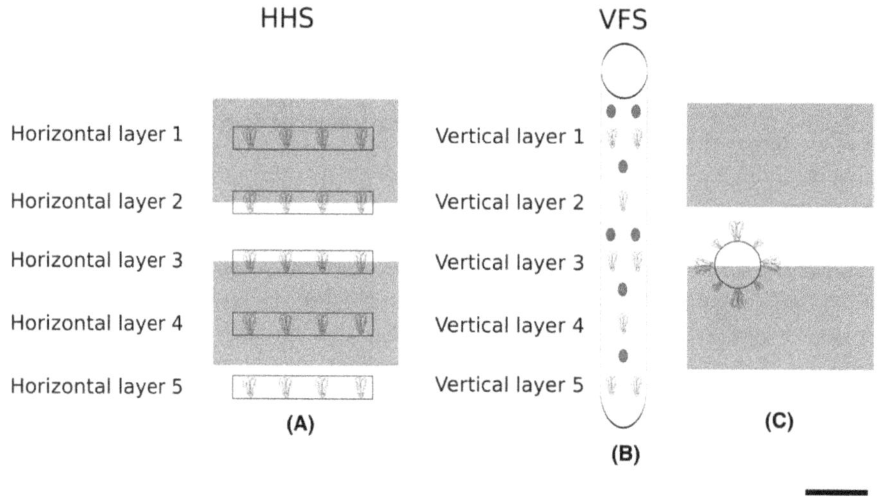

Figure 1. Schematic shows planting densities within the HHS and VFS. (A) Overhead view of HHS. (B) Side-view of VFS. (C) Overhead view of VFS. The HHS occupied 0.4 m^2 of growing floor area, whereas the VFS occupied 0.02 m^2 per column of floor area. The grey rectangles show the exact position of the 400 W metal halide lamps above the growth systems. The grey circles within the VFS show the exact measurement positions of the Macam Q203 Quantum radiometer quantum sensor. Light measurements in the HHS were obtained directly above the plants, at 20 cm distance from the PVC pipes. Scale bar is 20 cm. VFS, vertical farming system; HHS, horizontal hydroponic system.

$MnSO_4{\cdot}5H_2O$, 11.3 μmol·L^{-1} H_3BO_3, 0.3 μmol·L^{-1} $CuSO_4{\cdot}5H_2O$, 0.032 μmol·L^{-1} $(NH_4)_6Mo_7O_{24}{\cdot}4H_2O$, 1.04 μmol·L^{-1} $ZnSO_4{\cdot}7H_2O$, and 0.25 mmol·L^{-1} NaFe EDTA. Nutrient solution Electrical Conductivity (EC) was 1 ± 0.2 dS·m^{-1}. The nutrient solution in tanks was replaced weekly and 2 mol·L^{-1} H_3PO_4 was used to maintain a pH of 5.8 ± 0.2, which was checked daily. Hozelock 360° Micro Jet microsprinklers (Hozelock Limited, Aylesbury, UK) delivered the solution to the top layer of the VFS, allowing gravity-driven drip-irrigation of plants in growing modules below them in the column. In the HHS the nutrient solution was delivered to each pipe of the HHS by a microsprinkler. The effluent from the bottom layer of the VFS and from all PVC pipes of the HHS was subsequently returned to the tank and recirculated around the growing systems using a submersible aquarium water pump (All Pond Solutions Ltd, Middlesex, UK), capable of delivering a maximum of 3100 L·h^{-1}. Hozelock Coupling 13 mm hose connectors (Hozelock Limited, Aylesbury, UK) were used to connect all hoses and pumps. Each pump was programmed to operate for 1 min every hour using a multi purpose electronic digital programmable timer (JoJo Waterproof Digital Outdoor Electrical Timer; AuctionZ Ltd, Bradford, UK). Consequently, the microsprinklers sprayed nutrient solution within the top layers of the VFS and within the horizontal layers of the HHS for 1 min every hour.

Plant material

Romaine lettuce (*Lactuca sativa* L. cv. 'Little Gem') seeds were sown in 84-cell plug trays (tray dimensions: 53 cm × 31 cm × 5.5 cm) containing Levington M3 compost (Scotts UK, Ipswich, UK). Seedlings were watered daily with tap water and were grown at a PPFD of 200 μmol·m^{-2}·s^{-1} over a 16 h photoperiod. Plants were transplanted 20 days after sowing at the four true leaf stage.

Data collection and statistical analysis

Photosynthetic photon flux density measurements were obtained using a Macam Q203 Quantum radiometer (Macam Photometrics LTD, Livingstone, UK) 1 and 5 weeks after the plants were transferred to the growth systems. The quantum sensor was placed in the middle of the spacing collar, in a 10 cm radius zone around the vertical column in the VFS, and was placed 20 cm above the PVC pipes in the HHS. This approach ensured consistency in light intensity measurements and avoided plant damage during sampling. Shoot fresh weight was measured immediately after harvest on week 5 using a 2 decimal point scientific balance. To compare average shoot fresh weight per growth system using Student's *t*-test, the data were square root transformed, as they were not normally distributed (Table 1). Yield and number of plants per occupied growing floor area for growth systems was used to calculate the ratio of VFS to HHS (Table 1). Linear regression analysis was used to analyze the relationship between shoot fresh weight and vertical or horizontal layers within each growth system (Fig. 2). Significant differences in PPFD during week 1 within the VFS and the HHS were detected

Table 1. Comparison of the productivity of the vertical farming system (VFS) and horizontal hydroponic system (HHS).

Parameter	HHS	VFS	Result
Shoot fresh weight (g) Mean ± SE (n = 40)	138 ± 6	95 ± 6	$P < 0.001$*
Yield per occupied growing floor area[1] (kg FW·m^{-2})	6.9	95	VFS/HHS = 13.8
Number of plants per occupied growing floor area[1] (plant number m^{-2})	50	1000	VFS/HHS = 20

[1]HHS growing floor area: 0.4 m^2, VFS growing floor area: 0.02 m^2.
*Student's t-test on square root transformed data, t (78) = 5.656.

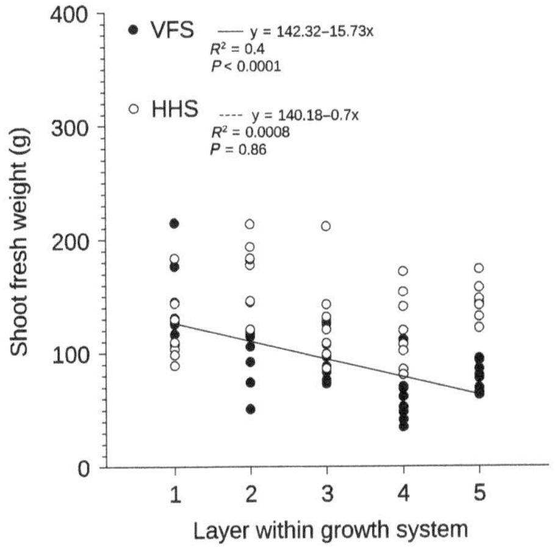

Figure 2. Linear regression analysis of shoot fresh weight versus layer in the VFS (solid line; closed symbols) and HHS (open symbols), respectively. When the linear regression was not significant the regression line was omitted. The regression equation, adjusted R^2 values and significance of the regression (P-value) are reported at the top of the panel. VFS, vertical farming system; HHS, horizontal hydroponic system.

using one-way ANOVA followed by Tukey post hoc analysis (Fig. 3). The relationship between PPFD during week 5 and shoot fresh weight was analyzed using linear regression analysis (Fig. 4), with P-value <0.05 considered to indicate a statistically significant difference. All statistical tests were performed using "R" version 3.1.2 software (R Development Core Team 2014).

Results

The VFS produced more crop per unit area compared to the HHS

The VFS produced 13.8 times more crop, calculated as a ratio of yield (kg FW) to occupied growing floor area (m^2). However, mean FW (g) for lettuce crops grown within the HHS was significantly higher than those grown within the VFS. Therefore, although the same number of plants was grown in each system the two HHS produced 1.7 kg more crop compared to the two VFS (5.5 and 3.8 kg of crop in total, respectively). Consequently, the higher productivity of the VFS, in terms of kg FW·m^{-2} growing floor area, can be attributed to the 20-fold higher number of plants per growing floor area (Table 1).

Yield decreased from top to base of VFS, whereas yield was uniform within the HHS

Shoot fresh weight decreased from top to base of vertical columns of the VFS, whereas no gradient in productivity was observed between horizontal layers in the HHS. Crop productivity was uniform in the HHS with a range of 133 g within a normal distribution, whereas in the VFS, crop productivity had a range of 180 g

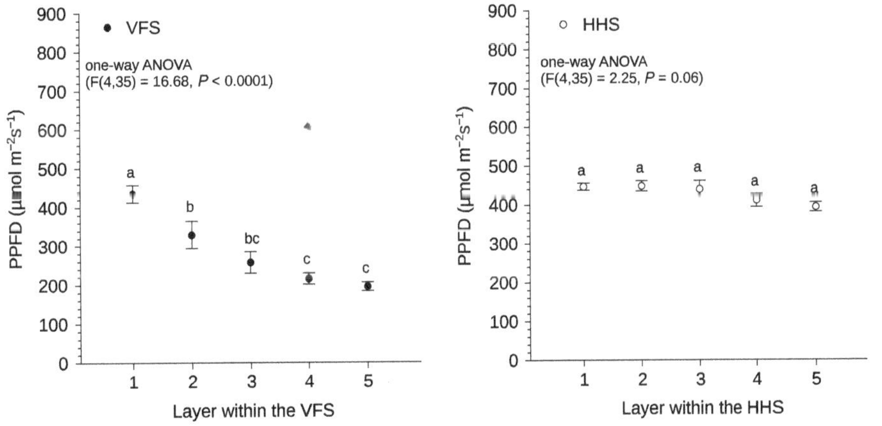

Figure 3. PPFD within the vertical farming system (VFS; closed symbols) and the horizontal hydroponic system (HHS; open symbols) plotted against layers in the growth systems. Values indicated with different letters indicate statistically significant differences, whereas those marked with the same letters show statistically similar values. Error bars represent SE (n = 8). VFS, vertical farming system; HHS, horizontal hydroponic system; PPFD, photosynthetic photon flux density.

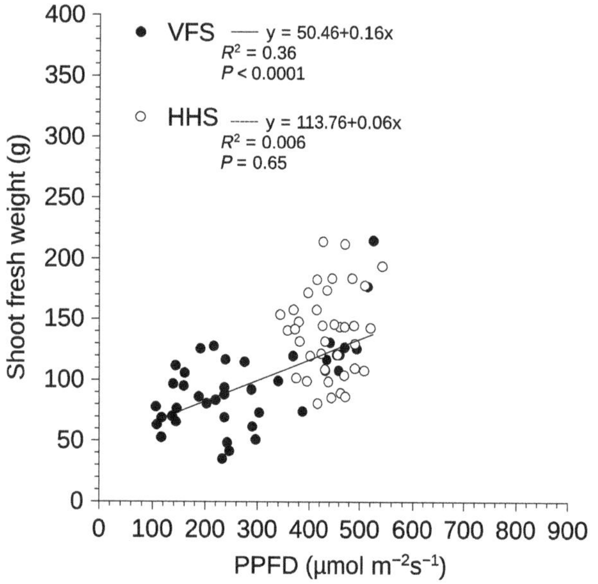

Figure 4. Linear regression analysis of shoot fresh weight versus PPFD in the vertical farming system (VFS; solid line; closed symbols) and horizontal hydroponic system (HHS; open symbols), respectively. When the linear regression was not significant the regression line was omitted. The regression equation, adjusted R^2 values, and significance of the regression (*P*-value) are reported at the top of the panel. VFS, vertical farming system; HHS, horizontal hydroponic system; PPFD, photosynthetic photon flux density.

within a nonnormal distribution with positive skewness (Sk = 1.035). Plants grown within the top layer of the VFS and within all layers of the HHS were of similar shoot fresh weight. However, in the middle and bottom layers (Layers 2–5) of the VFS, productivity decreased significantly. As a result, the bottom layer of the VFS produced 43% less shoot fresh weight than the top layer of the VFS (Fig. 2).

Light intensity decreased from top to base of VFS

Light intensity decreased significantly from top to base of vertical columns within the VFS, whereas no significant difference in PPFD was observed within the horizontal layers of the HHS. PPFD values varied between 491 and 134 $\mu mol \cdot m^{-2} \cdot s^{-1}$ from top to base of vertical column of the VFS and between 570 and 340 $\mu mol \cdot m^{-2} \cdot s^{-1}$ within the horizontal layers of the HHS. The top layer of the VFS received similar PPFD to all the horizontal layers of the HHS. However, within the vertical layers, as distance from the light source increased, there was a significant drop in PPFD values within the VFS. There was no significant difference in PPFD between layers 2 and 3 and between layers 3, 4, and 5 (Fig. 3).

Light intensity influenced growth in the VFS but not in the HHS

There was a significant positive relationship between shoot fresh weight and PPFD in the VFS, indicating that as light intensity increased so did crop productivity (Fig. 4). In contrast, there was no significant relationship between yield and PPFD within the HHS.

Discussion

Although it has been proposed that increases in yield per growing area can be achieved by extending plant cultivation into the vertical dimension using VFS (Eigenbrod and Gruda 2014), to date there is no conclusive evidence that this is indeed the case. However, our results show that crop productivity, defined as a ratio of yield to occupied growing floor area, is 13.8 times higher in VFS than the HHS. This is likely because by incorporating the vertical dimension into the growth environment, the VFS can grow 20-fold more plants per unit area than the HHS (Table 1). However, these calculations are based on independent vertical columns and do not consider the effect of column spacing on yield per occupied growing floor area. For example, in high wire crop training systems high planting densities impose intense competition for light within the growth system (Pettersen et al. 2010). This is an important factor that needs to be considered in future studies, as spacing between vertical columns influenced crop productivity in VFS glasshouse trials (Liu et al. 2004).

In contrast, the absolute yield of the HHS, in terms of shoot fresh weight, was higher than the VFS (Table 1). This can be explained by the significant decrease in PPFD from top to base of the vertical columns (Fig. 3) and significant causal relationship between shoot fresh weight and PPFD within the VFS (Fig. 4) that limited growth in the lower layers. Light intensity is one of the primary variables affecting lettuce yield and quality (Ferentinos et al. 2000; Son and Oh 2013; Ouzounis et al. 2015) and it is has been well documented that lettuce yield increases with increasing light intensity (Knight and Mitchell 1988; Frantz and Bugbee 2005). Therefore, since yield decreased from top to base of the vertical column, and yield was uniform within the HHS (Fig. 2), it was anticipated that the VFS would produce less crop in total than the HHS. Similarly, light intensity and shoot fresh weight were highly correlated and both decreased from top to base of vertical columns in a glasshouse (Liu et al. 2004). Light gradients from top to base of vertical column systems were also reported in glasshouse vertical strawberry cultivation (Ramírez-Gómez et al. 2012). Therefore, our data suggest that top to base gradients in light intensity and shoot fresh weight limit plant growth in vertical columns in both indoor and glasshouse settings.

Vertical light intensity gradients (e.g. Fig. 3) could be altered by natural illumination. In glasshouse trials with vertical columns, light intensity decreased from top to base of vertical columns, with lower PPFD values being recorded toward the northern side of columns compared to the southern side (Liu et al. 2004). Thus, natural illumination introduced an additional gradient in light distribution within the VFS. Future studies are therefore required to test whether natural illumination diminishes or exacerbates light intensity gradients.

Light intensity in growth chambers is known to decrease as distance from the light source increases (Poorter et al. 2012) and this phenomenon partially explained the large variance in PPFD observed in vertical layers 2–5 within the VFS (Fig. 3). In addition, a "shading effect" within the VFS (Linsley-Noakes et al. 2006) was due to higher positioned plants within the VFS obscuring lower positioned plants from the light source. Side-on rather than top-down illumination could potentially ameliorate the shading effect, consequently mitigating the gradient in crop productivity within the VFS. Side-on illumination, also known as interlighting has improved light distribution within tall canopies and, in some cases, increased crop yield and light use efficiency (Olle and Viršile 2013). Interlighting with light-emitting diodes (LEDs) ameliorated mutual shading within tomatoes at high planting density and increased tomato yield by 12–14% in comparison to the control (Lu et al. 2012). Overhead illumination combined with intracanopy lighting using HPS lamps increased cucumber yield in high-wire crop training system by 11% compared to traditional overhead illumination (Pettersen et al. 2010). In contrast, there were no differences in productivity when comparing LED interlighting against overhead HPS in high-wire tomato cultivation (Gómez et al. 2013). Similarly, interlighting by fluorescent tubes improved fruit quality but did not increase yield in high-wire cucumber production in the glasshouse (Heuvelink et al. 2006). This variation in crop responses to interlighting may be due to the different environmental conditions and crop management applied. However, since vertical column systems share similar light distribution properties within the vertical plane to plants grown in high wire crop training systems (Hovi et al. 2004), side-on illumination could potentially mitigate observed light gradients within the VFS.

Interestingly, only 36% of the variation in shoot fresh weight within the VFS was explained by the gradient in PPFD (Fig. 4) with the remaining 64% of variance being attributed to putative temperature gradients and putative nutrient concentration gradients along the vertical column (Jones 2014). Nutrient concentration gradients within the gully of Nutrient Film Technique (NFT) systems have been claimed to influence crop uniformity within the NFT (Puerta et al. 2007). An important difference between the VFS and the HHS of this study was that each layer of the HHS received nutrient solution directly from the tank whereas, in the VFS nutrient solution was delivered to the top layer and was gravity-driven drip-fed to vertical layers beneath creating the potential for marked gradients in nutrient availability within the VFS. Identifying the physiological effects of this putative nutrient concentration gradient to growth within VFS is therefore an important area for future studies.

To conclude, from a commercial point of view, the effects of gradients within the VFS on crop value will depend on how the crop is going to be processed and marketed. For example, if lettuce was grown to be sold as individual heads, then the nonuniform productivity of the VFS would be a potential weakness of the VFS over the HHS. However, if the crop was destined for precut salad bags then crop uniformity may be irrelevant while increased yield per unit area could be a significant advantage of the VFS. Therefore, crop utilization and marketability and an investigation of the cost-to-benefits ratio of these growing systems will be the ultimate criteria to decide whether VFS can provide an alternative to HHS.

Conclusions

Vertical column-based VFS presented a viable alternative to conventional horizontal growth systems by optimizing growing space use efficiency, thereby producing more crop per unit area. Further increases in yield could be achieved by incorporating artificial lighting within the VFS to mitigate the observed PPFD gradient. Future studies could investigate the influence of putative additional gradients (such as root and canopy zone temperature and nutrient concentration gradients) within the VFS and interactions between vertical columns in terms of competition for PPFD in large-scale vertical farming settings.

Conflict of Interest

None declared.

References

Chaplin-Kramer, R., R. P. Sharp, L. Mandle, S. Sim, J. Johnson, I. Butnar, et al. 2015. Spatial patterns of agricultural expansion determine impacts on biodiversity and carbon storage. Proc. Natl Acad. Sci. USA 112:7402–7407.

Corvalan, C., S. Hales, and A. J. McMichael. 2005. Ecosystems and human well-being: health synthesis. World Health Organization.

Despommier, D. 2011. The vertical farm: controlled environment agriculture carried out in tall buildings would create greater food safety and security for large urban populations. J. Für Verbraucherschutz Leb. 6:233–236.

Eigenbrod, C., and N. Gruda. 2014. Urban vegetable for food security in cities. A review. Agron. Sustain. Dev. 35:483–498.

Ferentinos, K. P., L. D. Albright, and D. V. Ramani. 2000. SE – structures and environment: optimal light integral and carbon dioxide concentration combinations for lettuce in ventilated greenhouses. J. Agric. Eng. Res. 77:309–315.

Foley, J. A., N. Ramankutty, K. A. Brauman, E. S. Cassidy, J. S. Gerber, M. Johnston, et al. 2011. Solutions for a cultivated planet. Nature 478:337–342.

Frantz, J. M., and B. Bugbee. 2005. Acclimation of plant populations to shade: photosynthesis, respiration, and carbon use efficiency. J. Am. Soc. Hortic. Sci. 130:918–927.

Gómez, C., R. C. Morrow, C. M. Bourget, G. D. Massa, and C. A. Mitchell. 2013. Comparison of intracanopy light-emitting diode towers and overhead high-pressure sodium lamps for supplemental lighting of greenhouse-grown tomatoes. HortTechnology 23:93–98.

Hayden, A. L. 2006. Aeroponic and hydroponic systems for medicinal herb, rhizome, and root crops. HortScience 41:536–538.

He, J., L. Qin, Y. Liu, and T. W. Choong. 2015. Photosynthetic capacities and productivity of indoor hydroponically grown *Brassica alboglabra* bailey under different light sources. Am. J. Plant Sci. 6:554.

Heller, H., A. Bar-Tal, S. Assouline, K. Narkis, S. Suryano, A. de la Forge, et al. 2014. The effects of container geometry on water and heat regimes in soilless culture: lettuce as a case study. Irrig. Sci. 33:53–65.

Heuvelink, E., M. J. Bakker, L. Hogendonk, J. Janse, R. Kaarsemaker, and R. Maaswinkel. 2006. Horticultural lighting in the Netherlands: new developments. In V International Symposium on Artificial Lighting in Horticulture. 711:25–34.

Hoagland, D. R., and D. I. Arnon. 1950. The water-culture method for growing plants without soil. Circ. Calif. Agric. Exp. Stn. 347:2nd edn.

Hochmuth, R., and G. J. Hochmuth. 2001. A decade of change in Florida's greenhouse vegetable industry: 1991–2001 Pp. 280–283 *in* Proc. Fla. State Hort. Soc.

Hovi, T., J. Näkkilä, and R. Tahvonen. 2004. Interlighting improves production of year-round cucumber. Sci. Hortic. 102:283–294.

Jones, Jr, J. B. 2014. Complete guide for growing plants hydroponically. CRC Press.

Kang, J. H., S. KrishnaKumar, S. L. S. Atulba, B. R. Jeong, and S. J. Hwang. 2014. Light intensity and photoperiod influence the growth and development of hydroponically grown leaf lettuce in a closed-type plant factory system. Hortic. Environ. Biotechnol. 54:501–509.

Kato, K., R. Yoshida, A. Kikuzaki, T. Hirai, H. Kuroda, K. Hiwasa-Tanase, et al. 2010. Molecular breeding of tomato lines for mass production of miraculin in a plant factory. J. Agric. Food Chem. 58:9505–9510.

Knight, S. L., and C. A. Mitchell. 1988. Effects of CO_2 and photosynthetic photon flux on yield, gas exchange and growth rate of *Lactuca sativa* L. "Waldmann"s Green'. J. Exp. Bot. 39:317–328.

Lambin, E. F., and P. Meyfroidt. 2011. Global land use change, economic globalization, and the looming land scarcity. Proc. Natl Acad. Sci. USA 108:3465–3472.

Lambin, E. F., H. K. Gibbs, L. Ferreira, R. Grau, P. Mayaux, P. Meyfroidt, et al. 2013. Estimating the world's potentially available cropland using a bottom-up approach. Glob. Environ. Change 23:892–901.

Linsley-Noakes, G., L. Wilken, and S. de Villiers. 2006. High density, vertical hydroponics growing system for strawberries. Acta Hortic. 708:365.

Liu, W., D. K. Chen, and Z. X. Liu. 2004. High efficiency column culture system in China. Acta Hortic. 691:495–500.

Lu, N., T. Maruo, M. Johkan, M. Hohjo, S. Tsukagoshi, Y. Ito, et al. 2012. Effects of Supplemental Lighting within the Canopy at Different Developing Stages on Tomato Yield and Quality of Single-Truss Tomato Plants Grown at High Density. Environ. Control. Biol. 50:1–11.

Mahdavi, S., M. Kafi, R. Naderi, and T. Sadat. 2012. Vertical mobile planting system consistent with the pattern of solar radiation and effects of system on light exposure and growth of Gerbera cut flowers (*Gerbera jamesonii* cv. Antibes), in greenhouse culture. J. Agric. Technol. 8:1461–1468.

Neocleous, D., C. Kaittanis, N. Seraphides, and P. Polycarpou. 2010. Horizontal and vertical soilless growing systems under Cyprus conditions. J. Appl. Hortic. 12:140–144.

Olle, M., and A. Viršile. 2013. The effects of light-emitting diode lighting on greenhouse plant growth and quality. Agric. Food Sci. 22:223–234.

Ouzounis, T., B. R. Parjikolaei, X. Fretté, E. Rosenqvist, and C. O. Ottosen. 2015. Predawn and high intensity application of supplemental blue light decreases the quantum yield of PSII and enhances the amount of phenolic acids, flavonoids, and pigments in *Lactuca sativa*. Front. Plant Sci. 6.

Pandey, B., and K. C. Seto. 2015. Urbanization and agricultural land loss in India: comparing satellite estimates with census data. J. Environ. Manage. 148:-53–66, Land Cover/Land Use Change (LC/LUC) and Environmental Impacts in South Asia.

Pettersen, R. I., S. Torre, and H. R. Gislerød. 2010. Effects of intracanopy lighting on photosynthetic characteristics in cucumber. Sci. Hortic. 125:77–81.

Poorter, H., F. Fiorani, M. Stitt, U. Schurr, A. Finck, Y. Gibon, et al. 2012. The art of growing plants for experimental purposes: a practical guide for the plant biologist. Funct. Plant Biol. 39:821–838.

Puerta, A. R., S. Sato, Y. Shinohara, and T. Maruo. 2007. A modified nutrient film technique system offers a more uniform nutrient supply to plants. Horttechnology 17:227–233.

R Core Team 2014. R: A language and environment for statistical computing. R Foundation for Statistical Computing, Vienna, Austria. Available at http://www.R-project.org/.

Ramírez-Gómez, H., M. Sandoval-Villa, A. Carrillo-Salazar, and A. Muratalla-Lúa. 2012. Comparison of hydroponic systems in the strawberry production. Acta Hortic. 947:165–172.

Resh, H. M. 2012. Hydroponic food production: a definitive guidebook for the advanced home gardener and the commercial hydroponic grower. CRC Press.

Safaei, M., J. Panahandeh, S. J. Tabatabaei, and A. R. M. Azar. 2015. Effects of different nutrients solutions on nutrients concentration and some qualitative traits of lettuce in hydroponics system. J. Sci. Technol. Greenh. Cult. 6:Pe1–Pe7, En8.

Son, K.-H., and M.-M. Oh. 2013. Leaf shape, growth, and antioxidant phenolic compounds of two lettuce cultivars grown under various combinations of blue and red light-emitting diodes. HortScience 48:988–995.

Tilman, D., C. Balzer, J. Hill, and B. L. Befort. 2011. Global food demand and the sustainable intensification of agriculture. Proc. Natl Acad. Sci. USA 108:20260–20264.

11

Genetically engineered bananas resistant to Xanthomonas wilt disease and nematodes

Leena Tripathi[1], Howard Atkinson[2], Hugh Roderick[2], Jerome Kubiriba[3] & Jaindra N. Tripathi[1]

[1]International Institute of Tropical Agriculture, Nairobi, Kenya
[2]Centre for Plant Sciences, University of Leeds, Leeds, UK
[3]National Agricultural Research Laboratories, PO Box 7084, Kampala, Uganda

Keywords
Banana, improvement, genetic engineering, Xanthomonas wilt, nematode, Africa

Correspondence
Leena Tripathi, International Institute of Tropical Agriculture, PO Box 30709, Nairobi, Kenya.

E-mail: l.tripathi@cgiar.org

Funding Information
The authors would like to thank United States Agency for International Development (USAID) for financial support to BXW research and the USAID ABSPII Program, the Biotechnology and Biological Sciences Research Council (BBSRC), Department for International Development (DFID), and CGIAR Research Program on Roots, Tubers and Bananas (RTB) for financial support to nematode research.

Abstract

Banana is an important staple food crop feeding more than 100 million Africans, but is subject to severe productivity constraints due to a range of pests and diseases. Banana Xanthomonas wilt caused by *Xanthomonas campestris* pv. *musacearum* is capable of entirely destroying a plantation while nematodes can cause losses up to 50% and increase susceptibility to other pests and diseases. Development of improved varieties of banana is fundamental in order to tackle these challenges. However, the sterile nature of the crop and the lack of resistance in *Musa* germplasm make improvement by traditional breeding techniques either impossible or extremely slow. Recent developments using genetic engineering have begun to address these problems. Transgenic banana expressing sweet pepper *Hrap* and *Pflp* genes have demonstrated complete resistance against *X. campestris* pv. *musacearum* in the field. Transgenic plantains expressing a cysteine proteinase inhibitors and/or synthetic peptide showed enhanced resistance to a mixed species population of nematodes in the field. Here, we review the genetic engineering technologies which have potential to improve agriculture and food security in Africa.

Introduction

Closing the yield gap of staple crops is a priority for ensuring future food security, especially in developing nations (Godfray et al. 2010). The population of Africa is projected to double between 2015 and 2050 to 2.5 billion and increase further to 4.4 billion by 2100 by which time 38% of the global population is projected to be African (UN 2015). In many parts of the world emphasis can be placed on cereals to address the demands of population growth (West et al. 2014) but not in some key areas in Sub-Saharan Africa (SSA).

Banana including plantain (*Musa* spp.) is an important staple crop in tropics. Annual global production of banana is about 145 million tons (FAOSTAT 2014). Approximately a third of that production is in Africa, and Africa accounts for about 72% of production of plantains (FAOSTAT 2014). Investment in banana improvement holds great potential for improving food security as these crops feed more people per unit area of production than other staple crops (West et al. 2014). For instance, Uganda produces 30% of the global production of cooking bananas and has the highest consumption per capita (FAOSTAT 2014). In southeastern Nigeria, smallholder farmers generate up to 30% of their income from plantain cultivation (Pasberg-Gauhl and Gauhl 1996). In Central and West Africa, plantains account for about 32% of total *Musa* production (Lescot 2008), which feed approximately 70 million people with >25% of their carbohydrates and 10% of their food energy (Ortiz and Vuylsteke 1996; Robinson 1996).

Most cultivated banana varieties are triploids with low to no fertility generated by hybridizations between two diploid species, *Musa acuminata* and *M. balbisiana*, which contribute to the A and B genomes, respectively (Ortiz et al. 1995). The sweet dessert banana that forms the bulk of the export market and East African highland bananas (EAHB) are AAA, plantains and East African dessert bananas are AAB, and most other cooking bananas are ABB (Simmonds 1987).

Banana production is severely hampered by several pests and diseases, particularly on low-input, subsistence farms. Banana Xanthomonas wilt (BXW) caused by *Xanthomonas campestris* pv. *musacearum* is seriously threatening the banana production in East Africa (Tripathi et al. 2009; Shimwela et al. 2016a). The disease starts with wilting of leaves or male bud and premature ripening of fruits leading to death of plant and rotting of fruits. Where it occurs, BXW causes acute infections that can lead to a complete loss of a plantation. It caused 30–50% decrease in banana yields in Uganda between 2001 and 2004 (Karamura et al. 2006; Shimwela et al. 2016a). Economic losses of about $2–8 billion have been reported over a decade in the East Africa (Tripathi et al. 2009; Nkuba et al. 2015; Shimwela et al. 2016a). BXW disease is transmitted mainly by insects, contaminated farming tools, infected planting materials, and probably rain splash (Shimwela et al. 2016a,b). It can be contained by the use of cultural practices such as removal of the male bud to prevent insect transmitted infection, using sterilized farming tools, destroying infected plants, and using clean pathogen-free planting materials. However, the adoption of these practices is inconsistent as these techniques are labor intensive and may enhance disease spread if cutting of plants occurs during rainy season (Shimwela et al. 2016a,b). The disease affects all banana varieties and no resistant source has been identified in *Musa* germplasm yet.

Nematodes cause losses globally to banana production. Analysis of data from experimental applications of nematicides across a range of African countries has demonstrated yield responses of 71 ± 16% over 3 years after nematicide application (Atkinson 2003). Losses of >50% have been confirmed in a field trial with plantain (Roderick et al. 2012a). Nematodes are often controlled in commercial banana plantations by periodic application of environmentally damaging pesticides, but they are not normally available or suitable for smallholders in Africa. Crop rotation is not often possible for such farmers, many of whom have insufficient land to accept the associated yield loss, given that plantains out produce all other staple crops in conditions that favor them.

Development of nematodes or banana Xanthomonas wilt-resistant cultivars by traditional crosspollination techniques is hampered by the sterility of the polyploid genomes of cultivated banana and plantains (Lorenzen et al. 2010). However, conventional breeding has produced hybrids with resistance to *Radopholus similis* though these tend to remain moderate hosts for *Pratylenchus* species (Quénéhervé et al. 2009). No hybrid has shown resistance to the concurrent infections by several nematode species (Pinochet 1988; De Waele and Elsen 2002; Lorenzen et al. 2010) as required to manage them on banana and plantain crops.

No resistant varieties of banana have been identified with both nematode and Xanthomonas wilt resistance, but transgenic plants with both of these resistance traits have been valued for Uganda alone at $962 m over a 30-year period (Kalyebara et al. 2007). Male and female sterility of most edible cultivars, lack of crossfertile wild relatives, and clonal propagation of banana all contribute to no risk of gene flow from transgenic banana plants to either wild or cultivated plants. Deployment of farmer preferred transgenic cultivars is unlikely to adversely affect the already very low genetic variability in the banana crop due to its perennial nature and a reliance on very few cultivars across large geographical areas. This review describes progress on developing transgenic banana resistance to both Xanthomonas wilt disease and nematodes and key issues to be resolved before their deployment to growers in Africa.

Xanthomonas wilt Resistant Banana

Genetic engineering is an important tool that facilitates transfer of genes for useful agronomic traits across species. It can complement conventional breeding of banana by allowing the bottlenecks of breeding for developing improved varieties to be overcome. In the absence of known host plant resistance among banana genotypes, genetic engineering provides a cost-effective alternative technique to develop Xanthomonas wilt resistant banana varieties. Host plant resistance against pathogens can be enhanced by expressing resistance (R) genes, antimicrobial genes, or defense genes (Tripathi et al. 2016; Table 1). The Hypersensitive Response Assisting Protein *(Hrap)* and Plant Ferredoxin Like Protein (*Pflp*) genes from sweet pepper (*Capsicum annuum*) are defense genes which can intensify the hypersensitive response (Lin et al. 1997; Chen et al. 2000). These genes have provided resistance against various bacterial pathogens such as *Erwinia, Pseudomonas, Ralstonia*, and *Xanthomonas* spp. in transgenic *Arabidopsis*, tobacco, tomato, orchids, calla lily, and rice (Tang et al.

2001; Ger et al. 2002; Liau et al. 2003; Huang et al. 2004, 2007; Pandey et al. 2005; Yip et al. 2007).

Transgenic bananas have been generated by inserting *Hrap* or *Pflp* gene in embryogenic cell suspensions of banana cultivars, the AAB sweet banana cultivar 'Sukali Ndiizi', and the AAA-EAHB cultivar 'Nakinyika', through *Agrobacterium*-mediated transformation (Tripathi et al. 2010; Namukwaya et al. 2012). The transgenic events were analyzed to confirm the presence of transgene by PCR and integration of transgene in banana genome by Southern blot analysis. Several of these transgenic events showed enhanced resistance under laboratory and glasshouse conditions (Tripathi et al. 2010; Namukwaya et al. 2012). The promising transgenic events showing 100% resistance against *Xcm* in glasshouse experiments were selected for confined field trial. Sixty-five transgenic events (40 *Hrap* gene lines and 25 *Pflp* gene lines) were evaluated for disease resistance in a confined field trial at the National Agricultural Research Laboratory (NARL), Kawanda, Uganda. Complete resistance to Xanthomonas wilt disease was demonstrated for 11 transgenic events (7 *Hrap* lines and 4 *Pflp* lines) for both mother and progeny crops (Tripathi et al. 2014a). Control nontransgenic plants developed disease symptoms and wilted completely. The results from field trial experiment confirmed the transfer of the disease resistance trait from mother to progeny. These 11 transgenic events, besides showing absolute resistance to Xanthomonas wilt disease, also showed agronomic characteristics (flowering and yield) similar to nontransgenic control varieties (Tripathi et al. 2014a). These transgenic events were further evaluated in a second confined trial to measure agronomic performance. As bacterial pathogens evolve fast, there is risk of breaking down of resistance in transgenic plants developed using single gene. To avoid or delay this situation, we are developing transgenic banana varieties using stacked genes (*Hrap-Pflp*). The transgenic banana expressing stacked *Hrap* and *Pflp* genes did not show higher or additive resistance against pathogen in comparison to individual genes; however, stacking might provide the benefit of durable resistance in case one transgene function is lost (Muwonge et al. 2016).

The rice pattern recognition receptor (PRR) XA21 has also been tested for resistance against *X. campestris pv. musacearum* in order to identify additional disease resistance genes for use in gene pyramiding strategies. The transgenic rice overexpressing *Xa21* gene confers resistance to the bacterial pathogen *X. oryzae* pv. *oryzae* (Ronald et al. 1992; Wang et al. 1996). Transgenic banana expressing rice *Xa21* gene were developed and tested for Xanthomonas wilt disease resistance. These transgenic plants demonstrated enhanced resistance against *X. campestris pv. musacearum* under glass house conditions (Tripathi et al. 2014b).

Several other potential transgenes have been identified that suppress development of disease lesions in other plants species or shown in vitro antibacterial effects against *Xanthamonas* sp. Transgenic tomato expressing the R-genes, *Pto* or *Bs2*, showed resistance against *X. campestris pv. vesicatoria* (Tai et al. 1999; Tang et al. 1999). The maize *Rxo1* gene provides resistance against *X. oryzae pv. oryzicola* causing bacterial streak disease in rice (Zhao et al. 2005). Overexpression of Arabidopsis *NPR1* or the rice *NH1* gene enhanced resistance to the rice bacterial blight pathogen *X. oryzae pv. oryzae* (Chern et al. 2005; Yuan et al. 2007). Expression of receptor EFR from *Arabidopsis thaliana* confers resistance against range of phytopathogenic bacteria in *Nicotiana benthamiana*, tomato, rice, and wheat (Lacombe et al. 2010; Schoobeek et al. 2015; Schwessinger et al. 2015). Cecropins derived from the Cecropia moth (*Hyalophora cecropia*) including native (cecropin B), synthetic (Shiva-1, D4E1), and mutant (SB-37, MB39) have shown antimicrobial activity against bacterial pathogens *X. campestris* and *X. populi* (Nordeen et al. 1992; Kaduno-Okuda et al. 1995; Rajasekaran et al. 2001; Mentag et al. 2003).

Nematode Resistant Bananas

The key nematode pests of banana in SSA are the migratory species *Radopholus similis, Pratylenchus goodeyi, P. coffeae, Helicotylenchus multicinctus,* and sedentary *Meloidogyne* spp. The migratory endoparasite *R. similis* is considered the most damaging where it occurs causing extensive root necrosis as the nematode migrates through the root feeding. This reduces root function and compromises plant anchorage leading to toppling during storms. *Pratylenchus* spp. are becoming increasingly prevalent pests of *Musa* across Africa, especially on plantain in West Africa, resulting in growing concern for their potential impact (Coyne 2009). They impose root pathology similar to *R. similis* (Bridge et al. 1997). *H. multicinctus* occurs in almost all banana-growing areas mainly in root cortex causing some necrosis. The sedentary root parasite *Meloidogyne* spp. differs in modifying plant cells into a feeding site at one locale (Gowen et al. 2005). Infestations of complexes of species are prevalent and the combination of nematode species present in banana plantations varies with the locality (Coyne et al. 2013).

Several transgenic defenses against nematodes are in different stages of development (Table 1). Cysteine proteinases are major digestive enzymes of many nematodes and can be inhibited by cysteine proteinase inhibitors (cystatins). The expression of plant cystatins by roots suppresses nematode growth and reproduction on several plants in containment including tomato (Urwin et al. 1995), *Arabidopsis* (Urwin et al. 1997, 2000), rice (Vain et al. 1998), dessert banana (Atkinson et al. 2004a),

Table 1. List of genes introduced to various crops for developing resistance to bacterial disease and nematodes.

Resistance Technology/ Target Gene	Origin	Target Organism	Crop	Mode of Action	Resistance: Green house	Resistance: Field	References
Bacterial Disease Resistance							
Hrap	Sweet pepper	*X. campestris pv. musacearum*	Banana	Hypersensitivity Response	Full	Full	Tripathi et al. (2010, 2014a)
Pflp	Sweet pepper	*X. campestris pv. musacearum*	Banana	Hypersensitivity Response	Full	Full	Namukwaya et al. (2012); Tripathi et al. (2014a)
Xa21	Rice	*X. campestris pv. musacearum*	Banana	Pathogen Recognition	Full	–	Tripathi et al. (2014b)
Pto	Tomato	*X. campestris pv. vesicatoria*	Tomato	Resistance (R) Gene	Enhanced	–	Tang et al. (1999)
Bs2	Sweet pepper	*X. campestris pv. vesicatoria*	Tomato	Resistance (R) Gene	Enhanced	–	Tai et al. (1999)
Rxo1	Maize	*X. oryzae pv. oryzicola*	Rice	Resistance (R) Gene	Enhanced	–	Zhao et al. (2005)
Npr1	Arabidopsis	*X. oryzae pv. oryzae*	Rice	Systemic Acquired Resistance	Enhanced	–	Chern et al. (2005)
NH1	Rice	*X. oryzae pv. oryzae*	Rice	Systemic Acquired Resistance	Enhanced	–	Yuan et al. (2007)
EFR	Arabidopsis	*Ralstonia solanacearum, Xanthomonas perforans*	Tomato	Pathogen Recognition	Enhanced	–	Lacombe et al. (2010)
D4E1	Synthetic	*X. populi pv. populi*	Popular	Cecropin Antimicrobial Peptide	Enhanced	–	Mentag et al. (2003)
Nematode Resistance							
CCII	Maize	*R. similis, H. multicinctus, Meloidogyne* sp.	Plantain	Antifeedant	84%	98%	Roderick et al. (2012b); Tripathi et al. (2015)
Peptide	Synthetic	*R. similis, H. multicinctus, Meloidogyne* sp.	Plantain	Behavioral Repellent	66%	99%	Roderick et al. (2012b); Tripathi et al. (2015)
CCII + Peptide	Synthetic	*R. similis, H. multicinctus, Meloidogyne* sp.	Plantain	As above	70%	95%	Roderick et al. (2012b); Tripathi et al. (2015)
OcIΔD86	Rice	*R. similis*	Banana	Antifeedant	70%	–	Atkinson et al. (2004a)
Cry5B	*B. thuringiensis*	*M. incognita*	Tomato	Bt Toxin	64%	–	Li et al. (2008)
16D10	*M. incognita*	*M. incognit, M. Javanica, M. arenaria, M. hapla*	*Arabidopsis*	RNAi	93%	–	Huang et al. (2006)
tp	*M. incognita*	*M. incognita*	Soybean	RNAi	82%	–	Ibrahim et al. (2010)
msp	*M. incognita*	*M. incognita*	Soybean	RNAi	85%	–	Ibrahim et al. (2010)
cb-1	*R. similis*	*R. similis*	Tobacco	RNAi	73%	–	Li et al. (2015a)
crt	*R. similis*	*R. similis*	Tomato	RNAi	75%	–	Li et al. (2015b)
Splicing Factor	*M. incognita*	*M. incognita*	Tobacco	RNAi	100%	–	Yadav et al. (2006)
Integrase	*M. incognita*	*M. incognita*	Tobacco	RNAi	99%	–	Yadav et al. (2006)
flp-14	*M. incognita*	*M. incognita*	Tobacco	RNAi	50%	–	Papolu et al. (2013)
flp-18	*M. incognita*	*M. incognita*	Tobacco	RNAi	58%	–	Papolu et al. (2013)

Full – transgenic lines identified with full resistance to bacterial pathogen, and Enhanced – transgenic lines identified with reduced disease symptoms. Best line percentage resistance to nematodes calculated from nematodes/100 g root relative to infected nontransgenic control plants.

aubergine (Papolu et al. 2016), and Easter Lily (Vieira et al. 2015). High levels of efficacy have also been established in confined field trials for both potato expressing an engineered rice grain cystatin (Urwin et al. 2001, 2003; Lilley et al. 2004) and plantain expressing a maize kernel cystatin (Roderick et al. 2012b).

A second well-developed transgenic resistance defense is based on expression of peptides that reduce invasion of roots without being lethal to nematodes. The peptides undergo retrograde transport along certain chemosensory dendrites to neuronal cell bodies of several nematodes including *R. similis* resulting in a loss of orientation to roots (Winter et al. 2002; Wang et al. 2011; Roderick et al. 2012b). The peptide that has been deployed in plantains is a disulfide-constrained 7-mer with the amino sequence CTTMHPRLC (Winter et al. 2002; Roderick et al. 2012b). It has provided resistance to *Globodera pallida* in the glasshouse (Lilley et al. 2011) and field (Green et al. 2012). Both the peptide and a cystatin provided a high level of resistance in plantain to both *R. similis* and *H. multicinctus* with evidence of an accumulative benefit as the crop advanced to harvest as the introduced nematodes failed to maintain their density on the growing root system (Tripathi et al. 2015; Fig. 1).

Bacillus thuringiensis derived Bt endotoxin genes, similar to the highly effective insecticidal genes deployed in several crops, have also demonstrated an ability to suppress *Meloidogyne* species (Li et al. 2008; Zhang et al. 2012; Yu et al. 2015). Plant lectins have also been shown to suppress *M. incognita* but are toxic to insects and mammals (Burrows and de Waele 1997). RNA interference (RNAi) based defenses are being developed and have shown promise against *Meloidogyne* species (Huang et al. 2006; Yadav et al. 2006; Ibrahim et al. 2011; Papolu et al. 2013), *R. similis* (Haegeman et al. 2009; Li et al. 2015a,b), and *Pratylenchus* species (Joseph et al. 2012; Tan et al. 2013). However, none of these resistance technologies have been deployed into banana as they lack the broad control required for concomitant infections typical in banana plantations (Wei et al. 2003).

Food and Environmental Biosafety

Bioinformatics approaches comparing the amino acid sequences of PFLP and HRAP proteins to known allergens (AllergenOnline.org and NCBI) and toxins (NCBI) confirmed that both the proteins are safe for human consumptions and do not have similarity with any toxin or allergen in database. The transgenic banana expressing *Hrap* or *Pflp* gene will be evaluated for food and environmental safety during next field trial.

There is a well-established case for the food and environmental safety of both the cystatin and peptide defenses. The rice and maize seed cystatins deployed in banana are not novel dietary proteins. They are consumed as part of the staple diet of many Africans and a similar protein is present in human saliva (Veerman et al. 1996). Experimental approaches concluded that they are neither toxins (Atkinson et al. 2004b) nor allergens (MAFF UK 2000). They also share no similarity with any known toxic or allergenic proteins in databases. The peptide expressed in plantains is destroyed by cooking and by simulated intestinal fluid. It is not recognized as a potential allergen by Allergenonline (http://www.allergenonline.com) or Allermatch (http://allermatch.org; Fiers et al. 2004), two tools that meet Food and Agriculture Organization/World Health Organization (FAO/WHO) Codex alimentarius guidelines for allergenicity assessment (http://bit.ly/CodexAlimentarius). The lack of allergenicity of the 1.16-kDa peptide is also consistent with the observation that proteins of less than 3 kDa do not normally elicit an allergic response in mammals (Van Beresteijn et al. 1994).

Expression of the engineered rice cystatin by potato plants does not pose a measurable environmental risk to aerial invertebrate associates of a transgenic potato crop (Cowgill et al. 2002a, 2004; Cowgill and Atkinson 2003), or perturb soil organism communities in the field (Cowgill et al. 2002b). Free-living soil nematodes are also unaffected by potato plants expressing the engineered rice cystatin (Green et al. 2012). The peptide is rapidly degraded in the soil, presumably being utilized by soil microorganisms. It is not lethal to nontarget invertebrates at levels above those produced by transgenic plants (Wang 2009). Its release from transgenic potato does not perturb nontarget communities of soil nematodes in the field (Green et al. 2012). Both food and environmental biosafety can be enhanced further by controlling expression under promoters that express preferentially in roots (Green et al. 2012). This strategy may also enhance the effectiveness of the peptide defense. Placing the peptide under control of a root-cap-specific promoter provided 94.9 ± 0.8% resistance to *G. pallida* in contained potato plant trial compared to 34.4 ± 8.4% resistance when the peptide was under control of the constitutive CaMV35S promoter (Lilley et al. 2011).

Future Perspectives

The progress of innovations to the market has been charted for a wide range of technologies using the Hype cycle (Fenn and Raskino 2008). Crucial to applying this approach is defining a metric of visibility. In the case of transgenic technologies citation number based on keyword searches, as used in a meta-analysis of the agronomic and economic impacts of GM crops (Klümper and Qaim 2014), provide a useful measure of the developmental state of specific genetic modifications. Citation frequency falls as workers with fundamental science interests disengage from the field leaving only those involved in translational and subsequent research for those applications showing commercial potential. Additionally, the years for citations to accumulate are less for more recent, translational research than the older publications on which that effort is based.

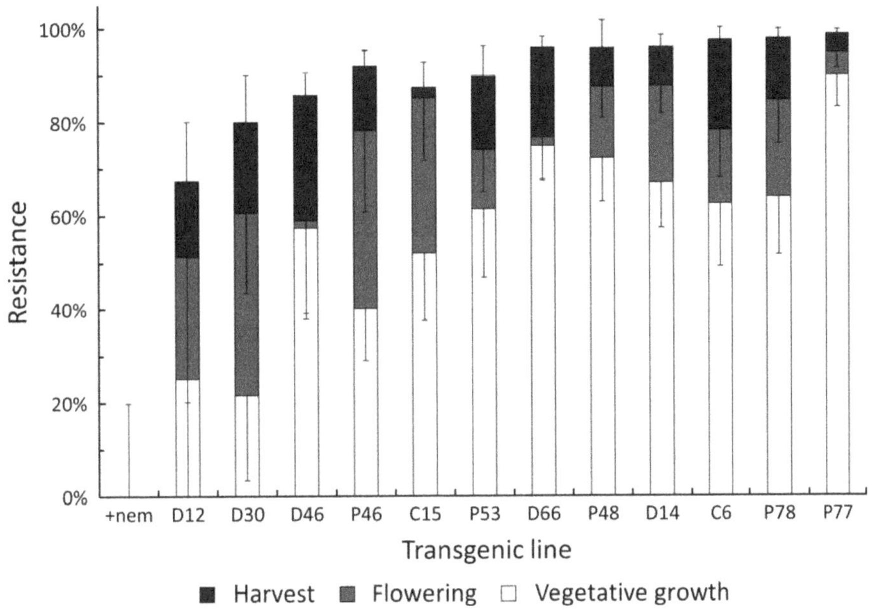

Figure 1. Stacked columns of cumulative percentage resistance (mean ± SEM) for the periods vegetative growth, flowering, and harvest (i.e., Line D30 had 22% resistance at vegetative sampling, 61% resistance at flowering sampling, and 80% resistance at harvest flowering) for 12 transgenic lines relative to the control plants to which nematodes were added before planting (+nem). Data are based on Tripathi et al. 2015. The expressed transgenes in the independent, transgenic events are as follow: C, cystatin; P, peptide; and D, both C and P.

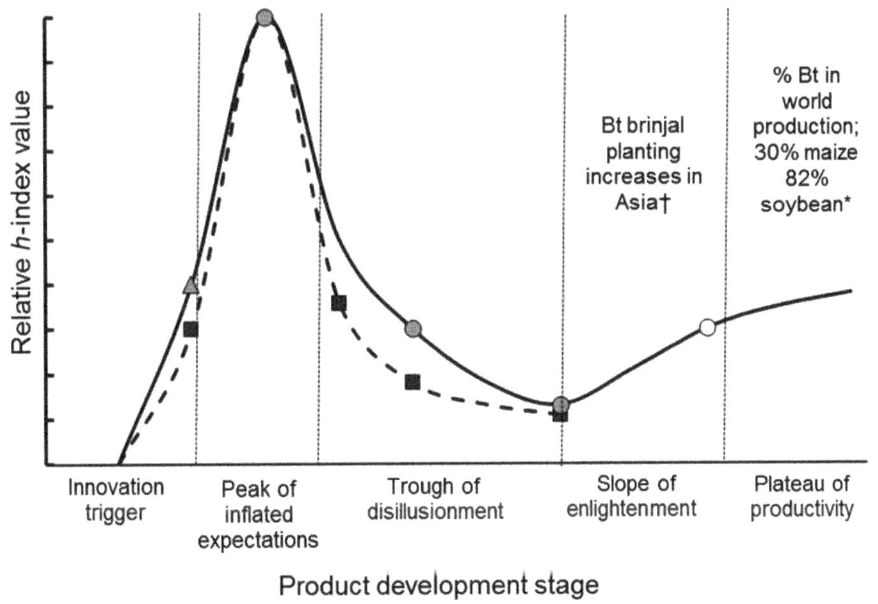

Figure 2. Gartner Hype cycle applied to development of Bt technology for transgenic insect control (solid line) and nematode resistance (dashed line and squares) expressed as percentage of the highest citation value in the peak of inflated expectations. The search terms used for Bt technology were "Bt + insect + crop" followed by stepped addition of "field", then "yield", then "benefit", and finally "Bt grower + society". For nematode resistance, these were "nematode + transgenic" followed by stepped addition of "crop", then "field", then "yield", and finally "improved". Estimated time to plateau of productivity for Bt technology; light-gray triangle, >10 years; gray circle <5–10 years; and open circles <2 years or on plateau. †, Herring 2015; *, James 2014.

The anti-insect protein from *Bacillus thuringiensis* (Bt) that confers resistance to certain insects can act as a comparator for nematode resistance technologies when applying the Hype cycle method (Fig. 2). Bt represents a mature resistance technology in food crops that has a stable market based on real benefits. A limitation of fitting the Hype cycle to development of nematode and particularly banana Xanthomonas wilt resistance is a shallower evidence base provided by a much smaller number of researchers involved in than those working on Bt. Despite this limitation, the analysis does suggest that nematode resistance technologies are within 5–10 years of achieving the plateau of productivity based on the similar stages of development in Bt insect resistance. A similar timeline is likely for the banana Xanthomonas wilt resistance technologies that are at a similar stage of development to the nematode resistance technologies.

The most substantial issue for the development of these public good technologies is maintaining a level of donor support required for the translation phases, by comparison the development of Bt in cotton and maize was more assured due to investment by biotechnology companies. Science-related factors include the need to demonstrate efficacy across all African regions where marketing is anticipated. Resistance breaking might eventually require management but is more likely to limit the duration of productivity rather emerge before widespread uptake. A more substantial issue is the capacity within Africa to produce the many millions of transgenic plantlets that would be required. Data compiled to date independently of the technology developers establish that neither of the new banana technologies poses toxicological or allergenic risk. Consequently, regulatory actions that resulted in withdrawal of soybeans expressing a novel protein that proved to be an allergen (Herman 2003) seem unlikely. However, regulatory processes have not yet optimized in Africa to support rapid and safe uptake of beneficial crops (Atkinson et al. 2015). A further major issue is political concerns that have hindered progress for aubergine (Brinjal) in India but not Bangladesh (Herring 2015). The future scientific objectives in addition to translational effort is to increase the benefits offered by transgenic banana by stacking traits such as resistance to nematode and banana Xanthomonas wilt in grower-preferred cultivars.

Acknowledgments

The authors would like to thank United States Agency for International Development (USAID) for financial support to BXW research and the USAID ABSPII program, the Biotechnology and Biological Sciences Research Council (BBSRC), Department for International Development (DFID), and CGIAR Research Program on Roots, Tubers and Bananas (RTB) for financial support to nematode research. We acknowledge Academia Sinica, Taiwan and the University of California, Davis for providing gene constructs for BXW resistance. We also acknowledge African Agricultural Technology Foundation (AATF) for negotiating license of *Pflp* and *Hrap* genes from Academia Sinica and for sublicensing the genes to International Institute of Tropical Agriculture (IITA).

Conflict of Interest

The authors declare no competing financial interests.

References

Atkinson, H. J. 2003. Strategies for resistance to nematodes in *Musa* spp. Pp. 74–107 *in* H. J. Atkinson, et al., eds. Genetic transformation strategies to address the major constraints to banana and plantain production in Africa. INIBAP, Montpellier, France.

Atkinson, H. J., S. Grimwood, K. A. Johnston, and J. Green. 2004a. Prototype demonstration of transgenic resistance to the nematode *Radopholus similis* conferred on banana by a cystatin. Transgenic Res. 13:135–142.

Atkinson, H. J., K. A. Johnston, and M. Robbins. 2004b. *Prima facie* evidence that a phytocystatin for transgenic plant resistance to nematodes is not a toxic risk in the human diet. J. Nutr. 134:431–434.

Atkinson, H. J., H. Roderick, and L. Tripathi. 2015. Africa needs streamlined regulation to support the deployment of GM crops. Trends Biotechnol. 33:433–435.

Bridge, J., R. Fogain, and S. Paul. 1997. The root lesion nematodes of banana. *Musa* Pest Fact Sheet No. 2. INIBAP, Montpellier, France.

Burrows, P. R., and D. de Waele. 1997. Engineering resistance against plant parasitic nematodes using anti-nematode genes. Pp. 217–236 *in* C. Fenoll, F. M. W. Grundler and S. A. Ohl, eds. Cellular and molecular aspects of plant-nematode interactions. Kluwer Academic Press, Dordrecht, Netherlands.

Chen, C. H., H. J. Lin, M. J. Ger, D. Chow, and T. Y. Feng. 2000. The cloning and characterization of a hypersensitive response assisting protein that may be associated with the harpin-mediated hypersensitive response. Plant Mol. Biol. 43:429–438.

Chern, M., H. A. Fitzgerald, P. E. Canlas, D. A. Navarre, and P. C. Ronald. 2005. Over-expression of a rice *NPR1* homolog leads to constitutive activation of defence response and hypersensitivity to light. Mol. Plant Microbe Interact. 18:511–520.

Cowgill, S. E., and H. J. Atkinson. 2003. A sequential approach to risk assessment of transgenic plants expressing protease inhibitors: effects on non-target herbivorous insects. Transgenic Res. 12:439–449.

Cowgill, S. E., C. Wright, and H. J. Atkinson. 2002a. Transgenic potatoes with enhanced levels of nematode resistance do not have altered susceptibility to non-target aphids. Mol. Ecol. 11:821–827.

Cowgill, S. E., R. D. Bardgett, D. T. Kiezebrink, and H. J. Atkinson. 2002b. The effect of transgenic nematode resistance on non-target organisms in the potato rhizosphere. J. Appl. Ecol. 39:915–923.

Cowgill, S. E., C. Danks, and H. J. Atkinson. 2004. Multitrophic interactions involving genetically modified potatoes, nontarget aphids, natural enemies and hyperparasitoids. Mol. Ecol. 13:639–647.

Coyne, D. 2009. Pre-empting plant-parasitic nematode losses on *Musa* spp. Acta Hortic. 828:227–236.

Coyne, D., A. Omowumi, I. Rotifa, and S. O. Afolami. 2013. Pathogenicity and damage potential of five plant-parasitic nematode species on plantain (*Musa* spp., AAB genome) cv. Agbagba. Nematol. 15:589–599.

De Waele, D., and A. Elsen. 2002. Migratory endoparasities: *Pratylenchus* and *Radopholus* species. Pp. 175–206 in J. J. Starr, R. Cook and J. Bridge, eds. Plant resistance to parasitic nematodes. CABI, Wallingford, UK.

FAOSTAT. 2014. Agriculture data. Available at http://faostat.fao.org. (accessed 27 September 2016).

Fenn, J., and M. Raskino. 2008. Mastering the hype cycle: how to choose the right innovation at the right time. Harvard Business Press, MA, USA.

Fiers, M. W. E. J., G. A. Kleter, H. Nijland, A. A. C. M. Peijnenburg, J. P. Nap, and R. C. H. J. van Ham. 2004. Allermatch™, a webtool for the prediction of potential allergenicity according to current FAO/WHO Codex alimentarius guidelines. BMC Bioinformatics 5:133.

Ger, M. J., C. H. Chen, S. Y. Hwang, H. E. Huang, A. R. Podile, B. V. Dayakar, et al. 2002. Constitutive expression of *Hrap* gene in transgenic tobacco plant enhances resistance against virulent bacterial pathogens by induction of a hypersensitive response. Mol. Plant Microbe Interact. 15:764–773.

Godfray, H. C., J. R. Beddington, I. R. Crute, L. Haddad, D. Lawrence, J. F. Muir, et al. 2010. Food security: the challenge of feeding 9 billion people. Science 327:812–818.

Gowen, S. C., P. Quénéherve, and R. Fogain. 2005. Nematode parasites of bananas and plantains. Pp. 611–643 in M. Luc, R. A. Sikora and J. Bridge, eds. Plant parasitic nematodes in subtropical and tropical agriculture, 2nd ed. CABI, Wallingford, UK.

Green, J., D. Wang, C. J. Lilley, P. E. Urwin, and H. J. Atkinson. 2012. Transgenic potatoes for potato cyst nematode control can replace pesticide use without impact on soil quality. PLoS ONE 7:e30973.

Haegeman, A., B. Vanholme, and G. Gheysen. 2009. Characterization of a putative endoxylanase in the migratory plant-parasitic nematode *Radopholus similis*. Mol. Plant Pathol. 10:389–401.

Herman, E. M. 2003. Genetically modified soybeans and food allergies. J. Exp. Bot. 54:1317–1319.

Herring, R. J. 2015. State science, risk and agricultural biotechnology: Bt cotton to Bt brinjal in India. J. Peasant Stud. 42:159–186.

Huang, S. N., C. H. Chen, H. J. Lin, M. J. Ger, Z. I. Chen, and T. Y. Feng. 2004. Plant ferredoxin-like protein AP1 enhances Erwinia induced hypersensitive response of tobacco. Physiol. Mol. Plant Pathol. 64:103–110.

Huang, G. Z., R. Allen, E. L. Davis, T. J. Baum, and R. S. Hussey. 2006. Engineering broad root-knot resistance in transgenic plants by RNAi silencing of a conserved and essential root-knot nematode parasitism gene. Proc. Natl Acad. Sci. USA 103:14302–14306.

Huang, H. E., M. J. Ger, C. Y. Chen, A. K. Pandey, M. K. Yip, H. W. Chou, et al. 2007. Disease resistance to bacterial pathogens affected by the amount of ferredoxin-I protein in plants. Mol. Plant Pathol. 8:129–137.

Ibrahim, H. M. M., N. W. Alkharouf, S. L. F. Meyer, M. A. M. Aly, A. Y. Gamal El-Din, E. H. A. Hussein, et al. 2011. Post-transcriptional gene silencing of root-knot nematode in transformed soybean roots. Exp. Parasitol. 127:90–99.

James, C. 2014. Global Status of Commercialized Biotech/GM Crops: 2014. ISAAA Brief No. 49. ISAAA, Ithaca, NY.

Joseph, S., G. Gheysen, and K. Subramaniam. 2012. RNA interference in *Pratylenchus coffeae*: knock down of Pc-pat-10 and Pc-unc-87 impedes migration. Mol. Biochem. Parasitol. 186:51–59.

Kaduno-Okuda, K., K. Taniai, Y. Kato, E. Kotani, and M. Yamakaula. 1995. Effects of synthetic *Bombyx mori* cecropin B on growth of plant pathogenic bacteria. J. Invertebr. Pathol. 65:309–319.

Kalyebara, R., S. Wood, and P. M. Abodi. 2007. Assessing the potential impact of selected technologies on the banana industry in Uganda. Pp. 141–153 in M. Smale and W. K. Tushemereirwe, eds. An economic assessment of banana genetic improvement and innovation in the Lake Victoria region of Uganda and Tanzania. IFPRI, Washington, DC.

Karamura, E., G. Kayobyo, G. Blomme, S. Benin, S. J. Eden-Green, and R. Markham. 2006. Impacts of BXW epidemic on the livelihoods of rural communities in Uganda. P. 57 in Proceedings of the 4th international bacterial wilt symposium, 17–20 July 2006. Central Science Laboratory, York, UK.

Klümper, W., and M. Qaim. 2014. A meta-analysis of the impacts of genetically modified crops. PLoS ONE 9:e111629.

Lacombe, S., A. Rougon-Cardoso, E. Sherwood, N. Peeters, D. Dahlbeck, H. P. van Esse, et al. 2010. Interfamily transfer of a plant pattern-recognition receptor confers broad-spectrum bacterial resistance. Nat. Biotechnol. 28:365–369.

Lescot, T. 2008. Genetic diversity of banana in figures. Fruitrop 155:29–33.

Li, X. Q., A. Tan, M. Voegtline, S. Bekele, C.-S. Chen, and R. V. Aroian. 2008. Expression of Cry5B protein from *Bacillus thuringiensis* in plant roots confers resistance to rootknot nematode. Biol. Control 47:97–102.

Li, Y., K. Wang, and H. Xie. 2015a. Cathepsin B cysteine proteinase is essential for the development and pathogenesis of the plant parasitic nematode *Radopholus similis*. Int. J. Biol. Sci. 11:1073–1087.

Li, Y., K. Wang, H. Xie, Y. T. Wang, and D. W. Wang. 2015b. A nematode calreticulin, Rs-CRT, is a key effector in reproduction and pathogenicity of *Radopholus similis*. PLoS ONE 10:e0129351.

Liau, C. H., J. C. Lu, V. Prasad, J. T. Lee, H. H. Hsiao, S. J. You, et al. 2003. The sweet pepper ferredoxin-like protein (*pflp*) conferred resistance against soft rot disease in Oncidium orchid. Transgenic Res. 12:329–336.

Lilley, C. J., P. E. Urwin, K. A. Johnston, and H. J. Atkinson. 2004. Preferential expression of a plant cystatin at nematode feeding sites confers resistance to *Meloidogyne incognita* and *Globodera pallida*. Plant Biotechnol. J. 2:3–12.

Lilley, C. J., D. Wang, H. J. Atkinson, and P. E. Urwin. 2011. Effective delivery of a nematode-repellent peptide using a root-cap-specific promoter. Plant Biotechnol. J. 9:151–161.

Lin, H. J., H. Y. Cheng, C. H. Chen, H. C. Huang, and T. Y. Feng. 1997. Plant amphipathic proteins delay the hypersensitive response caused by harpinPss and *Pseudomonas syringae* pv. *syringae*. Physiol. Mol. Plant Pathol. 51:367–376.

Lorenzen, J., A. Tenkouano, R. Bandyopadhyay, B. Vroh, D. Coyne, and L. Tripathi. 2010. Over view of banana and plantain (*Musa* spp.) improvement in Africa: past and future. Acta Hortic. 879:595–603.

Mentag, R., M. Lukevich, M. J. Morency, and A. Seguin. 2003. Bacterial disease resistance of transgenic hybrid poplar expressing the synthetic antimicrobial peptide D4E1. Tree Physiol. 23:405–411.

Ministry of Agriculture, Fisheries and Food, United Kingdom. 2000. *Final Project Report: Development of methods to predict the allergenic potential of genetically modified foods and new protein products.* MAFF project code FS 3023.

Muwonge, A., J. N. Tripathi, K. Kunert, and L. Tripathi. 2016. Expressing stacked *Hrap* and *Pflp* genes in transgenic banana has no synergistic effect on resistance to Xanthomonas wilt disease. S. Afr. J. Bot. 104:125–133.

Namukwaya, B., L. Tripathi, J. N. Tripathi, G. Arinaitwe, S. B. Mukasa, and W. K. Tushemereirwe. 2012. Transgenic banana expressing *Pflp* gene confers enhanced resistance to Xanthomonas wilt disease. Transgenic Res. 4:855–865.

Nkuba, J., W. Tinzaara, G. Night, N. Niko, W. Jogo, I. Ndyetabula, et al. 2015. Adverse impact of Banana Xanthomonas Wilt on farmers' livelihoods in Eastern and Central Africa. Afr. J. Plant Sci. 9:279–286.

Nordeen, R. O., S. L. Sinden, J. M. Jaynes, and L. D. Owens. 1992. Activity of cecropin SB37 against protoplasts from several plant species and their bacterial pathogens. Plant Sci. 82:101–107.

Ortiz, R., and D. Vuylsteke. 1996. Improving plantain and banana-based systems. Pp. 2–7 *in* R. Ortiz and M. O. Akoroda, eds. Plantain and banana production and research in West and Central Africa. IITA, Ibadan, Nigeria.

Ortiz, R., R. S. B. Ferris, and D. R. Vuylsteke. 1995. Banana and plantain breeding. Pp. 110–146 in S. Gowen, ed. Bananas and plantains. Chapman and Hall, London, UK.

Pandey, A. K., M. J. Ger, H. E. Huang, M. K. Yip, J. Zeng, and T. Y. Feng. 2005. Expression of the hypersensitive response-assisting protein in *Arabidopsis* results in harpin-dependent hypersensitive cell death in response to *Erwinia carotovora*. Plant Mol. Biol. 59:771–780.

Papolu, P. K., N. P. Gantasala, D. Kamaraju, P. Banakar, R. Sreevathsa, and U. Rao. 2013. Utility of host delivered RNAi of two FMRF amide like peptides, *flp*-14 and *flp*-18, for the management of root knot nematode, *Meloidogyne incognita*. PLoS ONE 8:e80603.

Papolu, P. K., T. K. Dutta, N. Tyagi, P. E. Urwin, C. J. Lilley, and U. Rao. 2016. Expression of a cystatin transgene in eggplant provides resistance to root-knot nematode, *Meloidogyne incognita*. Front. Plant Sci. 7:1122.

Pasberg-Gauhl, C., and F. Gauhl. 1996. *Musa* research in the plant health management division at IITA: activities at the high rainfall station, Onne in Nigeria. Pp. 7–14 *in* R. Ortiz and M. O. Akoroda, eds. Plantain and banana production and research in West and Central Africa. IITA, Ibadan, Nigeria.

Pinochet, J. 1988. Comments on the difficulty in breeding bananas and plantains for resistance to nematodes. Rev. Nematol. 11:3–5.

Quénéhervé, P., F. Salmon, P. Topart, and J. P. Horry. 2009. Nematode resistance in bananas: Screening results on some new *Mycosphaerella* resistant banana hybrids. Euphytica 165:123–136.

Rajasekaran, K., K. D. Stromberg, J. W. Cary, and T. E. Cleveland. 2001. Broad-spectrum antimicrobial activity in vitro of the synthetic peptide D4E1. J. Agric. Food Chem. 49:2799–2803.

Robinson, J. C. 1996. Bananas and plantain. CABI, Wallingford, UK.

Roderick, H., E. Mbiru, D. Coyne, L. Tripathi, and H. J. Atkinson. 2012a. Quantitative digital imaging of banana growth suppression by plant parasitic nematodes. PLoS ONE 7:e53355.

Roderick, H., L. Tripathi, A. Babirye, D. Wang, J. Tripathi, P. E. Urwin, et al. 2012b. Generation of transgenic plantain (*Musa* spp.) with resistance to plant pathogenic nematodes. Mol. Plant Pathol. 13:842–851.

Ronald, P. C., B. Albano, R. Tabien, L. Abenes, K. Wu, S. McCouch, et al. 1992. Genetic and physical analysis of the rice bacterial blight disease resistance locus, Xa21. Mol. Gen. Genet. 236:113–120.

Schoobeek, H. J., H. H. Wang, F. L. Stefanato, M. Craze, S. Bowden, E. Wallington, et al. 2015. Arabidopsis EF-Tu receptor enhances bacterial disease resistance in transgenic wheat. New Phytol. 206:606–613.

Schwessinger, B., O. Bahar, N. Thomas, N. Holton, V. Nekrasov, D. Ruan, et al. 2015. Transgenic expression of the dicotyledonous pattern recognition receptor EFR in rice leads to ligand-dependent activation of defense responses. PLoS Pathog. 11:e1004809.

Shimwela, M. M., R. C. Ploetz, F. D. Beed, J. B. Jones, J. K. Blackburn, S. I. Mkulila, et al. 2016a. Banana Xanthomonas wilt continues to spread in Tanzania despite an intensive symptomatic plant removal campaign: an impending socio-economic and ecological disaster. Food Secur. 8:939–951.

Shimwela, M. M., J. K. Blackburn, J. B. Jones, J. Nkuba, H. A. Narouei-Khandan, R. C. Ploetz, et al. 2016b. Local and regional spread of banana Xanthomonas wilt (BXW) in space and time in Kagera, Tanzania. Plant. Pathol. doi:10.1111/ppa.12637.

Simmonds, N. W. 1987. Classification and breeding of bananas. Pp. 69–73 *in* G. Persley, E. De Langhe, eds. Banana and plantain breeding strategies. ACIAR proceedings No. 21. ACIAR, Canberra, Australia.

Tai, T. H., D. Dahlbeck, E. T. Clark, P. Gajiwala, R. Pasion, M. C. Whalen, et al. 1999. Expression of the *Bs2* pepper gene confers resistance to bacterial spot disease in tomato. Proc. Natl Acad. Sci. USA 23:14153–14158.

Tan, J. A. C. H., M. G. K. Jones, and J. Fosu-Nyarko. 2013. Gene silencing in root lesion nematodes (*Pratylenchus* spp.) significantly reduces reproduction in a plant host. Exp. Parasitol. 133:166–178.

Tang, X., M. Xie, Y. J. Kim, J. Zhou, D. F. Klessig, and G. B. Martin. 1999. Overexpression of *Pto* activates defense responses and confers broad resistance. Plant Cell 11:15–29.

Tang, K., Y. Sun, Q. Hu, A. Wu, C. H. Lin, H. J. Lin, et al. 2001. Transgenic rice plants expressing the ferredoxin-like protein (AP1) from sweet pepper show enhanced resistance to *Xanthomonas oryzae* pv. *oryzae*. Plant Sci. 160:1035–1042.

Tripathi, L., M. Mwangi, S. Abele, V. Aritua, W. K. Tushemereirwe, and R. Bandyopadhyay. 2009. A threat to banana production in east and central Africa. Plant Dis. 93:440–451.

Tripathi, L., H. Mwaka, J. N. Tripathi, and W. K. Tushemereirwe. 2010. Expression of sweet pepper *Hrap* gene in banana enhances resistance to *Xanthomonas campestris* pv. *musacearum*. Mol. Plant Pathol. 11:721–731.

Tripathi, L., J. N. Tripathi, A. Kiggundu, S. Korie, F. Shotkoski, and W. K. Tushemereirwe. 2014a. Field trial of Xanthomonas wilt disease-resistant bananas in East Africa. Nat. Biotechnol. 32:868–870.

Tripathi, J. N., J. Lorenzen, O. Bahar, P. Ronald, and L. Tripathi. 2014b. Transgenic expression of the rice Xa21 pattern-recognition receptor in banana (*Musa* sp.) confers resistance to *Xanthomonas campestris* pv. *musacearum*. Plant Biotech. J. 12:663–673.

Tripathi, L., A. Babirye, H. Roderick, J. N. Tripathi, C. Changa, P. E. Urwin, et al. 2015. Field resistance of transgenic plantain to nematodes has potential for future African food security. Sci. Rep. 5:8127.

Tripathi, L., J. N. Tripathi, and J. Kubiriba. 2016. Transgenic Technologies for Bacterial Wilt Resistance. Pp. 197–209 in S. Mohandas and K. V. Ravishankar, eds. Banana: genomics and transgenic approaches for genetic improvement. Springer, Singapore.

United Nations, Department of Economic and Social Affairs, Population Division. 2015. *World Population Prospects: The 2015 Revision, Key Findings and Advance Tables.* Working Paper No. ESA/P/WP.241.

Urwin, P. E., H. J. Atkinson, D. A. Waller, and M. J. McPherson. 1995. Engineered oryzacystatin-I expressed in transgenic hairy roots confers resistance to *Globodera pallida*. Plant J. 8:121–131.

Urwin, P. E., C. J. Lilley, M. J. McPherson, and H. J. Atkinson. 1997. Resistance to both cyst and root-knot nematodes conferred by transgenic *Arabidopsis* expressing a modified plant cystatin. Plant J. 12:455–461.

Urwin, P. E., A. Levesley, M. J. McPherson, and H. J. Atkinson. 2000. Transgenic resistance to the nematode *Rotylenchulus reniformis* conferred by *Arabidopsis thaliana* plants expressing proteinase inhibitors. Mol. Breed. 6:257–264.

Urwin, P. E., K. M. Troth, E. I. Zubko, and H. J. Atkinson. 2001. Effective transgenic resistance to *Globodera pallida* in potato field trials. Mol. Breed. 8:95–101.

Urwin, P. E., J. Green, and H. J. Atkinson. 2003. Expression of a plant cystatin confers partial resistance to *Globodera*, full resistance is achieved by pyramiding a cystatin with natural resistance. Mol. Breed. 12:263–269.

Vain, P., B. Worland, M. C. Clarke, G. Richard, and M. Beavis. 1998. Expression if an engineered cysteine protease inhibitor (Oryzacystatin-I1D86) for nematode resistance in transgenic rice plants. Theor. Appl. Genet. 96:266–271.

Van Beresteijn, E. C. H., R. A. Peeters, J. Kaper, R. J. G. M. Meijer, A. J. P. M. Robben, and D. G. Schmidt. 1994. Molecular mass distribution, immunological properties and nutritive value of whey protein hydrolysates. J. Food Prot. 57:619–625.

Veerman, E. C., P. A. van den Keybus, A. Vissink, and A. V. Nieuw Amerongen. 1996. Human glandular salivas: their separate collection and analysis. Eur. J. Oral Sci. 104:346–352.

Vieira, P., S. Wantoch, C. J. Lilley, D. J. Chitwood, H. J. Atkinson, and K. Kamo. 2015. Expression of a cystatin transgene can confer resistance to root lesion nematodes in Lilium longiflorum cv'.Nellie White'. Transgenic Res. 24:421–432.

Wang, D. 2009. Reducing the environmental risks and hazards of crop production by biosafe use of transgenic crops. [Ph.D. thesis], Faculty of Biological Sciences, University of Leeds, Leeds.

Wang, G. L., W. Y. Song, D. L. Ruan, S. Sideris, and P. C. Ronald. 1996. The cloned gene Xa21 confers resistance to multiple Xanthomonas oryzae pv. oryzae isolates in transgenic plants. Mol. Plant Microbe Interact. 9:850–855.

Wang, D., L. M. Jones, P. E. Urwin, and H. J. Atkinson. 2011. A synthetic peptide shows retro- and anterograde neuronal transport before disrupting the chemosensation of plant-pathogenic nematodes. PLoS ONE 6:e17475.

Wei, J. Z., K. Hale, L. Cara, E. Platzer, C. Wong, S. C. Fang, et al. 2003. *Bacillus thuringiensis* crystal proteins that target nematodes. Proc. Natl Acad. Sci. USA 10:2760–2765.

West, P. C., J. S. Gerber, P. M. Engstrom, N. D. Mueller, K. A. Brauman, K. M. Carlson, et al. 2014. Leverage points for improving global food security and the environment. Science 345:325–328.

Winter, M. D., M. J. McPherson, and H. J. Atkinson. 2002. Neuronal uptake of pesticides disrupts chemosensory cells of nematodes. Parasitology 125:561–565.

Yadav, B. C., K. Veluthambi, and K. Subramaniam. 2006. Host generated double stranded RNA induces RNAi in plant-parasitic nematodes and protects the host from infection. Mol. Biochem. Parasitol. 148:219–222.

Yip, M. K., H. E. Huang, M. J. Ger, S. H. Chiu, Y. C. Tsai, C. I. Lin, et al. 2007. Production of soft rot resistant calla lily by expressing a ferredoxin-like protein gene (*pflp*) in transgenic plants. Plant Cell Rep. 26:449–457.

Yu, Z., J. Xiong, Q. Zhou, H. Luo, S. Hu, L. Xia, et al. 2015. The diverse nematicidal properties and biocontrol efficacy of *Bacillus thuringiensis* Cry6A against the root-knot nematode *Meloidogyne hapla*. J. Invertebr. Pathol. 125:73–80.

Yuan, Y., S. Zhong, Q. Li, Z. Zhu, Y. Lou, L. Wang, et al. 2007. Functional analysis of rice *NPR1* -like genes reveals that *OsNPR1 / NH1* is the rice orthologue conferring disease resistance with enhanced herbivore susceptibility. Plant Biotechnol. J. 5:313–324.

Zhang, F., D. Peng, X. Ye, Z. Yu, Z. Hu, L. Ruan, et al. 2012. In vitro uptake of 140 kDa *Bacillus thuringiensis* nematicidal crystal proteins by the second stage juvenile of *Meloidogyne hapla*. PLoS ONE 7:e38534.

Zhao, B., X. Lin, J. Poland, H. Trick, J. Leach, and S. Hulbet. 2005. A maize resistance gene functions against bacterial streak disease in rice. Proc. Natl Acad. Sci. USA 43:15383–15388.

12

Food security through translational biology between wheat and rice

Ravi Valluru[1], Matthew P. Reynolds[1] & Tanguy Lafarge[2]

[1]Global Wheat Program, International Maize and Wheat Improvement Center (CIMMYT), El Batan, Mexico
[2]Cirad, UMR AGAP, F-34398, Montpellier, France

Keywords
Comparative biology, rice, translational biology, wheat, yield potential

Correspondence
Ravi Valluru, Global Wheat Program, CIMMYT, El Batan, CP 56237, Mexico.

E-mail: r.valluru@cgiar.org

Abstract

Wheat and rice are the most important food crops in agriculture providing around 50% of all calories consumed in the human diet. While both are C_3 species, the evolution and domestication of wheat and rice occurred in very different environments, resulting in diverse anatomical and metabolic adaptation. This review focuses on the current understanding of their adaptation in an agronomic context. The similarities and differences between wheat and rice are discussed, focusing on traits related to phenology, photosynthesis, assimilate partitioning, and lodging resistance, these being the main abiotic drivers of yield expression in most agro-ecosystems. Currently, there are significant knowledge gaps in the major biological processes that account not only for differential adaption among cultivars within each species, but even between the two species. By addressing what is known as well as where gaps exist in a comparative context, this review aims to highlight translational research approaches that could provide insights into the genetic improvement of both crops.

Introduction

Wheat and rice are important cereal crops providing more than 50% of daily caloric intake worldwide. Taxonomically, wheat and rice represent different subfamilies of the Poaceae: Pooideae and Bambusoideae, respectively, and represent important models for comparative studies of the grasses because various aspects of their biology have been well studied (Sorrells et al. 2003; Shingaki-Wells et al. 2011; Cantu et al. 2013). Both wheat and rice have been domesticated in different climates, wheat is almost exclusively a dry-land winter crop, whereas rice is typically grown under tropical climates where most of the production comes from anaerobic/flooded soils (Nagai and Makino 2009; Shingaki-Wells et al. 2011). Such distinct evolutionary backgrounds triggered differential anatomical and developmental aspects resulting in various optimal growth conditions, yet they operate similar central metabolic processes (C_3 metabolism).

In the recent decades, the comparative biology between wheat and rice has been the subject of many studies (Sorrells et al. 2003; Shingaki-Wells et al. 2011; Cantu et al. 2013; for review, see Valluru et al. 2014). The study of two species may be useful in highlighting mechanisms of yield potential and stress adaptation in view of their domestication. Such distinct knowledge bases can lead to new knowledge, hypotheses, and predictions about related species and contribute to our knowledge of crop species evolution and adaptation to environment. Crop improvement programs can use comparative biology to transfer information on "traits" between relatively close species as well as more distant taxonomic groups. The genetic and molecular bases of yield-related traits were recently presented as a useful comparative biology approach between wheat and rice for application in translational research (Valluru et al. 2014). This review compares agronomic and physiological traits in two crop species.

Grain Yields of Wheat and Rice: Genetic Gains and Impacts

Genetic gains for wheat yields in the post Green Revolution period are widely reported to be in between 0.5% and 0.8% per year and globally the average annual absolute genetic gain of spring wheat reached about 40 kg ha^{-1} $year^{-1}$ (Morgounov et al. 2010) despite the similar relative gains in irrigated and rainfed environments (Rajaram and Braun, 2008). Genetic gains in spring wheat continue to date (1995–2010) in both irrigated (Sharma et al. 2012) and rainfed environments (Manès et al. 2012) at an average of 0.6% and 1.0% p.a., respectively. Genetic gains in winter wheat appear to be stagnating in many parts of the world including the great plains of North America and the Europe (Brisson et al. 2010; Graybosch and Peterson 2010).

In rice, absolute yield gain in the range of 16–80 kg ha^{-1} $year^{-1}$ were reported, being greater for Philippines (75–80 kg ha^{-1} $year^{-1}$) and smaller for Brazil (16–45 kg ha^{-1} $year^{-1}$) during the period 1966–1981 (Peng et al. 2000; Breseghello et al. 2011). The hybrid super rice shows 10–20% yield gain in China (Peng et al. 2008), while at IRRI tropical hybrids have showed 11–14% yield gain in Philippines (Fischer and Edmeades 2010). A recent comparison between hybrids and inbreds found 14–18% yield advantage for the hybrids (Bueno and Lafarge 2009). Overall, improvements in yield gain have been achieved through an ideotype approach (e.g., shorter plants, more fertile tillers, high early vigor, and CGR [crop growth rate], larger panicles) and the development of hybrid rice (which itself benefited from the ideotype approach).

Agronomic and Physiological Trait Associated With Yield Potential

With recent emphasis on food security, two major avenues have been highlighted to improve the genetic potential of yield in wheat: (1) increasing photosynthetic capacity and efficiency to raise the biological threshold in terms of total plant mass available for yield formation; and (2) optimizing partitioning to grain while maintaining lodging resistance such that increased genetic potential is realized as agronomic gains (Reynolds et al. 2009; Foulkes et al. 2011; Parry et al. 2011). Prebreeding to combine favorable expression of yield potential traits – focused on achieving a more optimal balance between assimilation capacity (source) and grain sites (sink) – has been successful in increasing yield levels in international trials (Reynolds et al. 2015).

In rice, trait-based selection was used to add specific traits into new plant types (e.g., NPTs) whereby NPTs served as a basis for the generation of high-yielding cultivars/hybrids with larger panicles and higher tillering fertility (Peng et al. 2008). However, yield-based selection has shown to be more effective in increasing grain yields in rice than trait-based selection; although some researchers question the relevance of the traits used (Yuan et al. 2011) like the higher tillering fertility that may be inversely correlated to the tillering capacity (Lafarge and Bueno 2009). Here, we briefly summarize important agronomic and physiological traits contributing to yield potential.

Crop phenology

For cereals, reproductive growth is clearly the most critical physiological function to understand and modify thereby permitting accumulation of biomass to be translated into maximal expression of yield. While wheat exhibits a great plasticity for phenology (Sadras et al. 2009), it is yet to achieve its theoretical maximum harvest index of 0.6, and typically expresses values between 0.4 and 0.5, suggesting major opportunities for yield improvement without even changing RUE (radiation use efficiency) or biomass (Foulkes et al. 2011). Breeding efforts during the 20th century have reduced sowing to anthesis duration in durum wheat in Spain (8 days) and Italy (2 days) (Álvaro et al. 2008; Isidro et al. 2011), although this trend has reversed in some recent CIMMYT cultivars (Fischer 2001; Underdahl et al. 2008). The duration between sowing to terminal spikelet decreased while the duration between booting to anthesis increased, which diminished floret abortion (24%), increased grain number (23%) and grain set (13%) from old to modern cultivars (Isidro et al. 2011). Furthermore, the stem elongation period has been shown to exhibit the greatest sensitivity to environment, independent of the major phenological genes (Garcia 2011). It is known that earliness per se affects the duration of preanthesis phases and spike development in cereals (Faricelli et al. 2009; Borràs-Gelonch et al. 2011, 2012). These studies suggest that earliness per se could be important to understand phenological phase plasticity to the environment, and it remains to be determined to what degree environmental effects confound or interact with genetic expression of phenological phases.

In rice, the genetic variation in postflowering period (from flowering to maturity) is relatively small for a given environment (Peng et al. 2000), while the preflowering phase (from sowing to flowering) varies greatly (Vergara and Chang 1985; Yin et al. 1997a). The cultivars released at IRRI since 1980s show slightly decreased crop duration (6 days) than the cultivars released before 1980s (Peng et al. 2000) due to shortened vegetative period, while the reproductive and grain-filling periods were largely unaffected. One of the objectives of the reduced crop cycle in the tropics is to grow three crops a year. In Texas, rice cultivars (1944–1992) showed a 12–18 days decrease in heading time (Tabien et al. 2008). The Brazilian cultivars (1984–2009) showed an average genetic loss of −0.25 day $years^{-1}$ in days to heading (Breseghello et al. 2011). Taken all together, both wheat and rice breeding programs have affected phenological pattern with crop duration and the length of the preanthesis period slightly decreased in both crops.

Photosynthesis and related traits

Light interception

LAI (leaf area index) beyond 3 is generally considered not an option for increasing light interception in wheat (Parry et al. 2011) because no apparent improvement in percent of light interception was observed in historical cultivars (Acreche et al. 2009). In contrast, LAI reaches 9 in rice depending on the developmental stage (Lü et al. 2007; Zheng et al. 2008; Zhang et al. 2009). Recent 3D models suggest that mean leaf inclination angles of rice cultivars range between 65° and 75° (70–90 in the upper canopy, while it is 30–60 in the lower canopy) (Zheng et al. 2008).

Modern rice and wheat cultivars have more erect and longer leaves in the upper canopy (flag-leaf). On average, rice plants receive 700–800 μmol photons m^{-2} sec^{-1} at the canopy top, with a total light interception of 80% of incident (Zheng et al. 2008). It has been estimated that leaves, panicles, and stems contribute about 80%, 15%, and 0.5% to the total light interception in rice (Saitoh et al. 2002). Reducing the panicle height to below top leaves or removing panicles (this allowed top leaves to access more light) have shown to increase leaf gas exchange rate about 15–50% (Setter et al. 1995; Saitoh et al. 2002). While more erect leaves/smaller leaves may allow light distribution to the lower canopy (Horton 2000), the genetic manipulation of leaf position, size, and thickness/density still needs to be explored since its manipulation is not straightforward.

Leaf photosynthesis

The relationship between the A_{max} (light-saturated only) and grain yield is not straightforward (Long et al. 2006) although the differences in A_{max} and its genetic variation can moderately contribute to variations in yields (Nelson 1988). Since modern hexaploid wheat species have lower A_{max} (postulated to be due to an inhibitory effect of D genome on gene expression in the A and/or B genomes and larger mesophyll cell sizes) than diploid and tetraploid species (Koç et al. 2003), these genetic pools may serve as sources for increasing A_{max}. For example, emmer wheat (*Triticum dicoccoides*) shows a large genetic diversity for A_{max}, which is independent of leaf anatomy (Carver and Nevo 1990). Genetic advances in wheat yield does not appear to be related to changes in photosynthesis rate on an area basis when measured in the flag-leaf or the spike, but only to a higher, whole spike photosynthesis rate (Zhou et al. 2014), suggesting that under potential yield conditions, the spike with respect to its size, may contribute more than the flag-leaf to yield formation. The awns of wheat are an important photosynthetic and transpiration organ on the spike. It has been reported that the pathway for assimilation movement from awns to the kernels is minimal (Evans et al. 1972). Furthermore, the activity of PEP carboxylase is much higher in awns than in flag leaves throughout ontogeny, and is particularly high at the late stages of grain filling (Li et al. 2002, 2006).

In rice, increasing A_{max} has been proposed to be a more effective strategy for increasing RUE than improving canopy architecture as modern rice varieties are believed to have an optimal canopy architecture in terms of light and N distribution (Peng and Ismail 2004). A large genotypic variation exists for A_{max} in rice and its relatives (Cook and Evans 1983) particularly in diploid wild species (Yeo et al. 1994). Such large variation in A_{max} is associated with variation in stomatal conductance (g_s) at mid-day conditions, Rubisco content, specific leaf area, and total leaf protein (Hubbart et al. 2007). A_{max} at each developmental stage was better explained by g_s, especially under mid-day conditions while N content explains changes in A_{max} associated with crop development (Ohsumi et al. 2007). Interestingly, since genotypic variation in N is unrelated to g_s; these two traits could be modified independently to increase A_{max}. Nevertheless, under light-saturated conditions, g_s generally decreases due to decreased leaf water potential, therefore increasing leaf hydraulic conductance under saturated light should be targeted in breeding (Taylaran et al. 2011).

The physiological regulation of two parameters, TPU (triose phosphate utilization) involved in inorganic phosphate (Pi) recycling, and g_m, the mesophyll conductance, has received far less attention. Limited sink strength results in the accumulation of photoassimilates in the leaf, as a signal for downregulation of photosynthesis, reducing TPU (Zhu et al. 2010; Nebauer et al. 2011). This points to the value of exploring genetic variability in TPU regulation. Also, the reduction in g_m, standing for the diffusive conductance of CO_2 from the stomata to the chloroplast, significantly limits photosynthesis (Flexas et al. 2008; Griffiths and Helliker 2013) with variation in leaf anatomy assumed to be a major factor (Tomas et al. 2013). However, despite substantial evidence for large genotypic variability of g_m, exactly how leaf anatomy interacts with g_m to determine leaf [CO_2] status remains unclear. Overall, these studies suggest that while both crops exhibit different limitations to increase net photosynthesis, manipulating specific trait dynamics may improve photosynthesis rate in both crops (Fig. 1).

Canopy photosynthesis and N distribution

Canopy photosynthesis is largely a function of the vertical distribution of light and nitrogen (light-harvesting proteins and Rubisco), as well as sink activity (Dreccer et al. 2000; Bertheloot et al. 2008). The relationship between plant height and photosynthetic capacity is always not consistent (Gent and Kiyomoto 1984; Gent 1995; Bishop and Bugbee 1998). Semi-dwarf wheat often shows higher canopy photosynthesis at anthesis, which may be linked to stronger sinks associated with more efficient assimilate partitioning at anthesis (Bishop and Bugbee 1998), suggesting that in tall genotypes, the rate of photosynthesis might be partly constrained by assimilate export from leaves. In addition, cultivars with greater capacity to store N in nonphotosynthetic tissues may improve canopy N distribution (Dreccer et al. 1998) particularly if it follows the light gradient (Dreccer et al. 2000; Bertheloot et al. 2008). Recently, the contribution of ear photosynthesis to grain filling was highlighted, especially in situations of source limitation (Maydup et al. 2012).

In rice, an improved canopy structure has increased canopy photosynthesis particularly during the grain-filling period (Zhang and Kokubun 2004). However, the relationships between canopy net photosynthetic rate and biomass, and yield were inconsistent (Peng 2000) probably due to a high genetic variation in canopy photosynthesis during postanthesis (Zhang and Kokubun 2004). With LAI of 5, erect leaves allow higher canopy photosynthesis, while with LAI lower than 3, droopy leaves can allow higher canopy photosynthesis (Peng and Ismail 2004). Therefore, an ideal phenotype should have a prostrate-leaf canopy in the very early growth stage to intercept PAR effectively, while more erect leaves at the later stages would allow light to penetrate more efficiently. Moreover, rice leaves can efficiently maintain lower canopy temperature via efficient transpiration cooling given sufficient water availability (Wassmann et al. 2009; Gao et al. 2011). Depending on the N concentration, either linear ((Makino et al. 1988) or curvilinear (Cook and Evans 1983) relationship with photosynthesis was observed (Shimoda 2012). Modeling studies suggest that if leaf N content can be increased above 1.8 g N m^{-2} in upper leaves, daily canopy photosynthesis rate can be increased regardless of plant density and light condition (Shiratsuchi et al. 2006). Such steeper gradient of leaf N in the canopy increases daily canopy photosynthesis by 20–25%. Both leaf area and tillering capacity should be large at early stages (Lafarge and Bueno 2009), but restricted during grain filling to avoid strong mutual shading and to allow radiation to reach deeper leaf layers to realize gains in canopy photosynthesis with increasing leaf N content. In this regard, it is particularly interesting that new plant types have very erect leaf orientation and relatively small number of productive tillers that could be useful to increase canopy photosynthesis (Horton 2000).

Radiation use efficiency

Until last century, RUE during preanthesis has not been significantly improved in modern wheat cultivars (Gutierrez-Rodriguez 2000). In irrigated environments, no

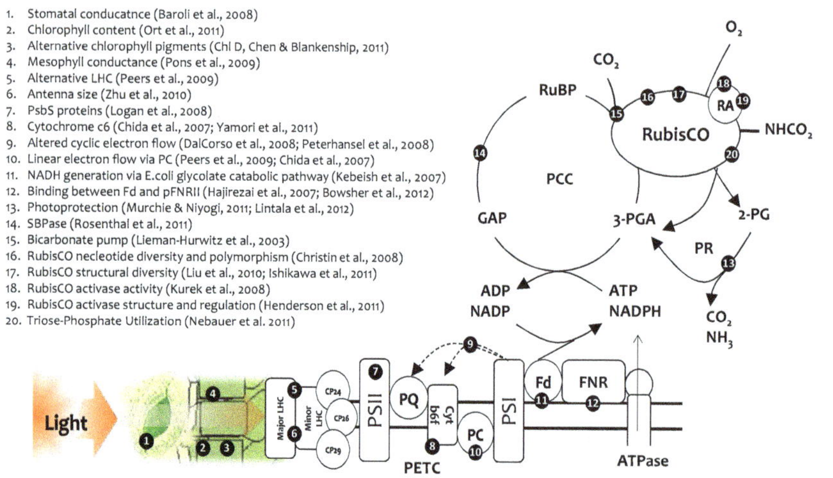

Figure 1. Proposed avenues for enhancing photosynthesis rate in wheat and rice.

differences in RUE or light interception between old and modern cultivars were found, while modern lines were shown to have higher RUE after anthesis (Calderini et al. 1997) largely driven by high sink. Cultivars with cooler canopy temperature exhibit an enhanced gas exchange rate (Ayeneh et al. 2002); hence, cooler canopy can increase crop yields especially under mild stress (Ayeneh et al. 2002). A 2.5°C range in canopy temperature depression at anthesis has been reported for Iranian wheat cultivars (Karimizadeh and Mohannadi 2011). Although diverse mechanisms may theoretically explain genetic variation in canopy temperature (Rebetzke et al. 2012), so far only root depth and capacity has been implicated through direct measurement (Lopes and Reynolds, 2010) and genetic studies in dry residual moisture and hot, irrigated environments (Pinto et al. 2010; Pinto et al. 2014). These studies suggest that RUE can still be improved in both crops and the underlying mechanisms of CTD differing between genotypes need to be explored.

In rice, RUE has reported to increase from old cultivars to modern cultivars. New indica varieties expressed more RUE (2.14 g MJ^{-1}) than traditional (1.8 g MJ^{-1}) and japonica varieties (1.8 g MJ^{-1}) (Boschetti et al. 2006) due to differences in morphophysiological characters (Kiniry et al. 2001). However, RUE in rice is considered to be lower (under aerobic conditions, ~1 g MJ^{-1}; in flooding, ~2.2 g MJ^{-1}) as compared to the average quantum requirement of C_3 species (2.9 g MJ^{-1})(Katsura et al. 2010). This may underlie the poor association between RUE and yield superiority of super hybrid rice varieties (Katsura et al. 2008, 2010). Nevertheless, a higher RUE especially during postanthesis can increase thousand-grain weight (Sadras and Lawson 2011). In wheat, photosynthetic capacity can be improved by increasing green leaf area duration and chlorophyll content of tissue, while stomatal conductance and a cooler canopy temperature are indicative of improved RUE (Sadras and Lawson 2011; Xiao et al. 2012).

Leaf senescence

In wheat, leaves contribute about 40% to the daily rate of grain N accumulation at mid grain filling (Simpson et al. 1983). Hence, a delayed leaf senescence or stay green, which maintains active photosynthesis for longer period may increase grain yields (Gregersen et al. 2008; Bogard et al. 2011). In contrast, accelerated senescence leads to low C but high N remobilization indicating the plasticity of senescence linked to carbon and N remobilization. Such senescence regulation of C and N could be a viable strategy to alter both C and N remobilization to the actively developing grains. While several patterns of senescence were proposed (Thomas and Howarth 2000), an ideal senescence phenotype in wheat, in general in cereals, still needs to be identified (Gregersen et al. 2008; Bogard et al. 2011). The advent of high-throughput spectral tools such as the Greenseeker has made it more feasible to study total canopy N remobilization dynamics (Lopes and Reynolds 2012).

In rice, a functional stay-green phenotype exhibits higher light conversion efficiency and net photosynthetic rate at saturated light throughout grain filling (Fu and Lee 2012). It also maintains mesophyll conductance due to high chlorophyll content and delays the degradation of photosystem II. Interestingly, the stay-green trait is associated with high root activity so higher capacity to accumulate and maintain high N (Fu et al. 2009) and with higher grain-filling rate under limited radiation conditions (Bueno and Lafarge, unpublished data). These studies suggest that the stay-green trait is beneficial to increase root activity and N accumulation during grain-filling period under favorable environments.

Nitrogen use efficiency and biological nitrification inhibition

Genetic gains in NUE under low N supply were associated with the improvements in NUpE (uptake efficiency) in spring wheat and to NUtE (utilization efficiency) in winter wheat (Brancourt-Hulmel 2003), while under high N, both components contribute approximately equally to NUE (Ortiz-Monasterio et al. 1997; Muurinen et al. 2006; Wang et al. 2011). These studies indicate that NUpE may be influenced by G × N interaction, while G × E interaction may influence NUtE. Although leaf senescence kinetics could explain the genetic variability associated with yield variation under low N (Gaju et al. 2011), the strong negative association between NUpE and NUtE highlights the necessity of understanding mechanisms influencing the uptake and utilization of N by crops. Modified root architecture associated with dwarfing genes is one parameter influencing NUE (Wojciechowski et al. 2009; Gooding et al. 2012). Alternatively, BNI (biological nitrification inhibition) may improve NUE. The wild wheat, *Leymus racemosus*, shows higher BNI release rates ranging from 20 to 30 allyl thiourea g^{-1} root dry wt. day^{-1} (Subbarao et al. 2007). A locus for high BNI was transferred to cultivated wheat using chromosome addition lines of wild wheat, however, these lines carry many undesirable traits compromising yield potential.

In rice, a large genotypic variation in NUE has been reported, ranging from 38 to 84 kg kg^{-1} for grain and 100 to 160 kg kg^{-1} for biomass, probably under aerobic and rainfed management (Koutroubas and Ntanos 2003). However, improving NUE for grain in favorable (irrigated) environments has been limited in breeding (Samonte et al. 2006). Indica cultivars generally have higher NUE for grain than japonica cultivars (Koutroubas and Ntanos 2003). Such large variations could be related to yield potential because high-yielding cultivars show less genetic variation in NUE for grain (Samonte et al. 2006). Moreover, NUE for grain is also influenced by plant height, crop growing duration (short and late maturing cultivars have high NUE for grain than tall and early maturing varieties; Koutroubas and Ntanos 2003), percentage of straw N, grain N concentration, and HI. At HI of 0.5, both straw and grain N equally contribute to NUE for grain; therefore, increasing HI would be associated with NUE for grain. However, genotypic variation in grain N concentration is low (0.95–1.2; Tirol-Padre et al. 1996), therefore, it may not be feasible to improve NUE for grain by reducing grain N concentration using conventional breeding as grain N concentration is affected more by environments than by genotypes (Ladha et al. 1998). Genetic variability for nitrogen metabolism in the developing ear of maize has been suggested to improve yields (Cañas et al. 2012). BNI improves N uptake and NUE due to its inhibitory effects on nitrification (Subbarao et al. 2006). A significant genotypic variation in BNI (<10% to >50%) was reported (Pariasca Tanaka et al. 2010). However, traditional and upland varieties exhibit high BNI activity than the modern irrigated lowland varieties. While nitrification rates in paddy soils are lower compared to typical upland soils, application of synthetic nitrification inhibitors has been shown to increase yields (Li et al. 2009) suggesting that BNI has the potential to improve yields in rice.

Source:sink regulation and lodging resistance

An approach to increase spike dry matter is to increase assimilate partitioning to spike at anthesis (Demotes-Mainard and Jeuffroy 2004). To some extent, dwarf and semi-dwarf cultivars have achieved this by increasing assimilates partitioning to developing spikes (Álvaro et al. 2008; Foulkes et al. 2011). For example, the dry matter partitioning to spikes at anthesis improved from 9% in old cultivars to 27% in modern semi-dwarf durum wheat in Italy and Spain (Álvaro et al. 2008). The genetic range for partitioning to the respective plant organs at anthesis (spikes, 0.12–0.29; leaf lamina, 0.19–0.31; stems and leaf sheath, 0.48–0.63; as a proportion of above-ground biomass; Reynolds et al. 2009) suggests that there is a possibility to increase assimilate partitioning to spikes (Foulkes et al. 2011). However, decreasing plant height further by developing double dwarfs may not provide any additional benefits because of reduced light interception and total biomass (Butler et al. 2005). Moreover, reduced cell elongation and smaller cells have reduced seedling vigor in resource-poor environments (Botwright et al. 2005).

AGB (above-ground biomass) was not systematically increased from old to modern cultivars until relatively recently (China [Zheng et al. 2011]; U.K. [Shearman et al. 2005]; Australia [Sadras and Lawson 2011]; and Mexico [Aisawi et al. 2015]). Increased AGB shows two common trends with yield components: thousand kernel weight or grain weight in cultivars of Australia (Sadras and Lawson 2011) and Henan Province of China (Zheng et al. 2011), and grains per unit area or spikes per unit area in cultivars of U.K. (Shearman et al. 2005) and Shandong of China (Xiao et al. 2012). This suggests that biomass partitioning to yield components has been greatly modified resulting in distinct wheat ideotypes (Álvaro et al. 2008). An increased AGB especially during postanthesis shows strong correlation with yield (Reynolds et al. 2009). Nevertheless, the association between yield and biomass depends heavily on environment. Under stress, there is usually a strong association (Reynolds et al. 2007) but less so often when conditions are favorable. In addition, higher WSC (water-soluble carbohydrates) at anthesis are associated with higher grain weight and lower WSC storage is associated with increased grain number (Rebetzke et al. 2008; Sadras and Lawson 2011). While this trend is consistent with most modern cultivars, often an opposite trend with higher WSC storage and grain number was reported (Shearman et al. 2005; Xiao et al. 2012). These studies did not report any changes in grain weight, hence the associations between WSCs and yield components are unclear.

In rice, there has been a marked increase in HI and grain production by modification of the reproductive structures in the modern cultivars (Evans et al. 1984; Mohapatra et al. 2011) and hybrids (Yang et al. 2007), suggesting an increased partitioning of assimilates to reproductive organs (Lafarge and Bueno 2009). Modern cultivars show increased biomass production over old cultivars. Yield improvement in cultivars released before 1980 was associated with increases in HI, while the yield improvement in the cultivars released after 1980 was associated with increases in biomass (Peng et al. 2000) although it largely depends on comparisons between either semi-dwarf and traditional or among semi-dwarfs (Evans et al. 1984; Yamauchi 1994; Peng et al. 1998). Also, higher yields of hybrids were attributed to both high HI (Bueno and Lafarge 2009) and biomass (Bueno and Lafarge 2009; Zhang et al. 2009) mainly through higher growth rate of priority sinks and higher sink regulation, that is the timing and magnitude of the successive sink activities over crop growth (Bueno et al. 2010), to the favor of stem biomass at the end of the vegetative phase, panicle biomass at the end of the reproductive phase, and remobilization from stem to panicle at the beginning of the filling phase (Lafarge and Bueno 2009; Bueno et al. 2010). Yet, it was reported that higher yield was attributed to higher biomass accumulation under favorable conditions and to higher HI under suboptimal conditions in the tropics (Peng et al. 2003). A recent study conducted on 32 high-yielding contrasted genotypes (hybrids and inbreds) from diverse geographical origin, and respective of their maturity groups, highlighted, however, that relevant traits were those favoring light capture in the favorable (dry) season and short plant height and delayed leaf senescence in the suboptimal (wet) season (Bueno and Lafarge submitted).

It has been estimated that during heading to maturity, an inbred rice allocates dry matter about 39.5% to panicles, 4.3% to leaves, 33.5% to stems, 6.5% to late tillers, and 6.7% to roots (Kato et al. 2006), suggesting that panicles have high demand for dry matter partitioning (Zhao et al. 2008). Indeed, in favorable conditions, hybrid rice shows a higher partitioning coefficient to culms in the late vegetative and early reproductive while lower partitioning coefficient to culms and higher partitioning coefficient to panicles in the later reproductive phases as compared to inbreds (Lafarge and Bueno 2009). This suggests a strong sink regulation and remobilization capacity with hybrids over the crop cycle (Bueno et al. 2010) through larger biomass partitioning to culm at late reproductive phase and HI. Indeed, in high-yielding recently bred varieties, equipped with large panicles (Fujita et al. 2013), grain filling of late flowering basal spikelets is particularly poor (Panda et al. 2009). Both NSC accumulation in the stems at heading (Fu et al. 2011) and reduction in ethylene concentration in the basal spikelets (Wang et al. 2012) enhances sink strength of the basal spikelets at early grain filling by increasing number of endosperm cells and activities of sucrose synthase and AGPase (adenosine diphosphoglucose pyrophosphorylase). These studies suggest that increasing NSC accumulation in the stems and reducing ethylene production in the basal spikelets will induce fast-synchronized filling within the panicle and higher yield potential. In addition, there is growing evidence that the T6P (trehalose-6-phosphate) plays a central role in regulating carbohydrate and starch metabolism in plants as it indirectly reflects sucrose concentration and has been widely accepted as an indicator of sucrose status (O'Hara et al. 2013; Yadav et al. 2014). By regulating sucrose export to the grain, T6P could then increase sink strength in case of its higher concentration within the grain and has been shown to increase grain yields (Nuccio et al. 2015).

Lodging resistance

Strategies to raise HI should consider assimilate partitioning to structural organs to avoid any stem and root lodging. Models describing lodging resistance in wheat have highlighted stem thickness, stem wall material strength, and area of root plate in the stem and root lodging resistance (Berry et al. 2003b, 2007) (Fig. 2). Interestingly, lodged plants exhibited stronger source limitation during grain filling than unlodged ones (Acreche and Slafer 2011), highlighting the importance of photosynthetic activity

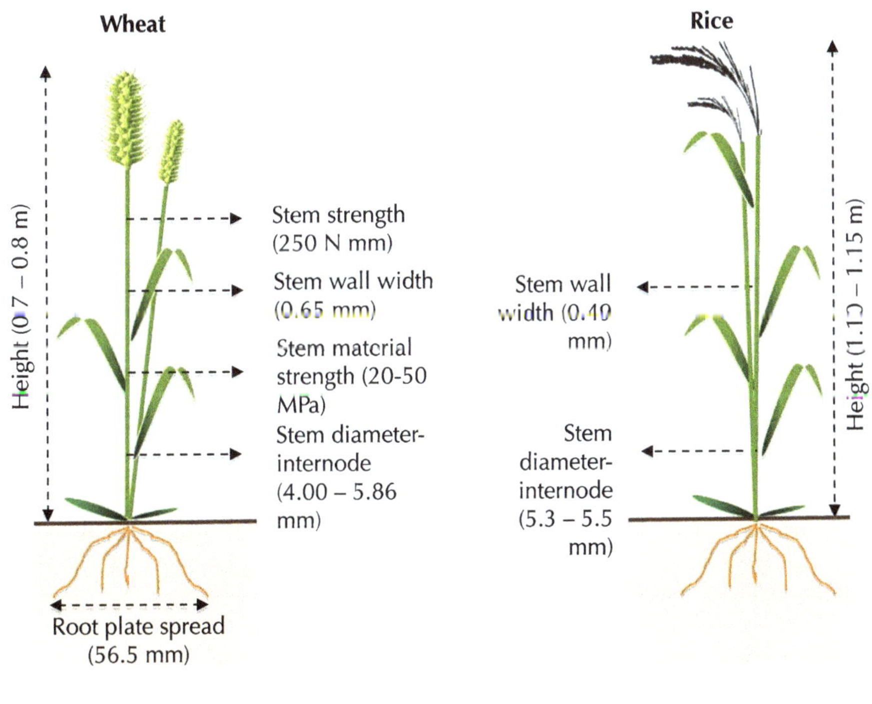

Figure 2. An ideotype design for lodging-resistant wheat and rice.

during grain filling to avoid any stem or root lodging. Reducing unproductive tiller number could also reduce lodging problems and increase grains per spike (9%) but decrease grain yields (8%) (Duggan et al. 2005; Gaju et al. 2011), as observed in China (Zheng et al. 2011) and Spain (Royo et al. 2006; Álvaro et al. 2008). The net redistribution of N and C to surviving shoots after their death is inconsistent being negative (Berry et al. 2003a) or positive (Thorne and Wood 1987). Although the efficiency of such transport is unknown, a high, early tiller production may enhance grain yields in cereals (Pasuquin et al. 2008).

Rice exhibits a large genetic variability in lodging resistance. Several lodging-related traits were identified (Islam et al. 2007). Lodging resistance was strongly correlated with the culm diameter and wall thickness of the basal internodes (Chuanren et al. 2004). A stem diameter of 5.3–5.5 mm (plant height of 114 cm under typhoon conditions) (Chuanren et al. 2004; Ishimaru et al. 2005) and stem wall thickness of 0.49 (Chuanren et al. 2004) were associated with lodging resistance. While plant height was not necessarily an important factor (Ishimaru et al. 2007), leaf sheath and the extent of wrapping (breaking strength of the shoot was increased by 30–60%) will be an important trait for selection against lodging at all developmental stages as shown in rice (Setter et al. 1994). In addition, mechanical properties such as tensile breaking point and the tensile elastic modulus are associated with lodging resistance (Chuanren et al. 2004). Clearly, while plant height is not necessarily being reduced, increasing stem wall thickness and mechanical properties can contribute to enhance lodging resistance.

Trait Differences and Similarities Between Wheat and Rice

Over the last two decades, the linear rate of yield change for wheat has been 25 kg ha^{-1} $year^{-1}$, while it is 38 kg ha^{-1} $year^{-1}$ for rice. However, the relative rates of yield increase seem to be the same (0.9%) for both crops (Fischer and Edmeades 2010). While reduction in plant height, increased biomass, and HI commonly resulted in yield gain in both crops, increased kernel number per unit area in wheat and increased CGR in rice are largely associated with yield gains (Fischer and Edmeades 2010; Pozo et al. 2014). The CGR between stem elongation and anthesis is increased in wheat particularly in the cultivars of Australia (Sadras and Lawson 2011), while CGR and early vigor (higher LAI) have been significantly increased in rice particularly in hybrids, which increased biomass accumulation and grain yield (Peng et al. 1998; Bueno and Lafarge 2009). In addition, net photosynthesis and RUE have been slightly increased in modern wheat while hybrid rice has a lower net photosynthetic rate during vegetative and grain-filling phases because of lower leaf N concentration (Peng et al. 1998).

Both wheat and rice crop phenophases are responsive to the environment. Breeding efforts have altered vegetative phase durations in both crops. The genetic factors such as earliness per se loci are absent in rice (Faricelli et al. 2009, 2010). Both crops were domesticated in different climates and so differ significantly in their growth conditions; wheat shows appreciably faster growth rate and allocates more biomass to leaves than rice; however, the optimal temperature for biomass production is higher for rice than wheat (Nagai and Makino 2009). While both species significantly contrast in photoperiod sensitivity, the response of developmental processes to optimum temperature is similar in both species and is not affected by either natural selection or artificial breeding (Parent and Tardieu 2012). This would suggest a strong interaction between developmental events and the environment in particular temperature per se.

Both species differ in LAI. Since both crops differ for leaf N content per unit area, they also differ for radiation conversion factor (Mitchell et al. 1998). Interestingly, the correlation between leaf N content and grain number vary with genotype and growth conditions in wheat, while it is reported to be independent in rice (Wada and Matsushima 1962; Makino 2011). Since the N concentration in the straw and grain in rice is lower than wheat, the NUE for biomass and grain is higher in rice (Ladha et al. 1998). Such differential N levels in the biomass and its allocation influence leaf senescence in both species. Furthermore, awns are absent or short in cultivated rice (Luo et al. 2013; Hua et al. 2015), while awns have been shown to contribute to net ear photosynthesis in wheat (Evans et al. 1972; Sanchez-Bragado et al. 2014).

The source–sink regulation is different between two crops. In rice (hybrids), the source capture is relatively higher than that of wheat, while the sink strength is more efficient during grain-filling period. For example, biomass production before stem elongation was higher in rice than in wheat due to high early vigor and CGR (Bueno and Lafarge 2009; Bueno et al. 2010). While assimilate partitioning to panicles over other organs is larger in rice after heading (Kato et al. 2006; Kumar et al. 2006), this has been shown to be limited in wheat. Although the remobilized carbon reserve percentage and its contribution to grain appears higher in wheat than rice, the HI is lower in wheat than in rice probably due to the existence of alternative sinks (Yang and Zhang 2006). Moreover, stem structural differences exist; the width of the stem wall in wheat (0.65 mm) is greater than that in rice (0.45 mm), while stem diameter in wheat (4.7–4.9) is less than the stem diameter of rice (5.3–5.5 mm) (Chuanren et al. 2004; Reynolds et al. 2009; Wu et al. 2011).

By virtue of the growth conditions, the crop management practices and environmental influences on crop growth and development differ between two species. While the conventional crop management practices in wheat have largely been unmodified, in rice, sound management practices have been in practice. Among others, planting patterns, planting density, and spacing have resulted in more open canopies, which increased the overall performance of the rice crop canopy in terms of leaf angle, tiller dynamics, light interception, and photosynthetic rate (Table 1).

A Research Gap Analysis for Future Research Priorities

Both, wheat and rice, crops have been explored in different ways. However, the genetic gains in yields of wheat and rice across the globe are less than 1% per annum. The following opportunities, but not limited to, may be promising to further increase in genetic gains in yields of both crops.

Yield components

Recent gains in crop yields has been achieved through larger panicle size with higher grain number per panicle in rice (Ying et al. 1998), while increased grain number predominantly enhanced grain yields in wheat (Pozo et al. 2014; Zhou et al. 2014). Historical data, however, support that grain number (either through extra panicles/spikes or extra grain per panicle/spike) needs to be increased in both crops. In addition, a number of genes important for yield formation have been identified in both crops (see review, Valluru et al. 2014, table 3). For example, genes controlling leaf and panicle size (SPIKE) and grain assimilate accumulation (*GIF1* and *Rg5*) have been identified in rice (Fujita et al. 2013; Ishimaru et al. 2005). In wheat, structural modification to AGPase gene has been proposed to increase starch accumulation in grain (Meyer et al. 2007). Furthermore, *GS3* and *Gn1* controlling grain number and weight have been identified in rice. These genes have largely not been explored in both crops (Valluru et al. 2014).

Harvest index

Increasing HI is not a strong candidate, as it is already high, above 0.5 in high-yielding varieties and sink regulation is also high and effectively expressed in hybrid rice (Ying et al. 1998; Lafarge et al. 2010). In wheat, an average HI reported is 0.45 (Zheng et al. 2011; Xiao et al. 2012). Increasing HI above the current values may lead to high risk of lodging unless plants have a lower center of gravity (i.e., are shorter), but then this would reduce light interception, consequently reducing biomass. More rigid stems involving anatomical changes like increasing concentrations of structural carbohydrates like lignin could be a long-term breeding objective.

Biomass

Increasing biomass accumulation may be a promising avenue in both crops. In rice, hybrids have significantly

Table 1. Trait similarities and differences between wheat and rice.

Trait	Wheat	References	Rice	References
Preanthesis period	Preanthesis period reduced in modern genotypes	Álvaro et al. (2008); Isidro et al. (2011)	Preanthesis period reduced in modern genotypes	Peng et al. (2000)
Earliness per se	Earliness per se locus are present in wheat	Gouis et al. (2011)	Earliness per se locus are reported to be absent in rice	Faricelli et al. (2009); (Faricelli et al. 2010)
LAI	LAI is lower	Acreche et al. (2009); Parry et al. (2011)	LAI is higher	Zheng et al. (2008); Zhang et al. (2009)
RUE	RUE increased in modern cultivars particularly during postanthesis period	Calderini et al. (1997)	RUE increased in modern cultivars	Boschetti et al. (2006)
Awns	Awns are reported to contribute to net ear photosynthesis	Evans et al. (1972); Sanchez-Bragado et al. (2014)	Awns are absent or short in the cultivated rice	Luo et al. (2013); Hua et al. (2015)
Leaf senescence	Leaf senescence is slightly delayed in modern cultivars due to the presence of nonfunctional grain protein content gene	Gregersen et al. (2008)	Indica and Japonica exhibit different leaf senescence patterns, early and late, respectively	Abdelkhalik et al. (2005)

LAI, leaf area index; RUE, radiation use efficiency.

higher aboveground biomass (10–12%) than inbred cultivars (Peng et al. 2008);(Bueno and Lafarge 2009). In wheat, biomass has, on average, a lower genetic gain (0.3), suggesting that biomass production may still be the main way forward to increase yields (Fischer and Edmeades 2010; Sylvester-Bradley et al. 2012). However, the key question that needs to be addressed would be whether the key driver of yield is, biomass accumulation or sink strength? Would higher biomass accumulation induce better sink strength or better sink strength support higher biomass accumulation? A crossing program at CIMMYT that combines high source with high sink traits is showing genetic gains for yield and biomass in international trials, suggesting that source and sink balance is important (Reynolds et al. 2015).

Leaf and crop photosynthesis

Improving TPU is promising in both crops. If sink strength is higher, photosynthate could be exported quicker from the source leaves to the sink to reduce downregulation of photosynthesis. Similarly, improving mesophyll conductance could be promising. Changes in leaf anatomy could facilitate CO_2 diffusion between the leaf compartments to also reduce downregulation of leaf photosynthesis. Further reducing night respiration could be important although few studies report limited diversity for this trait and limited effect on the overall performance (Peraudeau et al. 2015a,b). If the plant could uptake more N (via higher fertilization and/or better root system and/or improved water management for better uptake), with N being more available to fill grain, then so leaf senescence can be delayed and crop photosynthesis maintain for a longer duration. Numerous proposed avenues for enhancing photosynthesis rate in C_3 species are given in Figure 1.

Sink strength

Increasing sink strength through improving panicle/spike architecture could be important. Reducing hormonal (especially ethylene) regulation that delays the filling of basal spikelets could be promising for efficient grain filling. In addition, increasing the role of T6P, as sugar regulator, could facilitate the export of carbohydrates to the panicles/spikes.

Acknowledgments

The authors thank two reviewers for their useful comments.

Conflict of Interest

None declared.

References

Abdelkhalik, A. F., R. Shishido, K. Nomura, and H. Ikehashi. 2005. QTL-based analysis of leaf senescence in an indica/japonica hybrid in rice (*Oryza sativa* L.). Theor. Appl. Genet. 110:1226–1235.

Acreche, M. M., and G. A. Slafer. 2011. Lodging yield penalties as affected by breeding in Mediterranean wheats. Field Crops Res. 122:40–48.

Acreche, M. M., G. Briceño-Félix, J. A. Martín Sánchez, and G. A. Slafer. 2009. Radiation interception and use efficiency as affected by breeding in Mediterranean wheat. Field Crops Res. 110:91–97.

Aisawi, K. A. B., M. P. Reynolds, R. P. Singh, and M. J. Foulkes. 2015. The physiological basis of the genetic progress in yield potential of CIMMYT spring wheat cultivars from 1966 to 2009. Crop Sci. 55:1749–1764.

Álvaro, F., J. Isidro, D. Villegas, L. F. García del Moral, and C. Royo. 2008. Breeding effects on grain filling, biomass partitioning, and remobilization in Mediterranean durum wheat. Agron. J. 100:361.

Ayeneh, A., M. van Ginkel, M. Reynolds, and K. Ammar. 2002. Comparison of leaf, spike, peduncle and canopy temperature depression in wheat under heat stress. Field Crops Res. 79:173–184.

Berry, P., J. Spink, M. Foulkes, and A. Wade. 2003a. Quantifying the contributions and losses of dry matter from non-surviving shoots in four cultivars of winter wheat. Field Crops Res. 80:111–121.

Berry, P. M., M. Sterling, C. J. Baker, J. Spink, and D. L. Sparkes. 2003b. A calibrated model of wheat lodging compared with field measurements. Agric. For. Meteorol. 119:167–180.

Berry, P. M., R. Sylvester-Bradley, and S. Berry. 2007. Ideotype design for lodging-resistant wheat. Euphytica 154:165–179.

Bertheloot, J., P. Martre, and B. Andrieu. 2008. Dynamics of light and nitrogen distribution during grain filling within wheat canopy. Plant Physiol. 148:1707–1720.

Bishop, D. L., and B. G. Bugbee. 1998. Photosynthetic capacity and dry mass partitioning in dwarf and semi-dwarf wheat (*Triticum aestivum* L.). J. Plant Physiol. 153:558–565.

Bogard, M., M. Jourdan, V. Allard, P. Martre, M. R. Perretant, C. Ravel, et al. 2011. Anthesis date mainly explained correlations between post-anthesis leaf senescence, grain yield, and grain protein concentration in a winter wheat population segregating for flowering time QTLs. J. Exp. Bot. 62:3621–3636.

Borràs-Gelonch, G., M. Denti, W. T. B. Thomas, and I. Romagosa. 2011. Genetic control of pre-heading phases in the Steptoe × Morex barley population under different conditions of photoperiod and temperature. Euphy 183:303–321.

Borràs-Gelonch, G., G. J. Rebetzke, R. A. Richards, and I. Romagosa. 2012. Genetic control of duration of pre-anthesis phases in wheat (*Triticum aestivum* L.) and relationships to leaf appearance, tillering, and dry matter accumulation. J. Exp. Bot. 63:69–89.

Boschetti, M., S. Bocchi, D. Stroppiana, and P. Brivio. 2006. Estimation of parameters describing morpho-physiological features of Mediterranean rice varieties for modelling purposes. Ital. J. Agrometeorol. 49:3.

Botwright, T. L., G. J. Rebetzke, A. G. Condon, and R. A. Richards. 2005. Influence of the gibberellin-sensitive Rht8 dwarfing gene on leaf epidermal cell dimensions and early vigour in wheat (*Triticum aestivum* L.). Ann. Bot. 95:631–639.

Brancourt-Hulmel, M. 2003. Genetic improvement of agronomic traits of winter wheat cultivars released in France from 1946 to 1992. Crop Sci. 43:37–45.

Breseghello, F., O. P. de Morais, P. V. Pinheiro, A. C. S. Silva, E. Maia de Castro, A. P. de Castro, et al. 2011. Results of 25 years of upland rice breeding in Brazil. Crop Sci. 51:914.

Brisson, N., P. Gate, D. Gouache, G. Charmet, F.-X. Oury, and F. Huard. 2010. Why are wheat yields stagnating in Europe? A comprehensive data analysis for France. Field Crops Res. 119:201–212.

Bueno, C. S., and T. Lafarge. 2009. Higher crop performance of rice hybrids than of elite inbreds in the tropics: 1. Hybrids accumulate more biomass during each phenological phase. Field Crops Res. 112:229–237.

Bueno, C. S., and T. Lafarge. Submitted. Relevant traits for high yield in rice in the tropics as dependent on maturity groups and growing seasons.

Bueno, C. S., E. Pasuquin, B. Tubaña, and T. Lafarge. 2010. Improving sink regulation, and searching for promising traits associated with hybrids, as a key avenue to increase yield potential of the rice crop in the tropics. Field Crops Res. 118:199–207.

Butler, J. D., P. F. Byrne, V. Mohammadi, P. L. Chapman, and S. D. Haley. 2005. Agronomic performance of alleles in a spring wheat population across a range of moisture levels. Crop Sci. 45:939.

Calderini, D. F., M. F. Dreccer, and G. A. Slafer. 1997. Consequences of plant breeding on biomass growth, radiation interception and radiation use efficiency in wheat. Field Crops Res. 52:271–281.

Cañas, R. A., I. Quilleré, A. Gallais, and B. Hirel. 2012. Can genetic variability for nitrogen metabolism in the developing ear of maize be exploited to improve yield? New Phytol. 194(2):440–52.

Cantu, D., B. Yang, R. Ruan, K. Li, V. Menzo, D. Fu, et al. 2013. Comparative analysis of protein-protein interactions in the defense response of rice and wheat. BMC Genom. 14:166.

Carver, B. F., and E. Nevo. 1990. Genetic diversity of photosynthetic characters in native populations of *Triticum dicoccoides*. Photosynth. Res. 25:119–128.

Chuanren, D., W. Bochu, W. Pingqing, W. Daohong, and C. Shaoxi. 2004. Relationship between the minute structure and the lodging resistance of rice stems. Colloids Surf B Biointerfaces 35:155–158.

Cook, M. G., and L. T. Evans. 1983. Some physiological aspects of the domestication and improvement of rice (Oryza spp.). Field Crops Res. 6:219–238.

Demotes-Mainard, S., and M.-H. Jeuffroy. 2004. Effects of nitrogen and radiation on dry matter and nitrogen accumulation in the spike of winter wheat. Field Crops Res. 87:221–233.

Dreccer, M. F., G. A. Slafer, and R. Rabbinge. 1998. Optimization of vertical distribution of canopy nitrogen. J. Crop. Prod. 1:47–77.

Dreccer, M. F., M. van Oijen, A. H. C. M. Schapendonk, C. S. Pot, and R. Rabbinge. 2000. Dynamics of vertical leaf nitrogen distribution in a vegetative wheat canopy. Impact on canopy photosynthesis. Ann. Bot. 86:821–831.

Duggan, B. L., R. A. Richards, A. F. van Herwaarden, and N. A. Fettell. 2005. Agronomic evaluation of a tiller inhibition gene (tin) in wheat. I. Effect on yield, yield components, and grain protein. Aust. J. Agric. Res. 56:169–178.

Evans, L. T., J. Bingham, P. Jackson, and J. Sutherland. 1972. Effect of awns and drought on the supply of photosynthate and its distribution within wheat ears. Ann. Appl. Biol. 70:67–76.

Evans, L. T., R. M. Visperas, and B. S. Vergara. 1984. Morphological and physiological changes among rice varieties used in the Philippines over the last seventy years. Field Crops Res. 8:105–124.

Faricelli, M. E., M. Valárik, and J. Dubcovsky. 2009. Control of flowering time and spike development in cereals; the earliness per se Eps-1 region in wheat, rice, and Brachypodium. Funct. Integr. Genomics 10:293–306.

Faricelli, M. E., M. Valarik, and J. Dubcovsky. 2010. Control of flowering time and spike development in cereals: the earliness per se *Eps-1* region in wheat, rice, and *Brachypodium*. Funct. Integr. Genomics 10:293–306.

Fischer, R. 2001. Selection traits for improving yield potential. Appl. Physiol. Wheat Breed. 148–159.

Fischer, R. A., and G. O. Edmeades. 2010. Breeding and cereal yield progress. Crop Sci. 50:S85–S98.

Flexas, J., M. Ribas-Carbo, A. Diaz-Espejo, J. Galmes, and H. Medrano. 2008. Mesophyll conductance to CO_2: current knowledge and future prospects. Plant, Cell Environ. 31:602–621.

Foulkes, M. J., G. A. Slafer, W. J. Davies, P. M. Berry, R. Sylvester-Bradley, P. Martre, et al. 2011. Raising yield potential of wheat. III. Optimizing partitioning to grain while maintaining lodging resistance. J. Exp. Bot. 62:469–486.

Fu, J., and B. Lee. 2012. Changes in photosynthetic characteristics during grain filling of a functional stay-green rice SNU-SG1 and its F1 hybrids. J. Crop. Sci. Biotechnol. 11:75–82.

Fu, J., Y. Yan, and B. Lee. 2009. Physiological characteristics of a functional stay green rice SNU-SG1 during grain filling period. J. Crop. Sci. Biotechnol. 12:47–52.

Fu, J., Z. Huang, Z. Wang, J. Yang, and J. Zhang. 2011. Pre-anthesis non-structural carbohydrate reserve in the stem enhances the sink strength of inferior spikelets during grain filling of rice. Field Crops Res. 123:170–182.

Fujita, D., K. R. Trijatmiko, A. G. Tagle, M. V. Sapasap, Y. Koide, K. Sasaki, et al. 2013. NAL1 allele from a rice landrace greatly increases yield in modern indica cultivars. Proc. Natl Acad. Sci. USA 110: 20431–20436.

Gaju, O., V. Allard, P. Martre, J. W. Snape, E. Heumez, J. LeGouis, et al. 2011. Identification of traits to improve the nitrogen-use efficiency of wheat genotypes. Field Crops Res. 123:139–152.

Gao, J., Y. Han, X. Wang, X. Zhao, Z. Ma, H. Xu, et al. 2011. Canopy temperature difference and the related characteristics at heading stage in rice. J. Shenyang Agric. Univ. 4:5–7.

Garcia, G. A. 2011. Variability of duration of pre-anthesis phases as a strategy for increasing wheat grain yield. Field Crops Res. 124:408–416.

Gent, M. 1995. Canopy light interception, gas exchange, and biomass in reduced height isolines of winter wheat. Crop Sci. 35:1636–1642.

Gent, M., and R. Kiyomoto. 1984. Comparison of canopy and flag leaf net carbon dioxide exchange of 1920 and 1977 New York winter wheats. Crop Sci. 25:81–86.

Gooding, M. J., M. Addisu, R. K. Uppal, J. W. Snape, and H. E. Jones. 2012. Effect of wheat dwarfing genes on nitrogen-use efficiency. J. Agric. Sci. 150:3–22.

Gouis, J., J. Bordes, C. Ravel, S. Faure, S. Praud, N. Galic, et al. 2011. Genome-wide association analysis to identify chromosomal regions deter- mining components of earliness in wheat. Theor. Appl. Genet. 124:597–611.

Graybosch, R. A., and C. J. Peterson. 2010. Genetic improvement in winter wheat yields in the great plains of North America, 1959–2008. Crop Sci. 50:1882.

Gregersen, P. L., P. B. Holm, and K. Krupinska. 2008. Leaf senescence and nutrient remobilisation in barley and wheat. Plant Biol. 10:37–49.

Griffiths, H., and B. R. Helliker. 2013. Mesophyll conductance: internal insights of leaf carbon exchange. Plant, Cell Environ. 30:600–616.

Gutierrez-Rodriguez, M. 2000. Photosynthesis of wheat in a warm, irrigated environment: II. Traits associated with genetic gains in yield. Field Crops Res. 66:51–62.

Horton, P. 2000. Prospects for crop improvement through the genetic manipulation of photosynthesis: morphological and biochemical aspects of light capture. J. Exp. Bot. 51:475–485.

Hua, L., D. R. Wang, L. Tan, Y. Fu, F. Liu, L. Xiao, et al. 2015. LABA1, a domestication gene associated with long, barbed awns in wild rice. Plant Cell 27:1875–1888.

Hubbart, S., S. Peng, P. Horton, Y. Chen, and E. H. Murchie. 2007. Trends in leaf photosynthesis in historical rice varieties developed in the Philippines since 1966. J. Exp. Bot. 58:3429–3438.

Ishimaru, K., T. Kashiwagi, N. Hirotsu, and Y. Madoka. 2005. Identification and physiological analyses of a locus for rice yield potential across the genetic background. J. Exp. Bot. 56:2745–2753.

Ishimaru, K., E. Togawa, T. Ookawa, T. Kashiwagi, Y. Madoka, and N. Hirotsu. 2007. New target for rice lodging resistance and its effect in a typhoon. Planta 227:601–609.

Isidro, J., F. Álvaro, C. Royo, D. Villegas, D. J. Miralles, and L. F. García del Moral. 2011. Changes in duration of developmental phases of durum wheat caused by breeding in Spain and Italy during the 20th century and its impact on yield. Ann. Bot. 107:1355–1366.

Islam, M. S., S. Peng, R. M. Visperas, N. Ereful, M. S. U. Bhuiya, and A. W. Julfiquar. 2007. Lodging-related morphological traits of hybrid rice in a tropical irrigated ecosystem. Field Crops Res. 101:240–248.

Karimizadeh, R., and M. Mohannadi. 2011. Association of canopy temperature depression with yield of durum wheat genotypes under supplementary irrigated and rainfed conditions. Aust. J. Crop Sci. 5:138–146.

Kato, M., M. Yokoo, and S. Maruyama. 2006. Dry-matter partitioning and accumulation of carbon and nitrogen during ripening in a female-sterile line of rice. Plant Prod. Sci. 9:185–190.

Katsura, K., S. Maeda, I. Lubis, T. Horie, W. Cao, and T. Shiraiwa. 2008. The high yield of irrigated rice in Yunnan, China: "A cross-location analysis". Field Crops Res. 107:1–11.

Katsura, K., M. Okami, H. Mizunuma, and Y. Kato. 2010. Radiation use efficiency, N accumulation and biomass production of high-yielding rice in aerobic culture. Field Crops Res. 117:81–89.

Kiniry, J. R., G. McCauley, Y. Xie, and J. G. Arnold. 2001. Rice parameters describing crop performance of four U.S. cultivars. Agron. J. 93:1354.

Koç, M., C. Barutçular, and I. Genç. 2003. Photosynthesis and productivity of old and modern durum wheats in a mediterranean environment. Crop Sci. 43:2089.

Koutroubas, S., and D. Ntanos. 2003. Genotypic differences for grain yield and nitrogen utilization in Indica and Japonica rice under Mediterranean conditions. Field Crops Res. 83:251–260.

Kumar, R., A. K. Sarawgi, C. Ramos, S. T. Amarante, A. M. Ismail, and L. J. Wade. 2006. Partitioning of dry matter during drought stress in rainfed lowland rice. Field Crops Res. 98:1–11.

Ladha, J. K., G. J. D. Kirk, J. Bennett, S. Peng, C. K. Reddy, P. M. Reddy, et al. 1998. Opportunities for increased nitrogen-use efficiency from improved lowland rice germplasm. Field Crops Res. 56:41–71.

Lafarge, T., and C. S. Bueno. 2009. Higher crop performance of rice hybrids than of elite inbreds in the tropics: 2. Does sink regulation, rather than sink size, play a major role? Field Crops Res. 112:238–244.

Lafarge, T., C. Seassau, M. Martin, C. S. Bueno, A. Clément-Vidal, E. Schreck, et al. 2010. Regulation and recovery of sink strength in rice plants grown under changes in light intensity. Funct. Plant Biol. 37:413–428.

Li, H. B., Y. X. Hu, D. Z. Bai, T. Y. Kuang, F. Zhou, and J. X. Lin. 2002. Comparison of chloroplast ultrastructure and 77 Kfluorescence emission spec- tra between awns and flag leaves in wheat. J. Chin. Elect. Microsc. Soc. 21:97–101.

Li, X. J., H. G. Wang, H. B. Li, L. Y. Zhang, N. Teng, Q. Q. Lin, et al. 2006. Awns play a dominant role in carbohydrate production during the grain-filling stages in wheat (Triticum aestivum). Physiol. Plant. 127:701–709.

Li, H., Y. Chen, X. Liang, Y. Lian, and W. Li. 2009. Mineral-nitrogen leaching and ammonia volatilization from a rice-rapeseed system as affected by 3,4-dimethylpyrazole phosphate. J. Environ. Qual. 38:2131–2137.

Long, S. P., X. Zhu, S. L. Naidu, and D. R. Ort. 2006. Can improvement in photosynthesis increase crop yields? Plant, Cell Environ. 29:315–330.

Lopes, M. S., and M. P. Reynolds. 2010. Partitioning of assimilates to deeper roots is associated with cooler canopies and increased yield under drought in wheat. Funct. Plant Biol. 37:147–156.

Lopes, M. S., and M. P. Reynolds. 2012. Stay-green in spring wheat can be determined by spectral reflectance measurements (normalized difference vegetation index) independently from phenology. J. Exp. Bot. 63:3789–3798.

Lü, C. G., J. S. Zou, N. Hu, and K. M. Yao. 2007. Effects of leaf morphological factors on canopy structure and transmission of photosynthetically active radiation in rice. Jiangsu J. Agric. Sci. 23:501–508.

Luo, J., H. Liu, T. Gu, X. Huang, Y. Shangguan, J. Zhu, et al. 2013. An–1 encodes a basic helix-loop-helix protein that regulates awn development, grain size, and grain number in rice. Plant Cell 25:3360–3376.

Makino, A. 2011. Photosynthesis, grain yield, and nitrogen utilization in rice and wheat. Plant Physiol. 155:125–129.

Makino, A., T. Mae, and K. Ohira. 1988. Differences between wheat and rice in the enzymic properties of ribulose-1,5-bisphosphate carboxylase/oxygenase and the relationship to photosynthetic gas exchange. Planta 174:30–38.

Manès, Y., H. F. Gomez, L. Puhl, M. Reynolds, H. J. Braun, and R. Trethowan. 2012. Genetic yield gains of the CIMMYT international semi-arid wheat yield trials from 1994 to 2010. Crop Sci. 52:1543.

Maydup, M. L., M. Antonietta, J. J. Guiamet, and E. A. Tambussi. 2012. The contribution of green parts of the ear to grain filling in old and modern cultivars of bread wheat (*Triticum aestivum* L.): evidence for genetic gains over the past century. Field Crops Res. 134:208–215.

Meyer, R. C., M. Steinfath, J. Lisec, M. Becher, H. Witucka-Wall, O. Törjék, et al. 2007. The metabolic signature related to high plant growth rate in *Arabidopsis thaliana*. Proc. Natl Acad. Sci. USA 104:4759–4764.

Mitchell, P., J. Sheehy, and F. Woodward. 1998. potential yields and the efficiency of radiation use in rice. IRRI Discuss Pap SEr No 32 62

Mohapatra, P. K., R. Panigrahi, and N. C. Turner. 2011. Physiology of spikelet development on the rice panicle: is manipulation of apical dominance crucial for grain yield improvement?. Advances in Agronomy 110: 333–359.

Morgounov, A., V. Zykin, I. Belan, L. Roseeva, Y. Zelenskiy, H. F. Gomez-Becerra, et al. 2010. Genetic gains for grain yield in high latitude spring wheat grown in Western Siberia in 1900–2008. Field Crops Res. 117:101–112.

Muurinen, S., G. A. Slafer, and P. Peltonen-Sainio. 2006. Breeding effects on nitrogen use efficiency of spring cereals under northern conditions. Crop Sci. 46:561.

Nagai, T., and A. Makino. 2009. Differences between rice and wheat in temperature responses of photosynthesis and plant growth. Plant Cell Physiol. 50:744–755.

Nebauer, S. G., B. Renau-Morata, J. L. Guardiola, and R. V. Molina. 2011. Photosynthesis down-regulation precedes carbohydrate accumulation under sink limitation in Citrus. Tree Physiol. 31:169–177.

Nelson, C. 1988. Genetic associations between photosynthetic characteristics and yield: review of the evidence. Plant Physiol. Biochem. 26:543–554.

Nuccio, M. L., J. Wu, R. Mowers, H.-P. Zhou, M. Meghji, L. F. Primavesi, et al. 2015. Expression of trehalose-6-phosphate phosphatase in maize ears improves yield in well-watered and drought conditions. Nat. Biotechnol. 33:862–869.

O'Hara, L. E., M. J. Paul, and A. Wingler. 2013. How do sugars regulate plant growth and development? New insight into the role of trehalose-6-phosphate. Mol. Plant 6:261–274.

Ohsumi, A., A. Hamasaki, H. Nakagawa, H. Yoshida, T. Shiraiwa, and T. Horie. 2007. A model explaining genotypic and ontogenetic variation of leaf photosynthetic rate in rice (*Oryza sativa*) based on leaf nitrogen content and stomatal conductance. Ann. Bot. 99:265–273.

Ortiz-Monasterio, J., K. Sayre, S. Rajaram, and M. McMahon. 1997. Genetic progress in wheat yield and nitrogen use efficiency under four nitrogen rates. Crop Sci. 37:898–904.

Panda, B. B., E. Kariali, R. Panigrahi, and P. K. Mohapatra. 2009. High ethylene production slackens seed filling in compact panicled rice cultivar. Plant Growth Regul. 58:141–151.

Parent, B., and F. Tardieu. 2012. Temperature responses of developmental processes have not been affected by breeding in different ecological areas for 17 crop species. New Phytol. 194:760–774.

Pariasca Tanaka, J., P. Nardi, and M. Wissuwa. 2010. Nitrification inhibition activity, a novel trait in root exudates of rice. AoB Plants 2010:plq014.

Parry, M. A. J., M. Reynolds, M. E. Salvucci, C. Raines, P. J. Andralojc, X.-G. Zhu, et al. 2011. Raising yield potential of wheat. II. Increasing photosynthetic capacity and efficiency. J. Exp. Bot. 62:453–467.

Pasuquin, E., T. Lafarge, and B. Tubana. 2008. Transplanting young seedlings in irrigated rice fields: early and high tiller production enhanced grain yield. Field Crops Res. 105:141–155.

Peng, S. 2000. Single-leaf and canopy photosynthesis of rice. Redesigning Rice Photosynth. Increase Yield Proc. Workshop Quest Reduce Hunger. Rice Photosynth. Elsevier, Pp 213–228.

Peng, S., and A. M. Ismail. 2004. Physiological basis of yield and environmental adaptation in rice. Pp. 72–118 in H. T. Nguyen and A. Blum, eds. Physiology and biotechnology integration for plant breeding. CRC Press, New York, Basel.

Peng, S., K. Cassman, S. Virmani, J. Sheehy, and G. Khush. 1998. Yield potential trends of tropical rice since the release of IR8 and the challenge of increasing rice yield potential. Crop Sci. 39:1552–1559.

Peng, S., R. C. Laza, R. M. Visperas, A. L. Sanico, K. G. Cassman, and G. S. Khush. 2000. Grain yield of rice cultivars and lines developed in the Philippines since 1966. Crop Sci. 40:307.

Peng, S., J. Yang, M. R. C. Laza, A. Sanico, R. M. Visperas, and T. T. Son. 2003. Physiological basis of heterosis amd crop management strategies for hybrid rice in the tropics. Pp. 153–170 in S. S. Virmani, C. X. Mao and B. Hardy, eds. Hybrid rice for food security, poverty alleviation, and environmental protection. Proceedings of the 4th International Symposium on Hybrid Rice, Hanoi, Vietnam, 14–17 May 202. IRRI, Los Banos (Phillipines).

Peng, S., G. S. Khush, P. Virk, Q. Tang, and Y. Zou. 2008. Progress in ideotype breeding to increase rice yield potential. Field Crops Res. 108:32–38.

Peraudeau, S., T. Lafarge, S. Roques, C. O. Quinones, A. Clément-Vidal, P. B. F. Ouwerkerk, et al. 2015a. Effect of carbohydrates and night temperature on night respiration in rice. J. Exp. Bot. 66:3931–3944.

Peraudeau, S., S. Roques, C. O. Quinones, D. Fabre, J. Van Rie, P. B. F. Ouwerkerk, et al. 2015b. Increase in night temperature in rice enhances respiration rate without significant impact on biomass accumulation. Field. Crop. Res. 171:67–78.

Pinto, R. S., and M. P. Reynolds. 2015. Common genetic basis for canopy temperature depression under heat and drought stress associated with optimized root distribution in bread wheat. TAG Theor. Appl. Genet. Theor. Angew. Genet. 128:575–585. doi:10.1007/s00122-015-2453-9.

Pinto, R. S., M. P. Reynolds, K. L. Mathews, C. L. McIntyre, J.-J. Olivares-Villegas, and S. C. Chapman. 2010. Heat and drought adaptive QTL in a wheat population designed to minimize confounding agronomic effects. Theor. Appl. Genet. 121:1001–1021. doi:10.1007/s00122-010-1351-4.

Pinto, R. S., and M. P. Reynolds. 2015. Common genetic basis for canopy temperature depression under heat and drought stress associated with optimized root distribution in bread wheat. Theor. Appl. Genet. 128:575–585.

Pozo, A., I. Matus, M. D. Serret, and J. L. Araus. 2014. Agronomic and physiological traits associated with breeding advances of wheat under high-productive Mediterranean conditions. The case of Chile. Environ. Exp. Bot. 103:180–189.

Rajaram, S., H. J. Braun. 2008. Wheat yield potential. *in* eds. Reynolds et al. International symposium on wheat yield potential: challenges to international wheat breeding. Mexico DF, Mexico.

Rebetzke, G. J., A. F. van Herwaarden, C. Jenkins, M. Weiss, D. Lewis, S. Ruuska, et al. 2008. Quantitative trait loci for water-soluble carbohydrates and associations with agronomic traits in wheat. Aust. J. Agric. Res. 59:891–905.

Rebetzke, G. J., D. G. Bonnett, and M. H. Ellis. 2012. Combining gibberellic acid-sensitive and insensitive dwarfing genes in breeding of higher-yielding, semi-dwarf wheats. Field. Crop. Res. 127:17–25.

Reynolds, M., F. Dreccer, and R. Trethowan. 2007. Drought-adaptive traits derived from wheat wild relatives and landraces. J. Exp. Bot. 58:177–186.

Reynolds, M., M. J. Foulkes, G. A. Slafer, P. Berry, M. A. J. Parry, J. W. Snape, et al. 2009. Raising yield potential in wheat. J. Exp. Bot. 60:1899–1918.

Reynolds, M., et al. , 2015. Pre-breeding for yield potential traits: summary of international Data from 2nd WYCYT and performance of pipeline material. *in* M. Reynolds et al., eds. Proceedings of the International TRIGO (Wheat) Yield Potential Workshop 2015. CENEB, CIMMYT, Cd Obregon, Sonora, Mexico 24–26th March 2015. CIMMYT, Mexico DF.

Royo, C., F. Álvaro, V. Martos, A. Ramdani, J. Isidro, D. Villegas, et al. 2006. Genetic changes in durum wheat yield components and associated traits in Italian and Spanish varieties during the 20th century. Euphytica 155:259–270.

Sadras, V. O., and C. Lawson. 2011. Genetic gain in yield and associated changes in phenotype, trait plasticity and competitive ability of South Australian wheat varieties released between 1958 and 2007. Crop Pasture Sci. 62:533–549.

Sadras, V. O., M. P. Reynolds, A. J. de la Vega, P. R. Petrie, and R. Robinson. 2009. Phenotypic plasticity of yield and phenology in wheat, sunflower and grapevine. Field Crops Res. 110:242–250.

Saitoh, K., K. Yonetani, T. Murota, and T. Kuroda. 2002. Effects of flag leaves and panicles on light interception and canopy photosynthesis in high-yielding rice cultivars. Plant Prod. Sci. 5:275–280.

Samonte, S. O. P., L. T. Wilson, J. C. Medley, S. R. M. Pinson, A. M. McClung, and J. S. Lales. 2006. Nitrogen utilization efficiency. Agron. J. 98:168.

Sanchez-Bragado, R., G. Molero, M. P. Reynolds, and J. L. Araus. 2014. Relative contribution of shoot and ear photosynthesis to grain filling in wheat under good agronomical conditions assessed by differential organ δ13C. J. Exp. Bot. 65:5401–5413.

Setter, T. L., E. S. Ella, and A. P. Valdez. 1994. Relationship between coleoptile elongation and alcoholic fermentation in rice exposed to anoxia. II. Cultivar differences. Ann. Bot. 74:273–279.

Setter, T., E. Conocono, J. Egdane, and M. Kropff. 1995. Possibility of increasing yield potential of rice by reducing panicle height in the canopy. I. Effects of panicles on light interception and canopy photosynthesis. Funct. Plant Biol. 22:441–451.

Sharma, R. C., J. Crossa, G. Velu, J. Huerta-Espino, M. Vargas, T. S. Payne, et al. 2012. Genetic gains for grain yield in CIMMYT spring bread wheat across international environments. Crop Sci. 52:1522.

Shearman, V., R. Sylvester-bradley, R. Scott, and M. Foulkes. 2005. Physiological processes associated with wheat yield progress in the UK. Crop Sci. 45:175–185.

Shimoda, S. 2012. Contribution of leaf nitrogen to photosynthetic gas exchange in contrasting rice (*Oryza sativa* L.) cultivars during the grain-filling period. Photosynthetica. 50: 387–394.

Shingaki-Wells, R. N., S. Huang, N. L. Taylor, A. J. Carroll, W. Zhou, and A. H. Millar. 2011. Differential molecular responses of rice and wheat coleoptiles to anoxia reveal novel metabolic adaptations in amino acid metabolism for tissue tolerance. Plant Physiol. 156:1706–1724.

Shiratsuchi, H., T. Yamagishi, and R. Ishii. 2006. Leaf nitrogen distribution to maximize the canopy photosynthesis in rice. Field Crops Res. 95:291–304.

Simpson, R. J., H. Lambers, and M. J. Dalling. 1983. Nitrogen redistribution during grain growth in wheat (*Triticum aestivum* L.). Plant Physiol. 71:7–14.

Sorrells, M. E., M. La Rota, C. E. Bermudez-Kandianis, R. A. Greene, R. Kantety, J. D. Munkvold, et al. 2003. Comparative DNA sequence analysis of wheat and rice genomes. Genome Res. 13:1818–1827.

Subbarao, G. V., O. Ito, K. L. Sahrawat, W. L. Berry, K. Nakahara, T. Ishikawa, et al. 2006. Scope and strategies for regulation of nitrification in agricultural systems—challenges and opportunities. Crit. Rev. Plant Sci. 25:303–335.

Subbarao, G. V., B. Tomohiro, K. Masahiro, I. Osamu, H. Samejima, H. Y. Wang, et al. 2007. Can biological nitrification inhibition (BNI) genes from perennial *Leymus racemosus* (Triticeae) combat nitrification in wheat farming? Plant Soil 299:55–64.

Sylvester-Bradley, R., P. Riffkin, and G. O'Leary. 2012. Designing resource-efficient ideotypes for new cropping conditions: wheat (Triticum aestivum L.) in the High Rainfall Zone of southern Australia. Field Crops Res. 125:69–82.

Tabien, R. E., S. O. P. Samonte, and A. M. McClung. 2008. Forty-eight years of rice improvement in texas since the release of cultivar bluebonnet in 1944. Crop Sci. 48:2097.

Taylaran, R. D., S. Adachi, T. Ookawa, H. Usuda, and T. Hirasawa. 2011. Hydraulic conductance as well as nitrogen accumulation plays a role in the higher rate of leaf photosynthesis of the most productive variety of rice in Japan. J. Exp. Bot. 62:4067–4077.

Thomas, H., and C. Howarth. 2000. Five ways to stay green. J. Exp. Bot. 51:329–337.

Thorne, G. N., and D. W. Wood. 1987. Effects of radiation and temperature on tiller survival, grain number and grain yield in winter wheat. Ann. Bot. 59:413–426.

Tirol-Padre, A., J. K. Ladha, U. Singh, E. Laureles, G. Punzalan, and S. Akita. 1996. Grain yield performance of rice genotypes at suboptimal levels of soil N as affected by N uptake and utilization efficiency. Field Crops Res. 46:127–143.

Tomas, M., J. Flexas, L. Copolovici, J. Galmes, L. Hallik, H. Medrano, et al. 2013. Importance of leaf anatomy in determining mesophyll diffusion conductance to CO_2

across species: quantitative limitations and scaling up by models. J. Exp. Bot. 64:2269–2281.

Underdahl, J. L., M. Mergoum, J. K. Ransom, and B. G. Schatz. 2008. Agronomic traits improvement and associations in hard red spring wheat cultivars released in North Dakota from 1968 to 2006. Crop Sci. 48:158.

Valluru, R., M. P. Reynolds, and J. Salse. 2014. Genetic and molecular bases of yield-associated traits: a translational biology approach between rice and wheat. Theor. Appl. Genet. 127:1463–1489.

Vergara, B., and T. Chang. 1985. The flowering response of the rice plant to photoperiod. 4th ed. IRRI, Los Banos.

Wada, G., and S. Matsushima. 1962. Analysis of yield-determining process and its application to yield-production and culture improvement of lowland rice. LXII. On the mechanisms of determining the number of spikelets. Proc. Crop. Sci. Soc. Jpn. 31:23–25.

Wang, R. F., D. G. An, C. S. Hu, L. H. Li, Y. M. Zhang, Y. G. Jia, et al. 2011. Relationship between nitrogen uptake and use efficiency of winter wheat grown in the North China Plain. Crop Pasture Sci. 62:504–514.

Wang, Z., S. Kang, C. R. Jensen, and F. Liu. 2012. Alternate partial root-zone irrigation reduces bundle-sheath cell leakage to CO2 and enhances photosynthetic capacity in maize leaves. J. Exp. Bot. 63:1145–1153.

Wassmann, R., S. V. K. Jagadish, S. Heuer, A. Ismail, E. Redona, R. Serraj, et al. 2009. Climate change affecting rice production: the physiological and agronomic basis for possible adaptation strategies. Academic Press, Advances in Agronomy, Vol 101. Burlington: Academic Press, 2009, pp.59–122.

Wojciechowski, T., M. J. Gooding, L. Ramsay, and P. J. Gregory. 2009. The effects of dwarfing genes on seedling root growth of wheat. J. Exp. Bot. 60: 2565–2573.

Wu, L., Z. Liu, J. Wang, C. Zhou, and K. Chen. 2011. Morphological, anatomical, and physiological characteristics involved in development of the large culm trait in rice. Aust. J. Crop Sci. 5:1356–1363.

Xiao, Y. G., Z. G. Qian, K. Wu, J. J. Liu, X. C. Xia, W. Q. Ji, et al. 2012. Genetic gains in grain yield and physiological traits of winter wheat in Shandong Province, China, from 1969 to 2006. Crop Sci. 52:44.

Yadav, U. P., A. Ivakov, R. Feil, G. Y. Duan, D. Walther, P. Giavalisco, et al. 2014. The sucrose-trehalose 6-phosphate (Tre6P) nexus: specificity and mechanisms of sucrose signaling by Tre6P. J. Exp. Bot. 65:1051–1068.

Yamauchi, M. 1994. Physiological bases for higher yield potential in F1 hybrids. Pp. 71–80 in S. S. Virmani, ed. Hybrid rice technology: new developments and future prospects. International Rice Research Institute, Los Baños, Philippines.

Yang, J., and J. Zhang. 2006. Grain filling of cereals under soil drying. New Phytol. 169:223–236.

Yang, W., S. Peng, R. C. Laza, R. M. Visperas, and M. L. Dionisio-Sese. 2007. Grain yield and yield attributes of new plant type and hybrid rice. Crop Sci. 47:1393.

Yeo, M. E., A. R. Yeo, and T. J. Flowers. 1994. Photosynthesis and photorespiration in the genus Oryza. J. Exp. Bot. 45:553–560.

Yin, X., M. J. Kropff, P. K. Aggarwal, S. Peng, and T. Hone. 1997a. Optimal preflowering phenology of irrigated rice for high yield potential in three Asian environments: a simulation study. Field Crops Res. 51:19–27.

Ying, J., S. Peng, Q. He, H. Yang, C. Yang, R. M. Visperas, et al. 1998. Comparison of high-yield rice in tropical and subtropical environments: I. Determinants of grain and dry matter yields. Field Crops Res. 57:71–84.

Yuan, W., S. Peng, C. Cao, P. Virk, D. Xing, Y. Zhang, et al. 2011. Agronomic performance of rice breeding lines selected based on plant traits or grain yield. Field Crops Res. 121:168–174.

Zhang, W.-H., and M. Kokubun. 2004. Historical changes in grain yield and photosynthetic rate of rice cultivars released in the 20th century in Tohoku region. Plant Prod. Sci. 7:36–44.

Zhang, Y., Q. Tang, Y. Zou, D. Li, J. Qin, S. Yang, et al. 2009. Yield potential and radiation use efficiency of "super" hybrid rice grown under subtropical conditions. Field Crops Res. 114:91–98.

Zhao, M., T. L. B. Acuña, H. R. Lafitte, G. Dimayuga, and E. Sacks. 2008. Perennial hybrids of Oryza sativa × Oryza rufipogon Part II. Carbon exchange and assimilate partitioning. Field Crops Res. 106:214–223.

Zheng, B., L. Shi, Y. Ma, Q. Deng, B. Li, and Y. Guo. 2008. Comparison of architecture among different cultivars of hybrid rice using a spatial light model based on 3-D digitising. Funct. Plant Biol. 35:900–910.

Zheng, T. C., X. K. Zhang, G. H. Yin, L. N. Wang, Y. L. Han, L. Chen, et al. 2011. Genetic gains in grain yield, net photosynthesis and stomatal conductance achieved in Henan Province of China between 1981 and 2008. Field Crops Res. 122:225–233.

Zhou, B., Á. Sanz-Sáez, A. Elazab, T. Shen, R. Sánchez-Bragado, J. Bort, et al. 2014. Physiological traits contributed to the recent increase in yield potential of winter wheat from Henan Province, China. J. Integr. Plant Biol. 56:492–504.

Zhu, X. G., S. P. Long, and D. R. Ort. 2010. Improving photosynthetic efficiency for greater yield. Ann. Rev. Plant Biol. 61:235–261.

13

Design and characterization of spatial units for monitoring global impacts of environmental factors on major crops and food security

René Gommes, Bingfang Wu, Zhongyuan Li & Hongwei Zeng

CAS Key Laboratory of Digital Earth Science, Institute of Remote Sensing and Digital Earth (RADI), Chinese Academy of Sciences, Beijing 100101, China

Keywords

Crop distribution, crop monitoring, environmental variability, global agriculture, spatial units.

Correspondence

Bingfang Wu, CAS Key Laboratory of Digital Earth Science, Institute of Remote Sensing and Digital Earth (RADI), Chinese Academy of Sciences, Beijing 100101, China.

E-mail: wubf@radi.ac.cn

Funding Information

In addition to regular program funding, the Digital Agriculture Division, Remote Sensing and Digital Earth Institute (RADI), Chinese Academy of Sciences received support from the special Fund for Grains-scientific Research in the Public Interest (Grants No. 201313009-2 and No. 201413003-7), the National High Technology Research and Development Program of China (863 program), Grant No. 2012AA12A307. The first author is supported by Grant No. 2013T1Z0016 for Visiting Professorships for Senior International Scientists.

Abstract

This paper describes the design and characterization of a limited number of areas to be used as spatial monitoring and reporting units (MRU) by Crop-Watch, the global crop monitoring system (http://www.cropwatch.com.cn/) hosted by the Chinese Academy of Sciences. The MRUs are modified and adapted from an existing standard vegetation zoning map. They are designed to be areas of uniform vulnerability as assessed by three sets of variables: environmental (such as climate and NDVI, a satellite-based Normalized Difference Vegetation Index), seasonality (such as intra-annual climate variations and interannual NDVI variability) and agronomy (such as presence/absence of major crops and their yield). The paper pays due attention to within-zone spatial variability for each of the three groups of variables: in general, the variability measured as the average of the spatial coefficient of variation inside MRUs, is much larger for agronomic variables (330%) than for environmental variables (38%) and seasonality (55%). The MRUs provide a rather coherent picture of the variations of global agriculture, for instance, the links between the distribution of crops, agricultural production potential, and environmental variability (over space and time). They closely delineate the distribution of the major nonsugar food crops (barley, cassava, maize, potatoes, rice, soybean, and wheat). The discussion focuses on the relations between the size of the MRUs and the within-MRU variability (spatial heterogeneity) of cropping, and environmental conditions including seasonality. The conclusion stresses that spatial variability in agriculture is bound to be larger than the variability in the environmental variables used to define the units, regardless of the units' size. The size of the spatial units (and therefore their number) is not a very critical constraint for operational impact reporting, especially if impact indicators focus on agricultural areas inside MRUs.

Introduction

Most definitions of "monitoring" and "crop monitoring" refer to the regular and standardized collection and analysis of information over a specific area, often by an external and neutral observer, with a view to understanding the causes of, preventing or limiting damage (Babu and Quinn 1994; Schmid 1998; OECD 2007; Snodderly 2011). The information collected for impact assessments includes both impacting factors and impacted system. For global crop monitoring and reporting, some generalization of methods is required because of the gap between the scale of the observations (whether points or large pixels) and the scale at which impacts are assessed and decisions are taken, usually administrative or other large units (Dalgaard et al. 2003).

There exists a number of different global monitoring units, such as in the developed and developing countries, Southern and Northern Hemisphere, continents, countries at risk, etc. The most popular monitoring or reporting unit is the country since it is consistent with statistical data, economic, and policy conditions, even if inhomogeneous for impacting factors. Size also varies a lot from small countries such as Singapore to the largest one (Russia, more than 9000 times larger than Singapore).

When designing a reporting unit, adequate indicators and monitoring units are interlinked (Wu et al. 2015). Ideal indicators are both systemic and normative (Binder et al. 2010). Systemic: they organically relate to the impacted systems (reporting unit); normative: they can be used to intercompare different spatial units or the same spatial unit at different times. While impacting factors may vary a lot spatially, the impacted systems or reporting unit definition can be designed in such a way that they are homogeneous, that is, the same impacting factors will yield a similar impact over most of the impacted systems. In other terms, the vulnerability patterns are expected to be uniform over the monitoring units.

Impacts, which are the object of monitoring, are brought about by a variety of factors, individually or collectively (Zhang et al. 2013). They include economic conditions (cost of oil and fertilizer: Chen et al. 2010; Kelly 2010; Akpan et al. 2012), policies (Lu 2002; Wiggins and Brooks 2010) but mostly, directly or indirectly, weather (Gornall et al. 2010; Reynolds 2010; Iizumi et al. 2014; Leblois et al. 2014) and agronomy, especially mechanization, inputs, improved cultivars and management (Loyce et al. 2012; Yu et al. 2012; Zhang et al. 2013; George 2014b; Rozbicki et al. 2015). The economic and policy factors follow country units in a statistical sense; weather and agronomic factors don't: they mostly follow agroecological zones over long time periods.

There exists a number of different global classifications of agriculture, agricultural land use and agroecological zones. Some of them focus on the environment of specific crops, such as the wheat mega-environments developed by the CGIAR institutes (Hodson and White 2007). Others put more weight on functional, social or environmental aspects (management, biodiversity), exemplified by the vast literature on agroecosystems (Doré et al. 2011). All agroecological zones work is based on the implicit or explicit assumption that environmental variables and the human activities (such as vegetation and agriculture) that depend on them are correlated and behave coherently. The link between them is strong and they constitute a "complex": natural vegetation directly links to dominant crops, which in turn often constitute, of have constituted the basis of agro-economic zones (Sombroek and Gommes 1996; Haberle and Chepstow-Lusty 2000; Gommes et al. 2004).

This paper examines the validity of the concept and the feasibility of global monitoring units for identifying and characterizing a limited number of zones to be used by CropWatch for global monitoring and reporting (Wu et al. 2014), based on an existing standard agricultural zoning map. The zones are characterized by three sets of variables: agronomic (such as presence/absence of major crops and their yield), environmental (such as climate) and seasonality (such as intra-annual climate or NDVI variability). The paper also pays attention to within-zone spatial variability in order to identify those where impacts may be inhomogeneous.

Data

Spatial data

The CropWatch Mapping and Reporting Units (MRUs) are based on several existing global maps, starting with the FAO Global Ecological Zones map (GEZ), a map designed for reporting forest and forest change statistics in the ambit of the Forest Resources Assessment (FRA; FAO 1999, 2012, downloadable from GeoNetwork, FAO 2014a). The GEZ map is basically a map of natural vegetation types.

Next to a recent Köppen climate map (1976–2000 data; Grieser et al. 2006b), the MRUs were also assessed against global land use and ecosystem maps and especially the suitability for agriculture maps from the ongoing Atlas of the Biosphere project, established in 2002 (SAGE, 2002). It is also in order to mention several global products primarily developed for climate change impact studies (Ramankutty et al. 2002; Monfreda et al. 2008, 2009), as well as most of the global zoning schemes described by van Wart et al. (2013), and others (including an array of current and potential land use maps available in digital form as grids or polygons from FAO, 2014a).

For China, the standard ecological zones of Sun (1994; of which an English language version is available in Hu and Zhang 2006) were integrated into the MRU map.

Impact factors

Impact factors are classified into three categories for charactering MRUs: agronomic (such as presence/absence of major crops and their yield), environmental (such as climate) and seasonality (such as intra-annual climate or NDVI variability). Impacting factor values and statistics are referred to whole MRUs (e.g., wheat yield in the Pampas), even if they "exist" only in cropped areas. All variables as well as their sources and spatial resolution ("pixel size") are succinctly described in Table 1.

Agronomic variables

Agronomic variables are prefixed with A, for example, A1: Arland%, average of the percentage of each pixel that

Table 1. List of variables used to characterize MRUs. Type: A, agronomic; E: Environmental; S, seasonality. As a rule, the values in the table are spatial averages over all grid points inside the MRUs. Period: the period to which the data refer ("Current" means that the data are continually updated). "Pixel" is the grid size (in km) of the original rasters provided by the respective Sources.

Type	Name	Unit	Definition	Pixel	Period	Source
A1	Arland%	0–100	Percent of arable land per pixel	23	Current	IIASA
A2	Irr%	0–100	Percent of pixel area equipped for irrigation	8	1990–2013	GMIA
A3	Bar%	0–100	Percent of pixels where barley is cultivated	7	Current	JRC
A4	Baryld	t/ha	Average barley yield	57	2008–2012	CW-1
A5	Casyld	t/ha	Average cassava yield	57	2008–2012	CW-1
A6	Mz%	0–100	Percent of pixels where maize is cultivated	3	Current	JRC
A7	Mzyd	t/ha	Average maize yield	57	2008–2012	CW-1
A8	Pot%	0–100	Percent of pixels where potato is cultivated	7	Current	JRC
A9	Potyld	t/ha	Average potato yield	57	2008–2012	CW-1
A10	Rc%	0–100	Percent of pixels where rice is cultivated	7	Current	JRC
A11	Padyld	t/ha	Average paddy yield	57	2008–2012	CW-1
A12	Soy%	0–100	Percent of pixels where soybean is cultivated	7	Current	JRC
A13	Soyyld	t/ha	Average soybean yield	57	2008–2012	CW-1
A14	Wh%	0–100	Percent of pixels where wheat is cultivated	7	Current	JRC
A15	Whyld	t/ha	Average wheat yield	57	2008–2012	CW-1
E1	Area	1000 km^2	Area in thousands of Km^2	n.a.	n.a.	CW-2
E2	Z	m a.s.l.	Altitude	16	n.a.	Worldclim
E3	Rain	mm	Annual rainfall total in mm	16	1950–2000	Worldclim
E4	Tavg	°C	Average annual temperature	16	1961–1990	Climond
E5	Rady	W/m^2	Annual mean radiation total	16	1961–1990	Climond
E6	NDVIavg	NDVI	Average annual NDVI (2 + 5 + 8 + 11)/4	21	1999–2012	VITO
E7	Avm	mmH_2O/m	Easily available soil moisture	24	n.a.	Geo-1
E8	NPP	$gDM/(m^2$ year)	Net Primary production potential	24	1976–2000	Grieser
S1	Rn0510	0–1	Fraction of annual precipitation that falls from May to October	16	1950–2000	CW-3
S2	T02-08	°C	Difference between average February and average August temperature	23	1950–2000	CW-3
S3	TCV	CV	Temperature seasonality, coefficient of variation in monthly temperature	16	1961–1990	Climond
S4	Tamp	°C	Temperature of warmest week minus temperature of coldest week	16	1961–1990	Climond
S5	NDVI02-08	NDVI (no unit)	Average difference between February and August NDVI	21	1999–2012	CW-4
S6	NDVIStDev	NDVI (no unit)	Variability of NDVI over time (standard deviation)	8	1981–2003	Geo-2

Data sources. Climond: Kriticos et al. (2012), and Xu and Hutchinson (2011, 2013); CW-1: see text under 3.1; CW-2: computed based on the MRU polygon using QGIS 2.2.0, http://www.qgis.org/en/docs/; CW-3: computed based on WorldClim grids; CW-4: computed based on VITO (2014); GMIA: Global Map of Irrigation Agriculture, FAO (2013) or Siebert et al. (2013); IIASA: Geo-Wiki (2014); Geo-1: http://data.fao.org/map?entryId=61946540-bdbf-11db-a0f6-000d939bc5d8; Geo-2: http://www.fao.org/geonetwork/srv/en/metadata.show?id=37059; Grieser: Grieser et al. (2006a); JRC: data made available by the EC/JRC MARS unit http://mars.jrc.ec.europa.eu, based on the methodology described by Vancutsem et al. (2013); VITO: VITO (2014); Worldclim: Hijmans et al. (2005).
MRU, monitoring and reporting units; CV, coefficient of variation.

is cultivated; Irr%, average of the percentage of each pixel that is equipped for irrigation; Bar%, percentage of pixels in the MRU where barley is cultivated; Baryld, average yield achieved for barley in the pixels where the crop is cultivated; Casyld, average yield achieved for cassava in the pixels where the crop is cultivated. Similar variables are defined for maize (Mz% and Mzyd), white potatoes (Pot%, Potyld), soybean (Soy%, Soyyld), and wheat (Wh%, Whyld). For rice, the percentage of pixels in the MRU the crop is cultivated (Rc%) is complemented by yield (Padyld) expressed as paddy (husked grain).

Average MRU yields were computed by assigning average 2008–2012 yield from FAOSTAT (FAO 2014b) to country centroids and interpolating them using inverse distance weighting (exponent 2) to a coarse 57 km grid, as shown in Table 1. This is, admittedly, a crude method when applied to large areas with a low density of sample points, but deemed acceptable to estimate 5-year reference "zonal yields." The gridded yields were subsequently used to derive the average and the standard deviation by crop over each MRU. Values were retained only for the MRUs where the crop is actually grown, according to the JRC crop masks (Vancutsem et al. 2013).

In order to compare agronomic performance, a simple "yield index" (YldInd) was derived as follows: for each of the seven crops, yields were ranked from 1 to 52 (potatoes, which occur in 52 MRUs out of 65), 1 to 21 (both cassava and barley, which occur in 21 Crop MRUs) etc.; the highest rank was assigned the highest value. Ranks were then divided by the number of occurrences, resulting in the highest value becoming 1 and the lowest 1 divided by the number of occurrences. The final step was the calculation of the average across the seven crops for each MRU, ignoring crops that do not occur. In the case of New Zealand (only potatoes and barley), only two values were averaged.

Environmental variables

Table 1 lists eight environmental variables (E as prefix), of which the first (E1, area of the MRU) is conventionally assigned to this group. The seven variables E2 to E8 thus include altitude (Z), total annual rainfall (Rain), average annual temperature (Tavg), total annual radiation (Rady), average annual NDVI, easily available soil moisture, and net primary production potential (NPP). The pixel count, spatial average values, and spatial standard deviation were directly extracted from the original grids, using a Mollweide projection for the area.

Seasonality variables

There are six seasonality variables (S-variables): Rn0510, percentage of rainfall falling during the May to October semester (in combination with latitude, the variable indicates whether the MRUs enjoys winter or summer rainfall); T02-08, the difference between average February and August average monthly temperature; TCV, the coefficient of variation of 12 monthly average temperature values; Tamp, the difference of temperature between the warmest week and coldest week (annual thermal amplitude); NDVI02-08, the difference between February and August the average NDVI and finally NDVIStDev, and the standard deviation of average annual NDVI over time. NDVIStDev is a particularly relevant variable as it is the only one that assesses variability over time, that is, a major component of farming risk.

In order to derive a generic measure of environmental variability, some of the S-variables were modified in such a way that high values express high variability, regardless of the hemisphere where the MRU occurs. For instance, the percentage of rainfall during May to October was taken as the absolute departure from 50%, as 50% corresponds to well distributed rainfall over the year. Variables derived as the difference between February and August values were taken as the absolute value to make northern and southern hemisphere comparable. The values were then ranked and the average rank was rescaled in order to provide a variable between 0 (lowest variability) and 1 (highest variability), providing a variability/seasonality indicator (VSI).

Method

Generation of MRUs

The spatial units used for global crop monitoring and reporting by CropWatch are referred to as Crop Monitoring and reporting units (MRUs). They are essentially a simplification of the FAO GEZ map with additional adjustments made based on the recent 1976–2000 Köppen climate map by Grieser et al. (2006b). The following four "rules" were broadly followed. First, a rather low level of detail (equivalent to merging GEZs) was adopted for minor- or nonagricultural areas, such as the Central Northern Andes (MRU 21 in the map of Fig. 1) which actually cover the whole spectrum from hyperarid climate (BW in the Atacama desert) to polar (E) at the highest elevation. Second, inclusions which are "small" relative to the size of the MRU were ignored, for instance the Alps in "non-Mediterranean western Europe" (MRU 60) or the semiarid area (BS) on the border between Karnataka and Andhra Pradesh states in "Southern Asia" (MRU 45). Third, the borders of the GEZs were modified where the map by Grieser and colleagues showed similar climate conditions (due to recent shifts in climate) across GEZ borders. This usually involves climate warming in high latitude or high altitude areas, typically the north-east "appendix" of eastern- central Asia (MRU 52) or the Andean part of the semiarid southern cone (MRU 28). Fourth, some MRUs are "residual" areas, that is, very inhomogeneous hilly terrain where the "inhomogeneity" is actual a defining feature, typically the area between the Black Sea and the Caspian (Caucasus, MRU 29) and the Pamir area (MRU 30). In China, the authors adopted the ecological zones of Sun (1994) which are closer to the new Köppen map than the GEZ map. The GEZ was found to be very largely compatible with the cereal suitability map (FAO 2007) and the map of major crop types which is part of FAO (2010).

Agricultural and nonagricultural areas were included in order to provide global coverage and to include some agriculturally marginal areas, such as low-rainfall rangeland at the edge of deserts.

Characterization of MRUs

The 65 MRUs are identified by a code from M01 to M65. The naming is conventional and was chosen only for easy reference (Fig. 1). ESRI shapefiles as well as the

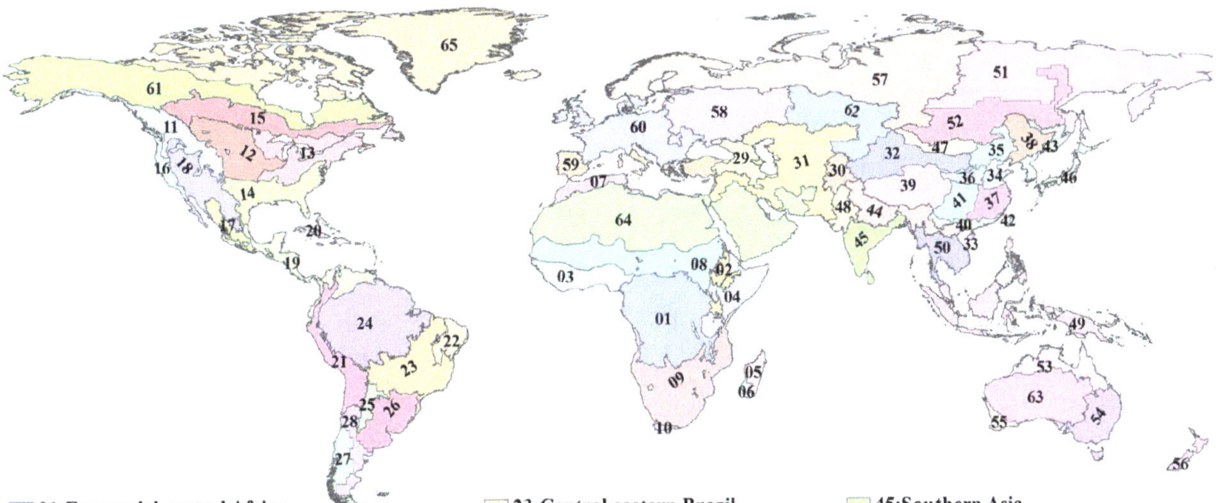

Figure 1. Spatial extent of 65 Crop MRUs designed for global crop monitoring, together with their numbers and conventional "long" names.

values of the variables for individual MRUs are available as supporting information (Tables S1 and S2) as well as from GeoNetwork (FAO 2014a). They can also be requested from the corresponding author.

Principal components

The environmental factors used to characterize MRUs (Table 1) are redundant because they are correlated. For instance, temperature is linked to elevation, and radiation is linked to rainfall through cloudiness. Principal components are therefore used below in several figures (starting with Fig. 3) to characterize the MRUs based on a limited number of synthetic factors.

The matrix of the coefficients of variation

In addition to average MRU values mentioned above, the spatial standard deviation for each variable was computed over the MRU polygons, and the spatial coefficient of variation (CV) was computed. As the CV is a standardized statistic, it can be used for intervariable and inter-MRU comparisons.

The matrices of the CV of MRUs × variables (15 A-variables, 6 S-variables and 8 E-variables of which E1 is area and E2 is elevation) were analyzed. As the matrix of A-variables is incomplete (not all crops are cultivated everywhere), a new matrix was prepared including the coefficients of spatial variation of A1, E2 to E8, S1 to S6 as well as the averages of A-variables (hereafter referred

Table 2. "Top 20" of MRUs for several agronomic variables. Number of crops: number of crops between barley, cassava, maize, potatoes, rice, soybean, wheat that are cultivated in the MRU; YldInd, Yield index (unit less, defined under 2.2.1), percentage of arable land and percentage of irrigation.

	No. of crops cultivated			Yield		Arable land			Irrigation		
	No.	MRU	YldInd	YldInd	MRU	Arland%	MRU	YldInd	Irr%	MRU	YldInd
1	7	M26	0.75	1.00	M56	100	M34	0.58	35	M34	0.58
2	6	M04	0.20	0.96	M60	100	M45	0.45	25	M48	0.40
3	6	M23	0.50	0.94	M46	99	M12	0.91	18	M44	0.24
4	6	M45	0.45	0.91	M12	99	M37	0.54	17	M37	0.54
5	5	M01	0.11	0.84	M15	99	M40	0.47	17	M45	0.45
6	5	M03	0.15	0.81	M28	99	M59	0.78	13	M42	0.58
7	5	M05	0.09	0.81	M61	98	M33	0.53	9	M16	0.77
8	5	M07	0.68	0.80	M18	98	M41	0.57	9	M36	0.65
9	5	M09	0.21	0.78	M59	98	M60	0.96	9	M46	0.94
10	5	M12	0.91	0.77	M16	97	M17	0.65	9	M50	0.51
11	5	M14	0.72	0.75	M26	97	M29	0.47	8	M30	0.58
12	5	M16	0.77	0.74	M58	96	M14	0.72	8	M40	0.47
13	5	M17	0.65	0.72	M14	96	M36	0.65	7	M29	0.47
14	5	M21	0.46	0.71	M27	96	M48	0.40	7	M33	0.53
15	5	M25	0.61	0.70	M13	90	M26	0.75	7	M59	0.78
16	5	M29	0.47	0.68	M54	89	M03	0.15	6	M20	0.06
17	5	M34	0.58	0.68	M07	88	M50	0.51	6	M38	0.49
18	5	M37	0.54	0.65	M36	87	M20	0.06	6	M41	0.57
19	5	M38	0.49	0.65	M17	87	M44	0.24	5	M07	0.68
20	5	M40	0.47	0.64	M57	86	M46	0.94	5	M14	0.72

MRU, monitoring and reporting units; YldInd, yield index; varA, A-variables; varE, E-variables; varS, S-variables; NPP, net primary production.

to as varA), E-variables (varE) and S-variables (varS), resulting in a 65 × 18 matrix.

Results

Map of MRUs

Figure 1 shows the extent and distribution of the 65 MRUs. The largest MRU correspond to mostly nonagricultural areas such as the deserts (M63, Australian desert; M64, old world deserts from the Sahara to the Afghan desert), boreal forest (tundra, taiga and permafrost areas: M51, Eastern Siberia; M57, boreal Eurasia and M65, subarctic America - which includes Greenland and Iceland). There are some other minor instances of transcontinental zones, although they are not easily visible in the map above: M53, North Australia also includes the southern fringe of the Western province in Papua New Guinea and areas in Indonesia and Timor Leste at the same latitude, as already mentioned above.

The smallest zones correspond to the southernmost islands of China (Hainan, M33; China Taiwan, M42). The next smallest area corresponds to Western Cape in South Africa (M10, a Mediterranean climate "enclave"); variables S1 and S6 show that seasons are inverted compared with the surrounding C9, Southern Africa MRU. The largest MRU is about 300 times larger than the smallest ones.

M53 ("North Australia") also includes the southern fringe of maritime South-east Asia; M59 ("Mediterranean Europe and Turkey") includes all of the Iberian Peninsula (including the Atlantic coast) and "Southern Africa" (M09) reaches as far as the coastal areas of southern Kenya.

Characteristics of crops in MRUs

Table 2 summarizes some of the main features about the crops grown in the various MRUs. It focuses on the seven major crops in terms of worldwide production and trade if sugar and oil crops are excluded: barley, cassava, maize, potatoes, rice, soybean and wheat. It is also stressed that the conclusions are only partly true in areas where other crops play an important part (e.g., coarse grains such as oats, or sunflower and oil palm, etc.). Additionally, the spatial variability inside MRUs could also depend on the farm size, field size (e.g., huge fields in USA, Kazakhstan or Russia), the applied agro-techniques (mixed dry farming and irrigation) etc.

Only one zone (M26, the humid Pampas) grows all seven crops, while three grow six: M04, the Horn of Africa; M23, Central-eastern Brazil and M45, southern Asia, including mainly India. The areas are identified by

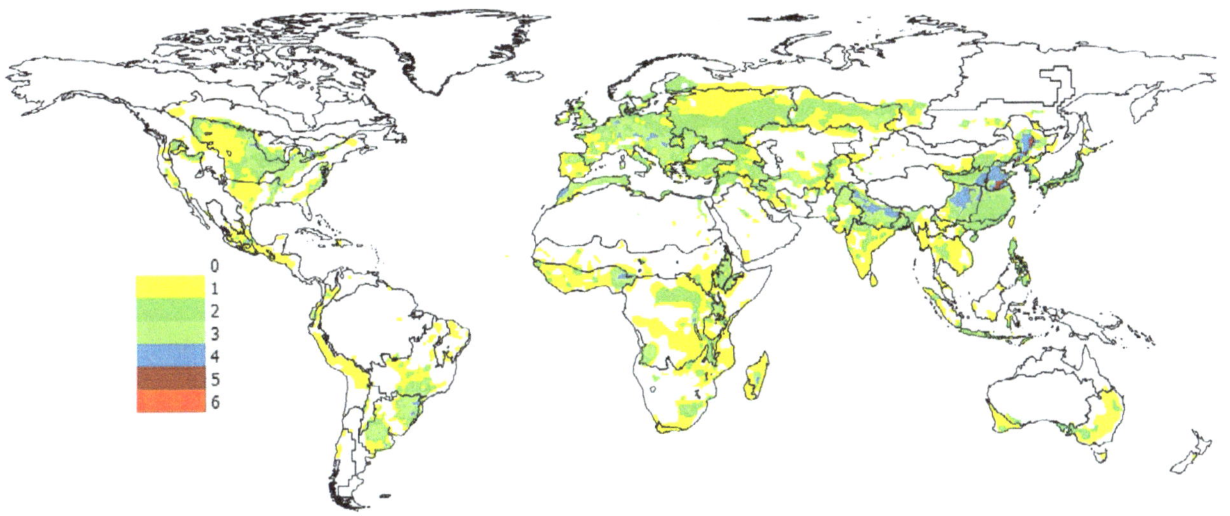

Figure 2. Number of crops (among barley, cassava, maize, potatoes, rice, soybean, wheat) that are gown in various areas. Seven crops occur in the Pampas (M26) but the number is not shown in the legend because the locations are not visible at the scale of the map.

the presence of ecosystems at relatively high elevations which can accommodate both tropical and temperate crops, sometimes with relatively low yields by world standards (e.g., the Horn of Africa).

A number of regions grow five crops, covering the yield spectrum from low (M5, north and central Madagascar; M1, equatorial central Africa and M3, the gulf of Guinea) to high (M12, American northern Great Plains). Figure 2 shows the distribution of the number of crops with additional detail. It is striking how the zones mentioned are all characterized by one or two dominant "background" crops with patches of ecologically diverse areas where additional crops occur.

The highest yields occur in New Zealand (M56), non-Mediterranean Western Europe (M60), Southern Japan and Korea (M46) and in the northern Great Plains in the USA. It is to be stressed that the number of crops in M56 is just two crops (barley and potatoes) which benefit from very long days, resulting in long periods of photosynthesis. The remaining areas cultivate between three and five crops. Lowest yields occur in Equatorial central Africa (M01), the east African highlands (M02), Madagascar (M05) and the Caribbean (M20). It is stressed that the listed crops are marginal in some of the areas, where yams, millets, teff, pigeon peas, plantain, and others play the greatest role as staple foods.

Table 2 also indicates how widespread arable land is in the various production zones. In 15 MRUs the arable land fraction exceeds 90%. Most of them occur in China and India (including M34, Huang Huaihai, southwest China M41 and M45, southern China), but also in western Europe and the northern Mediterranean (M59 and M60), Sierra Madre in Mexico and the US (M17) and the northern Great Plains (M12). Yields are mostly modest (in the YldInd range from 0.5 to 0.6) with the exception of the northern Mediterranean and Turkey (YldInd = 0.78) and Western Europe (0.96).

The observation about modest yields also applies to the main irrigated areas, with values of 0.58 in Huang Huaihai (M34, 35% of the area equipped for irrigation) and 0.40 in Punjab to Gujarat (M48, 25% irrigated).

Environmental features of MRUs

Based on the correlations between them, the seven environmental variables from E2 to E8 (refer to Table 1) can very clearly be subdivided into three groups of factors, corresponding to the three-first principal components. They account for a total of 92% of the variance of the variables, distributed among the components as follows: 49%, 28% and 14%. Figure 3 shows a plot of the production zones against the two-first principal components.

The first component correlates very strongly with average NDVI ($R = -0.91$), NPP ($R = -0.95$) and total rainfall ($R = -0.90$). It can thus be interpreted as "unproductivity" (MLA, 2015) a biological productivity variable, increasing from right to left. The second component PC2 (or "coldness") is clearly and negatively associated with radiation (Rady; $R = -0.93$) and with annual average temperature ($R = -0.71$). More surprisingly, PC2 is positively correlated with the soil moisture storage capacity (E7). The relation between E7, temperature and radiation is not direct, but it can be hypothesized that low temperature is associated slow soil organic matter degradation, as organic matter and other colloids play an important part in the soil moisture storage capacity.

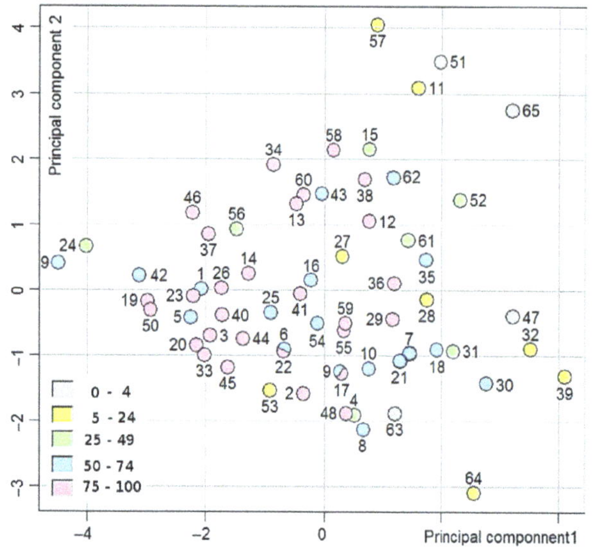

Figure 3. Plots of MRUs (numbers defined in Fig. 1) against the 1st and second principal components extracted from the six environmental variables E2–E7. The first component (unproductivity) corresponds to the production potential (decreasing from left to right) and the second (coldness) captures radiation and temperature (high at the bottom to low at the top). The colors indicate the percentage of arable land in five categories.

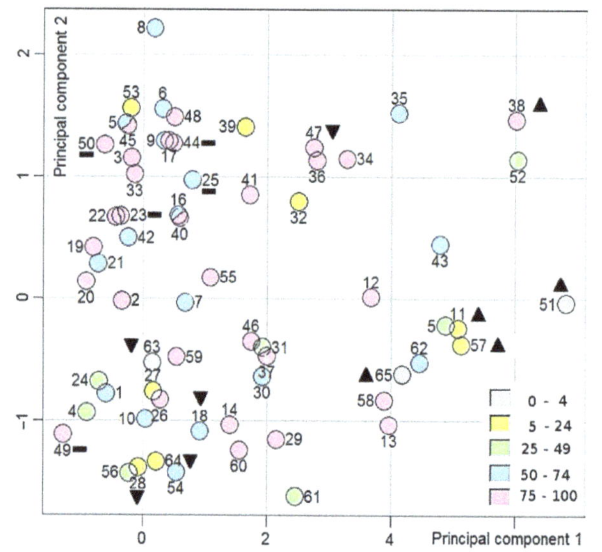

Figure 4. Seasonality in the 65 MRUs represented on the plane of the two-first principal components, where the first captures temperature and NDVI variability and the second represents rainfall seasonality. The colors indicate the percentage of arable land in five categories. Additional symbols (arrows) refer to the variability in NDVI over time: up for the five most variable MRUs, down for the least variable ones and a horizontal bar for average values.

The correlation between E7 and the factors which positively affect soil organic matter production is positive (NDVI: R = 0.42; NPP: R = 0.24 and Rain: R = 0.28) which is an additional argument in favor of the climatic determinants of E7.

An interesting observation refers to elevation, which is the variable most closely correlated with the third principal component (R = 0.73). This is not so surprising since, given the large geographic coverage of the production zones; rainfall and temperature are not correlated with altitude, as would happen for a spatially more detailed study. Therefore, the altitude variable is best regarded as a proxy for "type of landscape" as high elevations tend to be characterized by rugged terrain.

Unfavorable temperature and biomass conditions occur in the ecologically similar M51 (Eastern Siberia) and M65 (subarctic America), both characterized by little or no agriculture. Very favorable conditions of productivity occur in M24 (Amazon) and in C9 (Southern Africa) in spite of the presence of the semiarid Kalahari "desert".

M50 (mainland South-East Asia) and M19 (Central and Northern South America) are among the areas where a high fraction of arable land coincides with a high production potential. It is worth noting that some areas such as M42 (China Taiwan), M05 (northern and central Madagascar) M01 (Equatorial central Africa) do not, at present, fully exploit their potential, which may be due to a variety of factors, including low population densities and topography.

Seasonality and variability over time in MRUs

Based on the six S-variables in Table 1 modified as described under 2.2.3, the 6 × 65 data matrix was subjected to a principal components analysis and the MRUs were plotted against the two-first components (Fig. 4). The 67.8% of the variance is absorbed by the first component which correlates with the temperature-related variables and intra-annual NDVI amplitude (R between 0.89 and 0.96 for N = 65 MRUs); it can conveniently be referred to as "temperature variability". The second (16.8% of variance) is virtually equivalent to rainfall variability (the absolute value of February rain minus August rain, R = 0.99) while the third (13.1% of variance) is best correlated with interannual variability of NDVI, R = 0.62). Together, the three components account for 97.6% of the variance of the original variability/seasonality matrix. The above-mentioned VSI correlates well with temperature variability (R = 0.96). It is interesting to observe that, at the scale of the MRUs, rainfall seasonality, intra-annual temperature variability and interannual NDVI are largely uncorrelated variables.

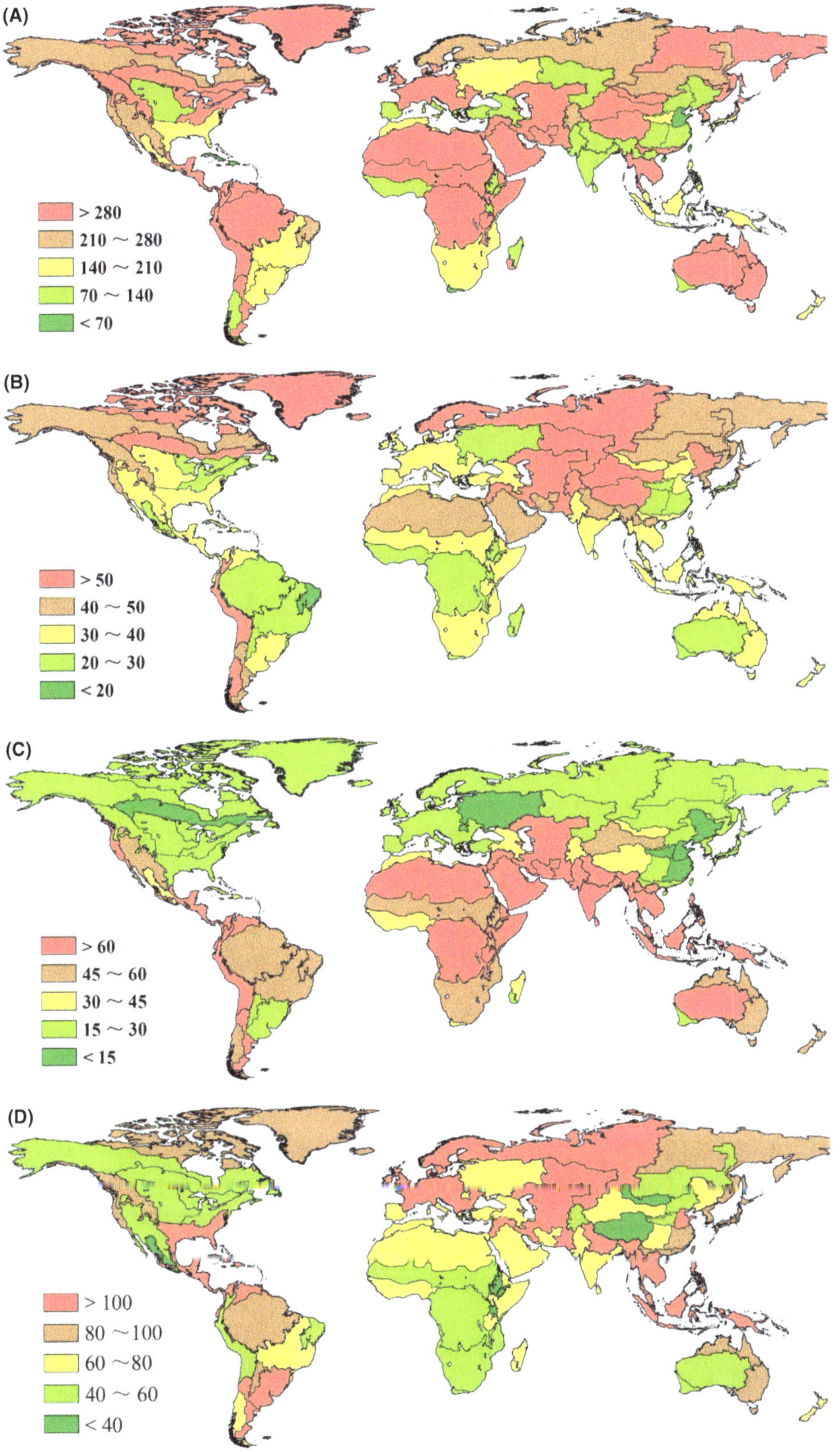

Figure 5. Within-MRU spatial variability (coefficients of variation) of agronomic (A), environmental (B), seasonality (C) variables, and elevation (D).

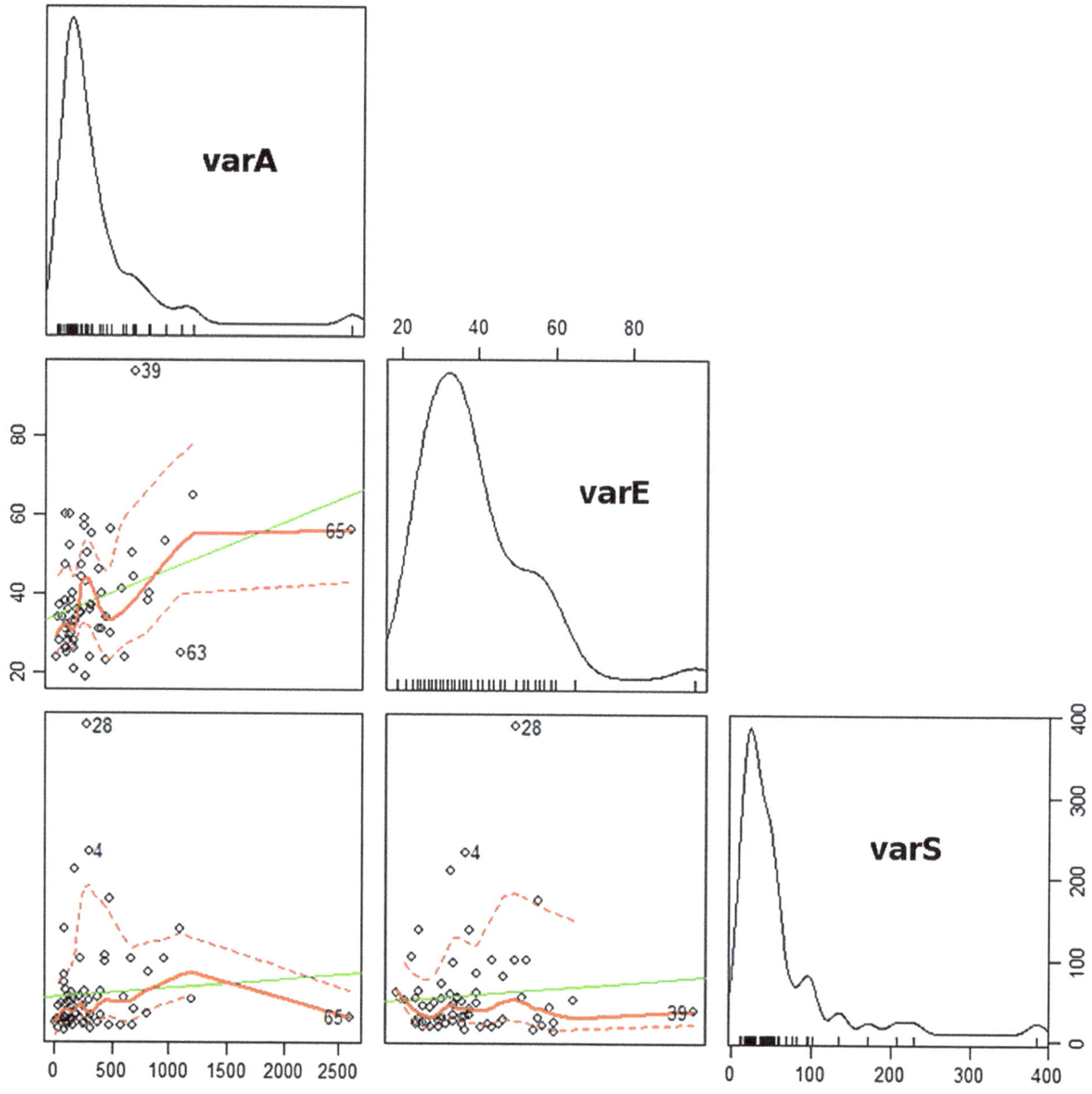

Figure 6. Plots of the average coefficients of spatial variation against each other. The indicated outliers and extremes are M28 (semiarid southern cone in Latin America), M39 (Qinghai-Tibet in China), M63 (Australian desert) and M65 (subarctic America).

rainfall is very low (M64, Sahara; M28, semiarid southern cone) and therefore cannot vary a lot, or rainfall is actually high but well distributed over the year (M56, New Zealand; M60, non-Mediterranean Western Europe). High rainfall seasonality characterizes the West African Sahel (M08), parts of China (North-East China, M38 and Inner Mongolia, M35), South-west Madagascar (M06) etc.

As expected, low temperature intra-annual variability affects equatorial areas (M19, Central America and northern South America; M20, Caribbean and south-east Asia, both continental and the islands, M49 and M50). Most variable temperatures are found in the most continental areas of north-east China (M38), central Asia (M52), and eastern Siberia (M51).

Spatial variability inside MRUs

While the A-, E-, and S-variables are relevant in describing the potential and actual farming in the CropWatch monitoring zones, spatial variability in the same variables needs to be looked at to understand how homogeneous the zones are, which is an important requisite for global monitoring.

A principal components analysis of the matrix of the CV shows that no <11 factors (components) are required to account for 95% of the variance of the matrix. If the original A, E and S variables are added (65 × 38), 18 components are needed to reach the same percentage. VarA is different from varE and varS in the sense that it is based largely on the "presence" (pixel value of 1) and absence (pixel value of 0) of crops. For a given crop, the average thus indicates the fraction of the area of the MRU where the crop is cultivated. As a result, the coefficient of variation is directly depends on the mentioned fraction. For values between 0.3 and 0.9, CV depends directly on the fraction and varies almost linearly from 150% (fraction = 0.3) to 17% (0.9).

The spatial variability in elevation (E2) can probably be regarded as a good yardstick when comparing spatial variability. The lowest values correspond to M39 (Qinghai-Tibet, 19%) and the east African Highlands (M02, 27%) while the five highest values from 132% to 146% correspond to M19 (central and northern South America), M20 (Caribbean), M44 (Southern Himalayas), M49 (maritime south-east Asia) and M62 (Ural to Altai Mountains). As expected, plateaus show low elevation variability while MRUs including lowlands and highlands display high values.

Figure 5 shows how varA, varE and varS vary among the MRUs, in comparison with the spatial variability in elevation, which is the first factor that disturbs zonality. The second is a weak E-W gradient that is visible in Fig. 5B (varE) at high northern latitudes. To interpret the spatial variability in elevation, it is necessary to keep in mind the definition of the coefficient of variation, which explains why highlands such as the East African Highlands (M2) and Qinghai-Tibet (M39) display relatively low values: the average elevation that divides standard deviation is high. On the other hand, the North China Plain (M34, Huang Huaihai) is at a low elevation, but includes some higher terrain in the west.

Western Asia (M31) and the central-northern Andes (M21) are among the most spatially inhomogeneous areas, as the heterogeneity affects all variables. The areas that display low variability for several groups of variables can fall in that category for several reasons. Hainan (M33), for instance, is a tropical island, but other areas (M34, Huang Huaihai) belong there because they are very extensively cropped, resulting in exceptionally low values of spatial variability (11%) for a variable that is very high in most other areas.

Large differences in coefficient of variation are observed between the categories of variables, 330%, 38%, and 55% on average for varA, varE, and varS, respectively. In general, the spatial variability in yield inside the MRUs is low, mostly between 10% and 15%, while the variability in pixels cultivated under various crops or irrigated pixels is very high, usually between 500% and 1000%.

It is also stressed that there is virtually no link between the spatial variability in the three variable types (Fig. 6, but also Fig. 5). In fact, the principal components of the average spatial variability carry 50%, 25%, and 25% of the variance. Figure 6 also shows the significant positive skew affecting all three variability variables.

Among the varA variables, the lowest spatial variability occurs, on average in M33 (11% in Hainan, China) while the most heterogeneous area, from an agronomic point of view, is M65 (2587%, subarctic America). The Sahara for instance, which belongs to M64 together with the Arabian Peninsula and hyperarid regions in Iran and Afghanistan displays a spatial variability of 668% as it is made up by low-variability desert with some irrigated pixels, mostly in the Nile valley.

E-variables, as mentioned, display limited variability, which results from the definition of the MRUs as environmentally homogeneous areas: the variation covers the range from 19% (M22, Brazilian Nordeste) to 96% (M39, Qinghai-Tibet in China), for an average of 38%. The spatial variability in the seasonality variables is larger: 72% on average, from a low value around 10% in M38 (north-east China) and M34 (Huang Huaihai) to values between 200% and 400% (207% in M49, maritime south-east Asia; 230% in M4, the Horn of Africa and 385% in M28, the semiarid southern cone in Latin America).

Discussion

Although primarily based on the natural vegetation GEZ map (FAO 2012), the MRUs shows a good agreement with the spatial distribution of major crops zones (Fig. 2). Both natural ecosystems and crops depend on natural resources, but crops benefit in addition from genetic improvements and management practices that take them to the "limit" of their ecophysiology into environments that would be marginal without human intervention. This is particularly visible in the areas where cropping is most intensive in terms of percentage of arable land and yields achieved.

It is also noted, as expected (Fig. 4), that high potential areas are also areas of low variability, which may result in high cropping intensities, as in continental and maritime SE Asia (M49 and M50) and in Central and northern South America (M19). Altogether, the MRUs thus retain the logic that underlies the GEZ, which is not surprising because agriculture remains a major component of the environment (Monfreda et al. 2008) and because arable land occupies a large fraction of land in most MRUs (Table 2). In fact, according to recent estimates, as much

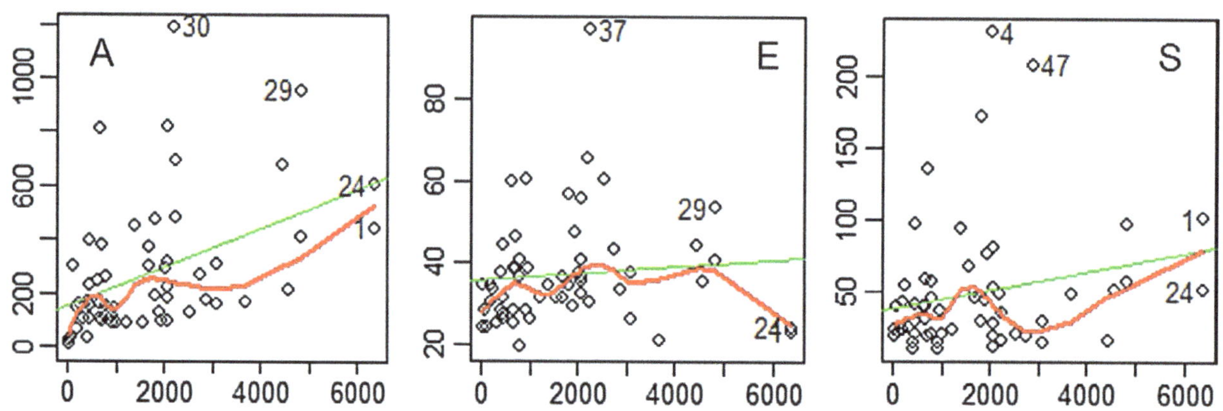

Figure 7. Plot of average coefficient of variation for agricultural (A for varA), environmental (E for varE) and seasonality (S for varS) variables against MRU area, after removal of mostly nonagricultural near-polar and/or arid areas (M27, M28, M51, M57, M61, M63, M64 and M65). The numbered points correspond to equatorial Africa (1, M01), the Amazon (24, M24), western Asia (29, M31), Gansu-Xinjiang (30, M32), Qinghai-Tibet (37, M39) and maritime south-east Asia (47, M49).

Table 3. Comparison of some MRU descriptors as a function of the fraction of arable land. <75%, less than 75% arable (36 MRUs); >75%, more than 75% arable (29 MRUs). R = coefficient of correlation of agricultural variability versus Area, S is the corresponding slope. Definition of variables is as in Table 1. Avg. is average and CV% is the coefficient of variation in % between the respective variables over MRUs belonging to each of the two groups. Reg. stands for the regression of varA against MRU area.

	Reg.		varA	varE	varS	Area	Arland%	Irr%	Z	Rain	Avm	NPP
<75%	R = 0.30	Avg.	446	43	63	2610	39	2	894	693	94	80
	S = 0.05	CV%	105	35	114	97	67	150	94	83	17	64
>75%	R = 0.19	Avg.	185	32	45	1380	91	7	556	1123	100	135
	S = 0.06	CV%	91	25	93	71	9	114	86	50	17	33

MRU, monitoring and reporting units; CV, coefficient of variation; NPP, net primary production.

as 23.8% of the potential global NPP is directly or indirectly managed by man (Haberl et al. 2007).

In addition, natural vegetation and crops are subjected to the same global environmental patterns of not only resources but variability as well, where latitude and topography play a major role. In general, a rather large number of agroecological classes are required to capture spatial variations of agriculture with sufficient detail. What does "sufficient" stand for? van Wart et al. (2013) found that, in general, all global zoning approaches that include less than 100 units retain significant within-unit variability. Indeed, all national agroecological zone studies go into significant detail, as amply documented by the national agro-ecological maps available in GeoNetwork (FAO, 2014a). To take just one example, for India, the agricultural area of which is smaller than China's, Pal et al. (2009) deem that about 130 zones are required for national work, while they encourage states to add additional level for more detailed local studies. In China, on the other hand, the large zones proposed by Sun are just nine, which are about the same size as the MRUs adopted for India. The 281 Food Production Units (FPU) proposed by Nelson et al. (2010) and Rosengrant et al. (2012) were originally designed to assess water-related issues. They result from the intersection of 115 geopolitical regions and 126 hydrographic basins and are generally found acceptable for water-related studies (Kummu et al. 2011). On the other hand, for statistical purposes, the United Nations geoscheme (UN, 2010) adopts just 21 global units, which are also used by OECD.

In reality, the issue of what amount of variability within zones is acceptable is bound to remain subjective as it depends on the intended application. In this study, we found that large differences (as measured by the average coefficient of variation varA, varE, and varS) are observed between the categories of variables: 330% for varA, which refers mostly to crop distribution within MRUs. The value of varA would drop to significantly lower values if the calculations were restricted to arable land (0% would be possible only in the case of one single crop). Spatial variability in yields, on the other hand, is in the range of 10%–15% (no doubt also partly due to the method used to estimate crop yields by MRU).

Environmental variables show a much lower range of spatial variability (average varE is 38%) and so do seasonality variables (average varS is 55%). It is stressed that seasonality is expected to be more variable than the environmental variables as it also includes NDVI variability over time, as a measure of cropping risk.

Figure 7 shows the plot of the three average spatial variability indicators against the MRUs' area, after exclusion of major cold and/or dry noncropped areas. Correlations are all positive, as expected, but significant only for VarA (agricultural variability, R = 0.45, significant below P = 0.01). As far as environmental and seasonality conditions are concerned, the MRUs may be considered homogeneous, which directly results from the derivation of MRUs from the FAO GEZs.

In Fig. 7A, the slope of varA against the area of the MRUs is 0.08 (the figure includes all but seven nonagricultural MRUs). Table 3 shows, among other descriptors from Table 1, that varA is highest in areas with low arable land, contrary to varE and varS, and some other variables. In Table 3 the arable land limit of 75% was chosen because it is close to the median (74%), so that the number of MRUs inside each category is approximately the same. Area with low varA, coincide with high potential areas (high rainfall, high NPP, widespread irrigation) while other variables, including elevation and soil moisture do not play a part that is commensurate with the role of the factors that directly affect NPP (e.g., rainfall).

It is also noted that the correlation between varA and MRU area is no longer significant (significance threshold for 35 observations is $R = 0.32$ at $P \leq 0.05$, and larger for 29 observations as well as for $P \leq 0.01$). In general, areas with high arable land fractions also correspond to higher average annual temperature and higher average NDVI, although this is not shown in Table 3 (17°C vs. 11°C and 0.54 vs. 0.39). On the other hand, the average standard deviation of NDVI over time (S6) and average annual NDVI amplitude (S5) are virtually identical between the groups (0.21–0.22). The fact that the CV between MRUs in the groups in Table 3 is comparable (in particular varA: 105% and 91%) indicates that the "ecological" logic behind the selection of the MRUs remains meaningful for agriculture, across different scales.

Conclusions

The current MRU subdivision by CropWatch adopts 65 units; it is derived from a GEZ map and used for operational monitoring, including areas with very limited or no crop or livestock production.

Spatial variability in environmental and seasonality variables inside MRUs only marginally depends on their size, which stresses their intrinsic homogeneity. They are spatially less homogeneous for agricultural variables.

This is a direct consequence of the low density of agriculture in the areas where the production potential is low; it does not seriously affect the validity of large agroecological zones as monitoring as well as reporting units. Regardless of the size of the MRUs, it is the prevailing environmental conditions that condition vulnerability patterns and the type of farming that can be practiced, an extreme case being irrigated crops in desert areas (e.g., in the Nile valley, or the semiarid areas in central Asia or the hills immediately bordering the Amazon). Even if the number of spatial units were increased, because of the fractal nature of landscapes, there would always remain some MRUs that include portions that are transitional to another MRU, and that attract or exclude agriculture more than neighboring homogeneous areas. Therefore, for the purpose of operational global monitoring and reporting, a limited but manageable number of spatial units defined based on general environmental variables is acceptable: the proposed CropWatch MRUs are meaningful reporting units at the global scale.

Acknowledgments

The authors specifically acknowledge the financial support through the Special Fund for Grains-scientific Research in the Public Interest (Grant No. 201313009-2 and 201413003-7) the National High Technology Research and Development Program of China (863 program), Grant No. 2012AA12A307; and the Visiting Professorships for Senior International Scientists, Grant No. 2013T1Z0016. The authors thank the reviewers for providing thorough comments that have improved the quality and readability of the paper.

Conflict of Interest

None declared.

References

Akpan, S. B., E. J. Udoh, and V. S. Nkanta. 2012. Factors influencing fertilizer use intensity among small holder crop farmers in Abak Agricultural Zone in Akwa Ibom State, Nigeria. J. Biol. Agric. Healthc. 2:54–65.

Babu, S. C., and V. J. Quinn. 1994. Food security and nutrition monitoring in Africa: introduction and historical background. Food Policy 19:211–217.

Binder, C. R., G. Feola, and J. K. Steinberger. 2010. Considering the normative, systemic and procedural dimensions in indicator-based sustainability assessments in agriculture. Environ. Impact Assess. Rev. 30:71–81.

Chen, S. T., H. I. Kuo, and C. C. Chen. 2010. Modeling the relationship between the oil price and global food prices. Appl. Energy 87:2517–2525.

Dalgaard, T., N. J. Hutchings, and J. R. Porter. 2003. Agroecology, scaling and interdisciplinarity. Agric. Ecosyst. Environ. 100:39–51.

Doré, T., D. Makowski, E. Malézieux, N. Munier-Jolain, M. Tchamitchian, and P. Tittonell. 2011. Facing up to the paradigm of ecological intensification in agronomy: revisiting methods, concepts and knowledge. Eur. J. Agron. 34:197–210.

FAO. 1999. FRA 2000, A concept and strategy for ecological zoning for the global forest resources assessment 2000. INTERIM REPORT. FAO, Forestry Dep. Working Paper 20. Rome. 28 pp.

FAO. 2007. Most suitable cereal. Digital map downloadable from FAO GeoNetwork web site http://www.fao.org/geonetwork/srv/en/main.home. Plate 48b in the FAO/IIASA GAEZ study http://www.fao.org/nr/gaez/en/. Methodology in 2002 document jointly published by FAO and IIASA and available from http://webarchive.iiasa.ac.at/Research/LUC/SAEZ/pdf/gaez2002.pdf (accessed September 2015).

FAO. 2010. Land Use Systems of the World. Digital map downloadable from FAO GeoNetwork web site http://www.fao.org/geonetwork/srv/en/main.home (accessed September 2015).

FAO. 2012. Global ecological zones for FAO forest reporting: 2010 Update. Forest Resources Assessment Working Paper N. 179. 42 pp. FAO, Rome.

FAO. 2013. Global map of irrigation areas - version 5. Digital map downloadable from FAO GeoNetwork web site http://www.fao.org/geonetwork/srv/en/main.home. The version used was downloaded on 20130901. Version 5 was published in October 2013 (accessed September 2015).

FAO. 2014a. GeoNetwork, a repository of spatial information about world agriculture. http://www.fao.org/geonetwork/srv/en/main.home (accessed September 2015).

FAO. 2014b. FOSTAT, the global statistical database of FAO. Available at: http://faostat3.fao.org/faostat-gateway/go/to/home/E (accessed September 2015).

George, T. 2014b. Why crop yields in developing countries have not kept pace with advances in agronomy. Glob. Food Sec. 3:49–58.

Geo-Wiki. 2014. Data downloaded from the Geo-wiki project. Available at http://agriculture.geo-wiki.org/login.php?menu=results. See also http://www.iiasa.ac.at/web/home/research/researchPrograms/EcosystemsServicesandManagement/Geo-Wiki.en.html. Dataset downloaded from Beta-hybrid.geo-wiki.org on 20130901 (accessed September 2015).

Gommes, R., J. du Guerny, M. H. Glantz, and L.-N. Hsu, 2004. Climate and HIV/AIDS: a hotspots analysis for early warning rapid response systems. UNDP/FAO/NCAR, UNDP SE Asia and Development Programme, Bangkok, 20 pp. Available at: http://www.fao.org/forestry/15532-0-0.pdf (accessed September 2015).

Gornall, J., R. Betts, E. Burke, R. Clark, J. Camp, K. Willett, et al. 2010. Implications of climate change for agricultural productivity in the early twenty-first century. Philos. Trans. R. Soc. Lond. B Biol. Sci. 365:2973–2989.

Grieser, J., R. Gommes, S. Cofield, and M. Bernardi, 2006a. World maps of climatological net primary production of biomass, NPP. FAO, 2005. 5 pp. http://www.juergen-grieser.de/downloads/NetPrimaryProduction/npp.pdf (accessed September 2015).

Grieser, J., R. Gommes, S. Cofield, and M. Bernardi, 2006b. New gridded maps of Koeppen's climate classification. Data, methodology and gridded data Available at: http://www.fao.org/nr/climpag/globgrids/KC_classification_en.asp. Methodology also downloadable from http://www.juergen-grieser.de/downloads/Koeppen-Climatology/Koeppen_Climatology.pdf (accessed September 2015).

Haberl, H., K. H. Erb, F. Krausmann, V. Gaube, A. Bondeau, C. Plutzar, et al. 2007. Quantifying and mapping the human appropriation of net primary production in earth's terrestrial ecosystems. Proc. Natl Acad. Sci. USA 104:12942–12947.

Haberle, S. G., and A. Chepstow-Lusty. 2000. Can climate influence cultural development? A view through time Environ. Hist., 34:9–369.

Hijmans, R. J., S. E. Cameron, J. L. Parra, P. G. Jones, and A. Jarvis. 2005. Very high resolution interpolated climate surfaces for global land areas. Int. J. Climatol. 25:1965–1978.

Hodson, D. P., and J. W. White. 2007. Use of spatial analyses for global characterization of wheat-based production systems. J. Agric. Sci. 145:115–125.

Hu, Z., and D. Zhang. 2006. China country pasture/forage resource profiles. Pp. 63. FAO, Rome.

Iizumi, T., J. J. Luo, A. J. Challinor, G. Sakurai, M. Yokozawa, H. Sakuma, et al. 2014. Impacts of El Niño Southern Oscillation on the global yields of major crops. Nat. Commun. 5:3712.

Kelly, V. A., 2010. Factors affecting demand for fertilizer in sub-Saharan Africa. Agriculture and Rural Development Discussion Paper N. 23. The World Bank, Washington, USA. 89 pp.

Kriticos, D. J., B. L. Webber, A. Leriche, N. Ota, I. Macadam, J. Bathols, et al. 2012. CliMond: global high-resolution historical and future scenario climate surfaces for bioclimatic modelling. Method Ecol. Evol. 3:53–64.

Kummu, M., H. de Moel, P. J. Ward, and O. Varis. 2011. How close do we live to water? A global analysis of population distance to freshwater bodies. PLoS One 6:e20578.

Leblois, A., P. Quirion, and B. Sultan. 2014. Price vs. weather shock hedging for cash crops: *ex ante* evaluation for cotton producers in Cameroon. Ecol. Econ. 101:67–80.

Loyce, C., J.M. Meynard, C. Bouchard, B. Rolland, P. Lonnet, P. Bataillon, et al. 2012. Growing winter wheat cultivars under different management intensities in France: a multicriteria assessment based on economic, energetic and environmental indicators. Field Crops Res. 125:167–178.

Lu, W. C. 2002. Effects of agricultural market policy on crop production in China. Food Policy 27:561–573.

MLA. 2015. Dictionary.com Unabridged. Random House, Inc. 22 Jul. 2015. Available at: http://dictionary.reference.com/browse/unproductivity.

Monfreda, C., N. Ramankutty, and J. A. Foley. 2008. Farming the planet: 2. Geographic distribution of crop areas, yields, physiological types, and net primary production in the year 2000. Global Biogeochem. Cycles, 22:GB1022.

Monfreda, C., N. Ramankutty, and T. W. Hertel. 2009. Global agricultural land use data for climate change analysis. Chap. 2 in Economic analysis of land use in global climate change policy (33–48) edited by Hertel, T.W., Rose, S., Tol, R. 2009. 348 pp. Routledge Press, London and New York. ISBN 978-0415619813.

Nelson, G. C., R. W. Rosegrant, A. Palazzo, I. Gray, C. Ingersoll, R. Robertson, et al. , 2010. Food Security, Farming, and Climate Change to 2050: Scenarios, Results, Policy Options. International Food Policy Research Institute IFPRI, Washington, D.C. 131 pp. Available at: http://www.ifpri.org/sites/default/files/publications/rr172.pdf (accessed September 2015).

OECD. 2007. Glossary of statistical term. 863 pp. OECD, Paris.

Pal, D. K., D. K. Mandal, T. Bhattacharyya, C. Mandal, and D. Sarkar. 2009. Revisiting the agro-ecological zones for crop evaluation. Indian J. Genet. Plant Breed. 69:315–318.

Ramankutty, N., J. A. Foley, J. Norman, and K. McSweeney. 2002. The global distribution of cultivable lands: current patterns and sensitivity to possible climate change. Glob. Ecol. Biogeogr. 11:377–392.

Reynolds, M. P., ed. 2010. Climate change and crop production. CABI climate change series 1. 292 pp. CABI, Wallingford, UK.

Rosengrant, M. W., and the Impact development team. 2012. International Model for Policy Analysis of Agricultural Commodities and Trade (IMPACT). Model Description. 50 pp. International Food Policy Research Institute IFPRI, Washington, DC. Available at: http://ebrary.ifpri.org/cdm/ref/collection/p15738coll2/id/12735 (accessed September 2015).

Rozbicki, J., A. Cegli, D. Gozdowski, M. Jakubczak, G. Cacak-Pietrzak, W. Madry, et al. 2015. Influence of the cultivar, environment and management on the grain yield and bread-making quality in winter wheat. J. Cereal Sci. 61:126–132.

SAGE. 2002. Available at: and http://library.mcmaster.ca/maps/geospatial/atlas-biosphere (accessed September 2015).

Schmid, A. P., 1998. Thesaurus and glossary of early warning and conflict prevention terms (abridged version). Synthesis Foundation. Erasmus University, Amsterdam. Edited for Forum on Early Warning and Early Response (FEWER) by S. B. Anderlini. FEWER Secretariat, London. 30 pp.

Siebert, S., V. Henrich, K. Frenken, and J. Burke. 2013. Update of the digital Global Map of Irrigation Areas (GMIA) to version 5. 171 pp. Institute of Crop Science and Resource Conservation, Rheinische Friedrich-Wilhelms-Universität Bonn, Bonn, Germany.

Snodderly, D., 2011. Peace Terms: Glossary of Terms for Conflict Management and Peacebuilding. United States Institute of Peace, Academy for International Conflict Management and Peacebuilding, Washington USA, 60 pp.

Sombroek, W. S., and R. Gommes, 1996. The climate change-agriculture conundrum, Pp. 1–14 *in* F. Bazzaz, W. Sombroek, eds. Global climate change and agricultural production. 345 pp. FAO and John Wiley, & Sons, Chichester, UK. ISBN 0 471 96927 3.

Sun, He., 1994. Agricultural natural resources and regional development of China. Jiangsu Science and Technology Press, Nanjing. (in Chinese)

UN. 2010. United Nations Statistics Division. Composition of Macro-Geographical (Continental) Regions, Geographical Sub-Regions, and Selected Economic and Other Groupings. Available at: http://millenniumindicators.un.org/unsd/methods/m49/m49regin.htm (accessed September 2015).

Vancutsem, C., E. Marinho, F. Kayitakire, L. See, and S. Fritz. 2013. Harmonizing and combining existing land cover/land use datasets for cropland area monitoring at the African continental scale. Remote Sens. 5:19–41.

VITO. 2014. NDVI Based on average 1999-2012 monthly SPOT VEGETATION NDVI, downloadable from https://earth.esa.int/web/guest/pi-community/apply-for-data and http://www.vito-eodata.be/PDF/portal/Application.html#Home. (accessed in September 2015).

van Wart, J., L. G. van Bussel, J. Wolf, R. Licker, P. Grassini, A. Nelson, et al. 2013. Use of agro-climatic zones to upscale simulated crop yield potential. Field Crops Res. 143:44–55.

Wiggins, S., and J. Brooks. 2010. The use of input subsidies in developing countries. 22 pp. OECD, Paris.

Wu, B., J. Meng, Q. Li, N. Yan, X. Du, and M. Zhang. 2014. Remote sensing-based global crop monitoring: experiences with China's CropWatch system. Int. J. Digital Earth 7:113–137.

Wu, B., R. Gommes, M. Zhang, H. Zeng, N. Yan, W. Zou, et al. 2015. Global crop monitoring: a satellite-based hierarchical approach. Remote Sens. 7:3907–3933.

Xu, T., and M. Hutchinson. 2011. ANUCLIM version 6.1 user guide. 85 pp. The Australian National University, Fenner School of Environment and Society, Canberra.

Xu, T., and M. F. Hutchinson. 2013. New developments and applications in the ANUCLIM spatial climatic and bioclimatic modelling package. Environ. Model. Softw. 40:267–279.

Yu, Y., Y. Huang, and W. Zhang. 2012. Changes in rice yields in China since 1980 associated with cultivar improvement, climate and crop management. Field Crops Res. 136:65–75.

Zhang, X., S. Wang, H. Sun, S. Chen, L. Shao, and X. Liu. 2013. Contribution of cultivar, fertilizer and weather to yield variation of winter wheat over three decades: a case study in the North China Plain. Eur. J. Agron. 50:52–59.

14

The case for evidence-based policy to support stress-resilient cropping systems

Matthew Gilliham[1], Scott Chapman[2], Lisa Martin[3], Sarah Jose[3] & Ruth Bastow[3]

[1]Plant Transport and Signalling Lab, ARC Centre of Excellence in Plant Energy Biology, School of Agriculture, Food and Wine, Waite Research Institute, University of Adelaide, Glen Osmond, South Australia 5064, Australia
[2]Commonwealth Scientific and Industrial Research Organisation (CSIRO) Agriculture and Food, The University of Queensland, Bioscience Precinct, St. Lucia, Queensland 4067, Australia
[3]Global Plant Council, Bow House, 1a Bow Lane, London EC4M 9EE, UK

Keywords
Crop productivity targets, food security, collaboration, research strategy.

Correspondence
Matthew Gilliham, Plant Transport and Signalling Lab, ARC Centre of Excellence in Plant Energy Biology, School of Agriculture, Food and Wine, Waite Research Institute, University of Adelaide, Glen Osmond, SA 5064, Australia.
E-mail: matthew.gilliham@adelaide.edu.au

Funding Information
M.G. funding – Grains Research and Development Corporation (UA00145), The International Wheat Yield Partnership (IWYP60FP/ANU00027), and the Australian Research Council (FT130100709 and CE140100008).

Abstract

Research and the dissemination of evidence-based guidelines for best practice in crop production are fundamental for the protection of our crop yields against biotic and abiotic threats, and for meeting ambitious food production targets by 2050. The advances in knowledge required for sustaining crop productivity targets will be gained through three research tracks: (1) basic strategic research in the field, for example, crop breeding, agronomy, and advanced phenotyping; (2) translational research involving the application of advances in fundamental science; and (3) pure fundamental research to fuel future translational research. We propose that policy and funding structures need to be improved to facilitate and encourage more interactions between scientists involved in all three research tracks, and also between researchers and farmers, to improve the effectiveness of delivering improvements in crop stress resilience. History illustrates that it is challenging for public researchers to "stretch across" all of these research tracks, with effective farm-level solutions being more likely when end-users and industry are directly engaged in the research pipeline. As research proceeds from fundamental through to applied research, the demand for experimental rigor and a wider understanding of appropriate methods and outcomes is paramount, that is, demonstrating value in yield at the field level requires the input of experienced practitioners from each research track. The development of evidence-based policies to support all funding structures and the engagement of producers with both the development of research, and with the findings of such research, will form an important capability in meeting food security targets. This commentary, concentrating on the development of policies to support research and its dissemination, is based on discussions held at the Stress Resilience Symposium organized by the Global Plant Council and Society of Experimental Biology in October 2015.

Introduction

Research into the development of stress-resilient plants and cropping systems has the potential to generate significant positive, worldwide impacts for dealing with climate change and ensuring global food security. Research alone cannot bring about fundamental change; rather research findings can be used to build evidence for reaching consensus, which in turn helps to exact a change in behavior or practice through policy development. The interface between science and policy is therefore important, and it is essential that research funding and policy development should be intrinsically linked. However, this is not always the case and more needs to be done to develop effective, global, evidence-based policies for plant science.

So how do these two, often disparate, areas of science and policy interact and influence each other? In terms of the effects of science on policy, the results of well-designed scientific research can provide the evidence on which to base consensus in the scientific community. In turn, scientific consensus provides the foundation on which to build adequate and robust policies for developing and researching efficient and sustainable practices that will support crop improvement and furnish food and agricultural products to the growing human population.

Policy impacts science in many different ways; for example, it affects the level of public and private investment in both research and development, and determines the strategic priorities that direct where this investment is targeted. Institutional, governmental, and national policies also impinge on and control how research data are collected, accessed, exchanged, and stored in both the public and private sectors. Policies and regulations related to the safe use and application of technologies also have significant impacts on the agricultural sector; for example, technologies that support decision-making (e.g., satellites and drones), and/or the implementation of solutions (e.g., genetically modified organisms [GMOs], chemicals, robots).

The potential for a disconnect between the science and policy arenas becomes amplified at the international level, with governments and funding bodies around the world varying in the ways in which they decide upon and develop funding strategies and regulatory and policy frameworks. These variations exist although all nations (developed and developing) face common challenges (as stated in the Millennium Development Goals and Sustainable Development Goals (United Nations, 2015)), common environments (e.g., rain-fed crops in Australia, and sub-Saharan Africa and central-western India frequently experience drought), grow common crops of interest (e.g., maize, wheat, rice, sorghum, barley, tubers, legumes), and are all inhabitants on a common planet (combating climate change is a universal issue).

Despite these universal links, it is the national agencies and bodies that are the major funders and regulators of research. As a result, national funders – quite reasonably – place restrictions on expenditure in other jurisdictions and wish to be in control of their own regulatory frameworks. This can result in fractured – and in some cases conflicting – policies across the globe that subsequently act as a barrier to collaboration, and limit the ability to collate and share data, resources, and intellectual property across national boundaries. The end point of this is the loss of added value that can be gained from linking parallel efforts or by concentrating research in priority areas of global significance.

Given the significant impact of policy on research, from a global perspective, it is vital that policy decisions take account of global scientific consensus, and that policy decisions impacting and influencing the plant and agricultural arena be fully informed by scientific method and results from the international research community. This in turn depends – to a large extent – on scientists summarizing and explaining their work in a manner that is accessible and understandable to nonspecialist audience so that others including funders and policy makers can understand current research efforts and what comprises high-quality scientific method in different tracks of research.

Learning from Past Experience

Biotic and abiotic stresses have negative impacts on crop productivity and thus are major limiting factors to global food and nutritional security, and to the production of agricultural products. Global wheat production, for example, is projected to decrease by 6% for each degree centigrade of global warming, together with an increased variability in yield across regions and seasons (Asseng et al. 2015). Yield losses are also predicted for other major crops (e.g., Challinor et al. 2014).

One of the predictions for a changing climate is an increasing incidence of extreme weather events. In some areas this will mean more prevalent drought, heat, and/or salinity events, and research into plant tolerance to these stresses will be paramount to improve crop suitability for such conditions and mitigate stress effects on crop yield. For example, Lobell et al. (2015) revealed that while drought will continue to impact yields of wheat and sorghum in Northeast Australia, breeders will need to increase the heat tolerance of these crops to mitigate the damage from increasing frequencies of extreme heat stress events in this region. Such research topics are not new endeavors, and the lessons learned from previous and current research can inform us how new policies might best support the development of stress-resilient cropping systems (Gilliham et al. 2017).

For instance, from the outcomes of previous research on drought and salinity research it is clear that tolerance to these stresses is complex requiring multiple traits, with tolerance to each stress composed of multiple traits (e.g., for salinity; exclusion of salt from the shoot, stomatal closure, detoxification of reactive oxygen species, the adaptation to low water potential in the soil); some of these traits are required for tolerance to both stresses (Chaves et al. 2009; Huang et al. 2009; Munns and Gilliham 2015). In extreme circumstances, crops can face multiple threats at the same time, for example, low water availability, high salinity, high temperatures and biological pests; therefore, crops must be well adapted to multiple threats.

Tolerance to abiotic stress requires both consideration of the crop's genetics (and its capacity to respond to the environment), and the best crop management to mitigate the impact of a stress. A holistic approach is needed whereby multiple research angles are deployed and integrated to achieve the best results, encompassing genetics, breeding, physiology, predictive modeling, agronomy, and extension to growers (Hammer et al. 2016). In combination, these approaches have led to research impact through translation in both the private and public sector, but have required a substantial cross-scale effort for delivery (Gilliham et al. 2017).

One such challenge is the agronomic management of water throughout the season. While it is sometimes viewed as an extension rather than a research challenge, new technologies have the potential to transform water management in agriculture (e.g., smart weather sensing and modeling for irrigation scheduling, or biodegradable plastic mulches; Clawson and Blad 1981; Lebourgeois et al. 2009; Li et al. 2013). In dry environments, the adoption of no-tillage systems has been a major factor in increasing soil organic matter by protecting it from erosion, and in conserving soil water through the season across much of the world's agricultural lands (Derpsch et al. 2010). These systems typically require mechanized management augmented by the careful use of biodegradable herbicides so that weeds do not develop resistance to chemical control. In addition to policies encouraging the use of these techniques, the development of herbicide-tolerant varieties (usually, but not necessarily, GMO) has facilitated the rapid adoption of no-till, especially in South America. Herbicide-tolerant GMO varieties have therefore had a significant impact on yield in dryland environments by encouraging the adoption of no-till and its water-conserving benefits.

There has been considerable investment in molecular and laboratory-based solutions to stress tolerance. These investments include the use of sequencing technologies, the deployment of molecular markers, "speed breeding", and – to a limited extent – genetic modification. GMOs (for herbicide and insect tolerance) have made substantial contributions to increased food production (Qaim 2009; Klümper and Qaim 2014), but genetically complex traits such as drought and salinity tolerance (much like the introduction of C4 photosynthesis into rice or nitrogen fixation into other cereals) require multigene solutions with significant lead times. These latter initiatives have received significant investment from NGOs; however, similar investment has not occurred for paradigm-shifting research in the stress resilience of cropping systems.

Assessment of impact and the evaluation of success and value can change and diverge substantially in the transition from fundamental through to applied research and to the field. This generates several challenges, particularly for research teams trying to span all levels of complexity to demonstrate and deliver new traits for stress resilience. For example, current funding and publication policies at the fundamental end of the research pipeline often require demonstration of impact, leading to numerous articles in plant journals containing statements such as "trait X has shown substantial yield improvement in glasshouse conditions". Critically, however, few of these effects will demonstrably translate to increases in yield under field conditions, despite the rigorous and repeatable demonstration of trait effects in controlled conditions. A good example of the large-sale yield monitoring technologies needed to demonstrate yield improvements in the field comes from DuPont researchers who showed a 6.9% yield advantage of highly water use-efficient corn hybrids across 2000 locations (Gaffney et al. 2015). We do not propose that such extensive approaches are required to demonstrate the value of potential new traits in the early part of discovery; rather that statements about "yield" require a rigorous evidence base, careful consideration of alternative hypotheses for the observed results, and a clear appreciation of the likelihood of the trait having an impact in the field, before inclusion in publications. For example, a yield increase of 50% is highly unlikely at a field level except in very specific circumstances. When such observations are included in publications, reviews, and funding policies without an appropriate field-based context, this can result in significant distortions in policy and investment strategies and outcomes.

It is important that fundamental research interrogating how crops survive stress is carried out so as to inform the development of new varieties. The development of water use-efficient wheat such as Drysdale, which has 10% higher yields in dry conditions, as well as the breeding of durum wheat with a 25% greater grain yield in saline conditions, were both the products of fundamental research discoveries applied to breeding programs (Rebetzke et al. 2002; Munns et al. 2012). These are good examples of why investment in both fundamental and applied solutions to stress tolerance is needed. Furthermore, investment in translational research is essential to establish the robustness of fundamental research solutions prior to their implementation in farming systems (Gilliham et al. 2017). It should also be noted that these are long-term investments; it took 20 years for Drysdale to be released following elucidation of the trait underlying its increased water use efficiency.

To most effectively deploy research aimed at improving the stress resilience of cropping systems, an integrated science, policy, and society approach is needed to ensure that disparate skillsets and expertise come together. Although it is not exclusively the case, scientists who perform most of their stress tolerance research in the laboratory, and those based primarily in the field, operate in different arenas and rarely interact or combine efforts. The involvement of primary producers in research strategy or projects can also be lacking. Future research cannot afford to ignore the potential synergies gained by involving all three stakeholders. Examples of projects that have benefited from broad involvement include those listed above, that is, the development of water use-efficient wheat or salt-tolerant durum wheat, but more are needed. Unfortunately, funding is a currently one barrier to progress in this area. Few funding structures can take on promising advances made in fundamental science, validate them, and assess whether they can be translated into the yield gains in the field (Gilliham et al., 2017). A recent report by the Australian Academy of Science identified a specific fund for translational science as a key priority for the future (Australian Academy of Sciences, 2017).

Funding is not the only issue. Stress-tolerance traits related to survival identified in laboratory-based studies are often not relevant for maintaining yield in the field (e.g., Chapman et al. 2002; Hammer et al. 2016); therefore, policies must be created to fund structures that work toward applying, translating, and researching yield improvement in the field over the mid-to-long-term (Gilliham et al., 2016). This is another reason why encouraging entities that have not commonly worked together (e.g., laboratory-based fundamental scientists and field-based researchers) to join forces is important to ensure that traits relevant to stress resilience in the field are examined and rigorous translational studies are conducted. The private sector designs research flows to ensure that links between research tracks function to deliver improved germplasm, but these are more challenging to develop within the public sector research system. Policy that encourages public–private partnerships is increasingly utilized in sponsoring agronomy research, but less so in plant breeding.

A more detailed assessment of traits relevant to stress tolerance in the field can be found elsewhere and are often informed by predictive modeling of agricultural systems (e.g., Cooper et al. 2014; *Hammer* et al. 2016). Borrell & Reynolds (2017) discuss the need to join and maximize islands of isolated knowledge to maximize potential outcomes. This applies to both isolated concurrent research being conducted on similar topics through forming effective collaborations. A balance needs to be struck to avoid "tipping the scales" and creating situations in which new initiatives are funded at the expense of pure fundamental research that feeds the innovations of tomorrow, or purely field-based studies that inform practice and lead to important gains.

The Bigger Picture

For multifactorial traits such as drought and salinity tolerance, research investment and policy decisions must take a "big picture" approach ranging from the gene, to the field, to the ecosystem level, to develop crops and cropping systems that will meet future needs. Such a holistic approach will require the intelligent use of "big data," and the development of new technologies to support research studies and the application of research findings in the field.

Agriculture, like many other areas, is undergoing a data revolution; examples range from real-time monitoring of livestock health and condition; automated glasshouse control of vegetable quality; and the application of satellites, unmanned aerial vehicles (drones), machine yield monitors, and farmer information crowdsourcing platforms. These new technologies and associated data generate a wealth of opportunities for new discoveries and innovative solutions, but at the same time create numerous challenges for policy development.

There is growing pressure from governments and funders across the globe to make public data more open and accessible. However, what open data mean in practice, and how it will contribute toward increasing food security, improving human health and nutrition, and ensuring the more sustainable management of natural resources, is still being assessed and the associated policies are in development. For example, how do you balance "data ownership" – which may encourage business development and competition – with the concept of "open data?"

Initiatives such as Global Open Data for Agriculture and Nutrition (GODAN) are grappling with this and many other issues associated with the open data concept. GODAN supports the proactive sharing of open data to deal with the urgent challenge of ensuring world food security. By bringing national governments together with nongovernmental, international, and private sector organizations, GODAN seeks to support global efforts to make agricultural and nutritionally relevant data available, accessible, and usable for unrestricted use worldwide. The initiative focuses on building high-level policy and public and private institutional support for open data. The existence and popularity of GODAN (450 partners) illustrates how far-reaching the issue of open data is, and that if progress is to be made, solutions and policies must be developed and adopted at a global scale.

For policies to truly have maximum impact, the conversation must move beyond the science and governance arena. Widening the conversation to the broader global community is therefore vital, but does require initiatives to train, equip, and encourage scientists and science communicators to engage members of the public with research of broader media interest. Although many scientists are well trained to perform their laboratory or field-based research, they often lack the skills required to communicate their research and its implications to a general audience in an easily accessible and approachable manner. Several organizations and programs across the globe provide training in communication and outreach; for example, Sense About Science, a UK-based charity with new EU and US-based counterparts, has established the Plant Science Panel, a group of expert plant scientists that answers questions (online) posed by members of the public to help promote understanding and address misconceptions on any plant science-related topic. The same charity has also set up the Voices of Young Science network, and an associated series of successful "Standing up for Science" workshops across the UK, that helps to better equip early career scientists to talk about their research, and provides advice on how to get involved in discussions in the public arena. Projects like this are essential to help provide researchers with the skills, confidence, and opportunities to contribute to science and policy-based discussions.

Forming evidence-based policies that will have the greatest impact will require a full consideration of all viewpoints and surrounding issues. Policies cannot be developed, much less implemented, unless there is consensus, and consensus cannot be reached without being fully informed of the research, funding, and policy landscape. Developing an informed "community" requires effective communication between all stakeholders. Such a dialog should take place in an international context and involve not just those embedded in the research, policy, and regulatory arenas but also the wider public.

Actions

Trying to help bridge the gap between science and policy is logistically, culturally, and politically challenging but, as has been demonstrated with "big issues" such as climate change and the work of the Intergovernmental Panel on Climate Change, it is possible. The question is: can the plant/agricultural community develop a similar global effort?

As a global organization with members across six continents, the Global Plant Council (GPC) (http://globalplantcouncil.org/) is well placed to help facilitate better integration between the research, funding, regulatory, and policy domains, as outlined by the mechanisms proposed below.

Building consensus view

Short term

- Develop an understanding of the current landscape by:
  - a. Gathering evidence on the current scientific consensus regarding what is required at the international level to develop stress-resilient crops and cropping systems; for example, reports, papers, and position statements.
  - b. Collating information on current projects across the globe and,
  - c. On current regulatory frameworks of relevance to stress resilience research.
- Assess current community needs by:
  - d. Undertaking electronic survey(s) of the GPC's membership, and the wider community, to understand the major bottlenecks to developing stress-resilient crops and cropping systems with respect to policy and regulatory frameworks.

Medium term

- Help build a consensus by:
  - a. Developing position statements based on the findings of the landscape study and survey of needs.
  - b. Working with other key organizations to develop international consensuses, taking key political issues into account, which outline current bottlenecks and potential solutions to improving the policy environment of relevance to developing stress-resilient crops and cropping systems.

Long term

- Advocate for the inclusion and implementation of scientific consensus in international policy and regulatory frameworks.
- Advise international bodies, funders, and other forums to inform funding, regulatory, and policy decisions.

Developing an informed community – promoting a global conversation

Short term

- Exchange research and policy knowledge by:
  - a. Using the GPC's social media channels to help those working on similar topics in the areas of stress-resilient crops and cropping systems to become more aware of each other's efforts and activities.
  - b. Raising awareness of current open access and data-sharing policies, as well as examples of appropriate and inappropriate experimental methods in different research tracks.
  - c. Generating an online database of the information gathered in the landscape study.

Short-to-medium term

- Develop viewpoint articles from a range of stakeholders on the current challenges and solutions in building adequate and robust frameworks at the science/policy interface.

- Publicize these articles via social and traditional media, with accompanying, engaging materials (e.g., videos, leaflets, case studies) to raise awareness of global issues in local contexts, and start a conversation with the public, stakeholders, and politicians.

Medium-to-long term

- Collate articles and viewpoints into an annual online publication.

Training a new generation of plant science communicators to "stand up" for evidence-based policies

Short term

- Collate information about existing science communication and science policy courses and advertise them to a wider audience.
- Provide discussion points about what constitutes rigor in demonstrating physiological (to scientists and funders) and field-level (to farmers) value of stress resilience research

Longer term

- Work with others to assess which approaches and courses are productive/successful, and for which audiences (school, community group, government, international). Based on this, the GPC could help to develop future online courses (e.g., massive open online courses) to enthuse and inspire a new generation of communicators.

Establishing consensus is the first step toward realizing effective, global, evidence-based policies for plant science and allied research and development. If multiple, diverse stakeholders from around the world can be brought together under the banner of a single, global organization, such as the Global Plant Council, then our combined voices will be much louder.

Acknowledgments

This paper is based on outcomes from a Stress Resilience Symposium held in Brazil in October 2015 organized by the Global Plant Council and Society for Experimental Biology. The authors would like to thank the Society for Experimental Biology for funding support for this symposium.

Conflict of Interest

The authors of this article are either affiliated to the Global Plant Council through membership of their Society or are employed by the Global Plant Council.

References

Asseng, S., F. Ewert, P. Martre, et al. 2015. Rising temperatures reduce global wheat production. Nat. Clim. Chang. 5:143–147.

Australian Academy of Sciences. 2017. Decadal plan for agriculture. Available at https://www.science.org.au/support/analysis/decadal-plans-science/decadal-plan-agriculture (accessed 10 February 2017).

Borrell, A., and M. Reynolds. 2017. Integrating islands of knowledge for greater synergy and efficiency in crop research. Food and Energy Secur. 6:26–36.

Challinor, A. J., J. Watson, D. B. Lobell, S. M. Howden, D. R. Smith, and N. Chhetri. 2014. A meta-analysis of crop yield under climate change and adaptation. Nat. Clim. Chang. 4:287–291.

Chapman, S. C., G. L. Hammer, D. W. Podlich, and M. Cooper. 2002. Linking bio-physical and genetic models to integrate physiology, molecular biology and plant breeding. Pp. 167–187 *in* M. Kang, ed. "Quantitative genetics, genomics, and plant breeding" (Invited paper at Symposium on Quantitative Genetics for the 21st Century, Baton Rouge, Louisiana, March 2001). CAB International, Wallingford, UK.

Chaves, M. M., J. Flexas, and C. Pinheiro. 2009. Photosynthesis under drought and salt stress: regulation mechanisms from whole plant to cell. Ann. Bot. 103:551–560.

Clawson, K. L., and B. L. Blad. 1981. Infrared thermometry for scheduling irrigation of corn. Agron. J. 74:311–316.

Cooper, M., C. Gho, R. Leafgren, T. Tang, and C. Messina. 2014. Breeding drought-tolerant maize hybrids for the US corn-belt: discovery to product. J. Exp. Bot. 65:6191–6204.

Derpsch, R., T. Friedrich, A. Kassam, and L. Hongwen. 2010. Current status of adoption of no-till farming in the world and some of its main benefits. Int. J. Agric. Biol. Eng. 3:1–25.

Gaffney, J., J. Schussler, C. Löffler, W. Cai, S. Paszkiewicz, C. Messina, et al. 2015. Industry scale evaluation of maize hybrids selected for increased yield in drought stress conditions of the U.S. Corn Belt. Crop Sci. 55:1608–1618.

Gilliham, M., J. A. Able, and S. J. Roy. 2017. Translating knowledge about abiotic stress tolerance to breeding programmes. Plant J. Online 8th February 2017. doi:10.1111/tpj.13456.

Hammer, G. L., G. McLean, A. Doherty, van Oosterom E., and S. C. Chapman. 2016. Sorghum crop modeling and its utility in agronomy and breeding. *In* I. Ciampitti, V. Prasad, eds. Sorghum: state of the art and future perspectives, agronomy monograph 58. ASA and CSSA, Madison, WI. doi:10.2134/agronmonogr58.2014.0064

Huang, X. Y., D. Y. Chao, J. P. Gao, M. Z. Zhu, M. Shi, and H. X. Lin. 2009. A previously unknown zinc finger protein, DST, regulates drought and salt tolerance in rice via stomatal aperture control. Genes Dev. 23:1805–1817.

Klümper, W., and M. Qaim. 2014. A meta-analysis of the impacts of genetically modified crops. PLoS ONE 9:e111629.

Lebourgeois, V., J.-L. Chopart, A. Bégue, and L. Le Mézo. 2009. Towards using a thermal infrared index combined with water balance modelling to monitor sugarcane irrigation in a tropical environment. Agric. Water Manag. 97:75–82.

Li, S. X., Z. H. Wang, S. Q. Li, Y. J. Gao, and X. H. Tian. 2013. Effect of plastic sheet mulch, wheat straw mulch, and maize growth on water loss by evaporation in dryland areas of China. Agric. Water Manag. 116:39–49.

Lobell, D. B., G. L. Hammer, K. Chenu, B. Zheng, G. McLean, and S. C. Chapman. 2015. The shifting influence of drought and heat stress for crops in northeast Australia. Glob. Change Biol. 21:4115–4127.

Munns, R., and M. Gilliham. 2015. Salinity tolerance of crops – what is the cost? New Phytol. 208:668–673.

Munns, R., R. A. James, B. Xu, et al. 2012. Wheat grain yield on saline soils is improved by an ancestral Na^+ transporter gene. Nat. Biotechnol. 30:360–364.

Qaim, M. 2009. The economics of genetically modified crops. Annu. Rev. Resour. Economics 1:665–694.

Rebetzke, G. J., A. G. Condon, R. A. Richards, and G. D. Farquhar. 2002. Selection for reduced carbon isotope discrimination increases aerial biomass and grain yield of rainfed bread wheat. Crop Sci. 42:739–745.

United Nations. 2015. Sustainable development goals [online]. Available at http://www.un.org/sustainabledevelopment/sustainable-development-goals/ (accessed 10 February 2017).

15

The CROPROTECT project and wider opportunities to improve farm productivity through web-based knowledge exchange

Toby J. A. Bruce

Rothamsted Research, Harpenden, Hertfordshire AL5 2JQ, UK

Keywords
Farmer research networks, impact, innovation, research policy, sustainable intensification, translational research

Correspondence
Toby J.A. Bruce, Rothamsted Research, Harpenden, Hertfordshire AL5 2JQ, UK.

E-mail: toby.bruce@rothamsted.ac.uk

Funding Information
Natural Environment Research Council (Grant/Award Number: "NE/M016676/1")

Abstract

A key global 21st century challenge is to maximize agricultural production while minimizing use of resources such as land, water, and energy to meet rising demand for produce. To meet this challenge, while also adapting to climate change, agriculture will have to become more knowledge intensive and deploy smarter farming techniques. The intention of this study was to: (1) Highlight the opportunity for web-based knowledge exchange to increase farm productivity and thus contribute to achieving food and energy security, (2) Give some examples of online farming information services such as the "CROPROTECT" tool I am developing in the UK, the CABI "Plantwise" Knowledge Bank and the IRRI "Rice Doctor," and (3) Consider lessons learnt so far. There are huge opportunities to facilitate knowledge exchange through online systems for farmers and people who advise farmers. CROPROTECT is interacting with users to determine priorities in terms of the pests, weeds, and diseases covered and is providing key information to assist with their management. Knowledge is a critical input for farming systems. Crop protection in particular is becoming more difficult due to evolution of pest resistance and changes in legislation. Up to date information can be made rapidly available and shared online through websites and smartphone Apps. Agricultural extension no longer relies solely on physical meetings and printed documents. The capacity to share information via the Internet is tremendous with its potential to reach a wide audience in the farming community, to provide rapid updates and to interact more with the users. However, in an era of information deluge, accessing relevant information and ensuring reliability are essential considerations. There is also a need to bring science and farming communities together to turn information into relevant farming knowledge.

Introduction

Agriculture is facing huge pressures in the 21st century to produce more as demand rises but, in an increasingly resource constrained world, there is less scope for it to use more resources such as land, water, energy, and nutrients (Baulcombe et al. 2009). The challenges of meeting surging demand for food, water, and energy while also adapting to climate change were clearly explained by Sir John Beddington in his "Perfect Storm" speech (Beddington 2009). The improvement of advisory services was identified as a priority in the UK government Foresight report, "The Future of Food and Farming" (Foresight 2011) and lack of information flow between scientists, practitioners, and policymakers is mentioned by Pretty et al. (2010) as a constraint to achieving growth in food production. Farm businesses need to find smart ways of producing and protecting high value and high yield produce. To achieve sustainable intensification of agriculture, there is a need to improve access to information to allow farmers and others in the farming community to make better decisions.

Effective knowledge exchange allows farmers and agronomists to benefit from scientific and technical advances. It can also provide invaluable feedback from farmers to the research community that will help improve the research and make it better targeted to farmers' needs and thus increase its impact (Phillipson et al. 2012; Kindred 2015). However, unfortunately, the prospect of stakeholder engagement is sometimes viewed as a distraction by the scientific community (Phillipson et al. 2012) and linkages between the advisory system and the scientific system may need improvement (Klerkx and Proctor 2013). Sometimes the solutions that are appropriate may not be immediately apparent either to the land managers or to the researchers and interaction can help to generate novel solutions. In terms of information about potential solutions, farmers and agronomists require decision support but not decision making because they are the ones to decide what is most appropriate for their local conditions (Wood et al. 2014).

For exploitation of research findings, current thinking is moving from a model of knowledge transfer (Garforth et al. 2004) toward a model of knowledge exchange (Phillipson et al. 2012; Ward et al. 2012; Wood et al. 2014). The implication of "knowledge transfer" is that it is separate from knowledge production but if this separation is not accepted, and networks allow transfer in both directions, then it is possible to have a more interactive model of knowledge production (Phillipson et al. 2012). This means that farmers and agronomists are more involved and provide input and feedback into the process. The book "Farmer First" by Chambers et al. (1989) radically prioritized farmer participation in development of knowledge but since then there has been growing realization that a wider innovation system is needed and the farmer cannot be considered in isolation (Pretty and Chambers 1994; Scoones and Thompson 1994, 2009; Klerkx et al. 2012; Wood et al. 2014). Thus, two-way flow of information can benefit all parties involved.

There is increasing awareness of the important role played by farmer-to-farmer dissemination pathways for effective sharing. Scientists create knowledge but farmers decide what is best practice (Kindred 2015). Farmers and agronomists are more likely to be convinced if they see a neighboring farmer successfully using a new technique or seeing it demonstrated at a trial site than reading about it. For example, farmer to farmer networks have played a key role in successful adoption and upscaling of a push-pull companion cropping system developed by *icipe* and Rothamsted, which is now used by approximately 100,000 smallholder farmers in East Africa (Amudavi et al. 2009). Although farming systems are different in different parts of the world, they all need information to optimize farm productivity. Whether in the developing or developed world, growers value real farming experience and

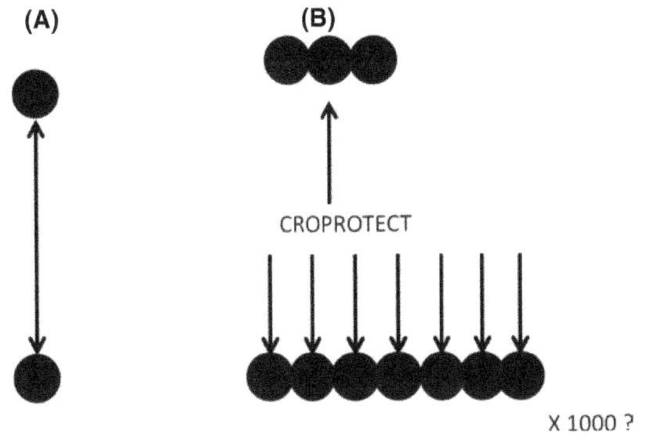

Figure 1. Interaction between knowledge brokers and users. (A) Driving to a farm to meet in person allows an in depth conversation but means farms have to be visited one at a time. (B) Providing online information is a more faceless form of communication but has the advantage of being able to reach a large number of users.

knowledge of the particular over knowledge in general (Wood et al. 2014). Wood et al. (2014) found that New Zealand farmers seek knowledge that can be applied to their individual farm by contacting individuals who can share the experiences of other equally individual farms. Mechanisms need to be found to couple these dissemination pathways with access to science based knowledge and data to improve farming operations. There is potential to do this through online knowledge sharing provided that there are people and structures set up to provide the relevant content.

Online Knowledge Sharing Platforms

There are opportunities to develop online systems for knowledge exchange. Although they will never be a complete substitute for face-to-face meetings, they can be highly complementary, especially in situations where face-to-face meeting do not occur often. Indeed, rather than replacing meetings in person they can open lines of communication that lead to subsequent new contacts and face-to-face meetings. Online systems have major advantages of improved accessibility, and mean that there can be a wider reach while saving time and travel costs. Internet access is widespread and increasingly via mobile devices which mean that information posted can be consumed at a time and place convenient for the user. Furthermore, electronic materials can be rapidly updated and allow interaction with users. Exciting developments in online publishing and social media have occurred but for improved knowledge exchange better systems integrating formal scientific reporting and informal farmer knowledge networks are needed.

There is immense potential to share information about "what works" via the Internet because of its tremendous reach and capacity for sharing information (Fig. 1). The

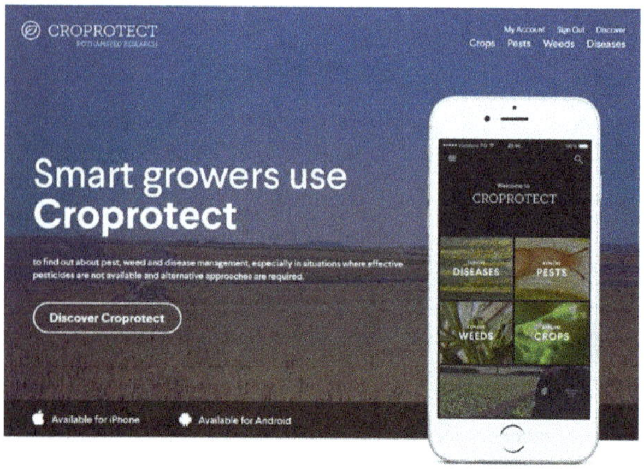

Figure 2. The CROPROTECT homepage (from https://croprotect.com/).

way information is structured, organized, and delivered needs consideration so that it can be optimized. Systems need to be developed to translate research findings into summaries of what they mean for the farmer. There is extensive information available online but it is scattered and of varying quality. The challenge is to find relevant information and provide it in a format where the key points are clearly explained. Some examples of online farming information services include:

Plantwise Knowledge Bank

The CABI online Plantwise Knowledge Bank, www.plantwise.org/KnowledgeBank, provides pest identification tools and a database of online factsheets about plant health (Leach and Hobbs 2013). It is used by "Plant Doctors" who provide extension advice to farmers in developing countries. This is a large scale operation with 3500 plant doctors in 1500 plant clinics in 34 countries. There is a Plantwise Factsheets Library App for Android smartphones as well as the website. The Plantwise program provides actionable knowledge to farmers through community based plant clinics, encouraging exchange of information between farmers and other stakeholders and by drawing together information and making it available in a central resource – the Knowledge Bank. The combination of online information with large numbers of plant doctors to advise farmers perhaps gives the best of both online and face-to-face meetings, however, even the large number of plant doctors may not meet demand. The knowledge bank also contains "Factsheets for Farmers," written by extension workers in-country to ensure they are locally relevant and these could allow direct access by farmers who have a smartphone but not a local plant doctor, Feedback from a user survey indicated that they found there was too much information and they wanted it filtered more, they wanted more images for identification and less technical language and better links to management information (Leach and Hobbs 2013), Internet access varied from country to country and in some countries printed factsheets were used due to lack of Internet connectivity. The Plantwise Knowledge Bank continues to grow and is focusing on adding more locally relevant content (Leach and Hobbs 2013). Some factsheets may require updating. CABI is also developing "Direct2Farm" – a service that turns factsheets into short SMS and voice messages that are delivered straight to smallholder farmers via mobile phones.

IRRI rice knowledge bank

"Rice Doctor" from the IRRI rice knowledge bank (http://www.knowledgebank.irri.org/) provides diagnostic tools and factsheets for management of rice diseases. It is available as a smartphone App for Android or iOS. "Crop Manager" provides rice, maize, and wheat farmers with personalized crop and nutrient management guidelines. Weed identification tools and information about rice crop management at different growth stages are also provided.

PlantVillage

PlantVillage Images, https://www.plantvillage.org/, is an open access database of 50,000+ images of healthy and diseased crops. PlantVillage are launching and growing this database in order to enable the development of open access machine learning algorithms that can accurately classify crop diseases on a smartphone. It also has a function where users can ask questions via the website.

Agriculture and Horticulture Development Board

This UK levy board has a website, http://www.ahdb.org.uk/, containing many valuable factsheets. These are in the crop management section of its website. Information is provided about field drainage, crop diseases, weed management, pest management, nutrient management, soil management, grain storage and sampling, precision farming, and stewardship.

Please note the above examples are not an exhaustive list of all online agricultural knowledge exchange systems – they are only intended to give an idea of the existing state of the art. I now provide a description of the CROPROTECT system I am developing.

CROPROTECT

CROPROTECT is a new UK-based website, www.croprotect.com, which we are developing to provide information about pest, weed, and disease management (Fig. 2). It is a new and growing resource for farmers and agronomists designed to provide easy access to management recommendations for key pests. From the onset we have been engaged in two-way communication with users. This way we can build a system which is relevant to them and meets their requirements, raise awareness and grow the user base at the same time as developing the system. The first version of the website was launched when the project began in November 2014. On this, pioneer users could register and say what their pest, weed, and disease priorities were. Thus, the target pests, for which management information is being compiled, are the ones specified by the users. Not surprisingly, the most frequently mentioned targets are ones which have evolved resistance to pesticides and for which alternative or supplementary management is required. Agriculture is becoming more difficult as the era of cheap effective chemical pesticides is ending and new biological solutions and information about how to implement those solutions is required (Hillocks 2012; Andersons Centre 2014). To address these issues, CROPROTECT aims to draw together existing and new information about crop protection as there is a general problem of information sources being fragmented and disconnected (Klerkx and Proctor 2013). Alternative approaches are often more complicated relying on a combination of resistant cultivars, biocontrol, agronomic practices and rationalized, better targeted pesticide use. Information about integrated pest, weed, and disease management is scattered in disparate places which are hard for busy farmers to track down for every pest, weed, and disease threat they face.

The first pest management information pages on the website were released in February 2015. Information pages were only available to registered users to provide an incentive for users to register and specify their crop protection targets. However, we realized that registration and login was a barrier to entry and after more information pages were compiled we decided to make the information pages available to nonregistered users. At the time of writing we now have over 640 users and in the last month there were 379 visitors to the website. The system is evolving and has only been running for 1 year so far. Electronic documents can be easily updated and tweaked as new information, for example, about a new crop variety or treatment becomes available. This is a major advantage compared to printed documents which are fixed once printed.

In response to user demand for even easier access, a CROPROTECT smartphone App (for Android and iOS) was developed and launched in December 2015. Part of the inspiration for CROPROTECT was the active community of farmers, agronomists and other agricultural professionals who use social media, particularly Twitter, for farming discussions. AgriChatUK, an online farming forum, hold regular weekly discussion sessions every Thursday evening and CROPROTECT has featured in several of them. The discussions are run on Twitter using the #agrichatuk hashtag. They reach many people in the agricultural sector and the AgriChatUK Twitter account has over 18,000 followers. We are using these channels to raise awareness of CROPROTECT and have a representative of, AgriChatUK, on the steering committee for CROPROTECT. Guidance is also being obtained from the Agriculture and Horticulture Development Board (AHDB, a UK levy board), the National Farmers Union and three agronomist organizations (Hutchinsons, Agrii, and AICC).

While the primary aim of CROPROTECT is to provide management recommendations, it also contains a reporting function through which users can say what their pest targets are. User reports are mapped on a GIS (geographical information system) using Google maps which is used to create pest incidence maps. These provide crowdsourced information on which pest, weed, and disease problems occur in different parts of the country. As the number of users grows there may be further opportunities to use these data to predict the spread of pests, weeds, and diseases and could be particularly powerful if coupled with meteorological data and soil data. User data are kept are not put in the public domain unless they are anonymized to protect user privacy. The resolution of pest, weed, and disease incidence maps provided online is restricted so that individual farms are not identified.

Feedback from users has been very positive. They appreciate having straightforward information focusing on key management recommendations. We have emphasized quality and relevance of information over quantity because we know that users' time is limited and they require key pieces of information rather than to have to sift through many pages. The ability to link to other information resources is expected to make CROPROTECT even more useful. We are currently linking extensively to AHDB resources. The capacity to share information via the Internet is tremendous and access is increasingly via mobile devices. These have the potential to reach a wide audience in the farming community, to provide rapid updates and to interact more with the users. In the Internet age, availability of information is not the main constraint, there is more of an issue of accessing relevant information.

The UK, as a member state of the EU, is obliged to implement EU directive 2009-128, the "Sustainable Use Directive" (European Parliament 2009), which not only places restrictions on the use of conventional toxic pesticides but also has a requirement for "Promoting the use of IPM (integrated pest management) and of alternative approaches". IPM is where a combination of control tactics is used against pests such as biocontrol or resistant crop varieties so that reliance on pesticides is reduced (Kogan 1998). For example, resistant varieties and a decision support system using data from pheromone traps have allowed reductions in use of chlorpyrifos against orange wheat blossom midge (Bruce and Smart 2009). CROPROTECT is designed to make information about IPM more readily accessible either through the website or through the App. As such, it will put the UK in a better position for achieving the intention of EU directive 2009-128 which is supposed to provide IPM solutions.

A major advantage of electronic systems is that they can be easily and rapidly updated as new information becomes available. They do not have to be reprinted each time there is an update and this means small changes can be made frequently. However, to ensure the information is kept up to date it is important that there is someone to maintain the site and that regular review points are set.

Closing Yield Gaps

Scientific and technical advances need to be taken to the farmer. In addition to ensuring farmers and agronomists have access to information about crop protection to minimize crop losses in the UK, there are wider opportunities to reach the farming community globally. There are large parts of the world where average crop yields are below their climatic potential which is known a "yield gap" and where yields could improve with different land management practices (Licker et al. 2010). There are opportunities to increase yields across many parts of Africa, Latin America and Eastern Europe (Foley et al. 2011). Part of the reason for comparatively low agricultural productivity is that knowledge about improved farming practices has not yet been delivered and web-based knowledge exchange systems could help reach areas where conventional agricultural advisory services have not reached. For example, the remote locations, small sizes and very large number of smallholder farms in Africa mean that knowledge exchange is an immense challenge. Even if there were large numbers of extension officers it would be difficult to reach all the people requiring help. The reality is that extension services are minimal and in many places nonexistent. Therefore, new methods of reaching farmers need to be found. Being successful in agriculture is challenging even at the best of times but smallholder farmers in sub-Saharan Africa face severe constraints to their production systems due to economic disadvantage, increasingly unpredictable weather patterns, lack of inputs, and lack of extension support. Nevertheless, there is a vast potential for improvement with typical cereal yields of only 1 ton per hectare. Given the right support there is tremendous potential for yield improvement which can make big changes to the livelihoods of smallholder farmers.

In the developed world there may not be such large yield gaps but farmers are coming under increasing pressure to comply with new legislation. Knowledge sharing networks are particularly important in the area of crop protection where pesticides are no longer a reliable solution as pests are evolving resistance to old products and fewer new products are coming through the regulatory system. This is discussed in the next section.

How to Increase Participation of Farmers, Agronomists and Researchers in Online Knowledge Sharing?

For online knowledge sharing to happen the people involved need to actively participate. Access to information technology is a prerequisite but is not as much as of a limiting factor as it was 10 or even 5 years ago. Internet use is becoming almost ubiquitous although a minority of mostly older people in developed countries or poorer people in developing countries may not take part because they do not have the technology. A Defra survey conducted in 2012 (Defra, 2013) found that 86% of UK farms had access to a computer and 29% of UK farmers had a smartphone. Even smallholder farmers in developing countries have widespread mobile phone use (Kikulwe et al. 2014) and they are increasingly using basic version of smartphones that have Internet access. The reasons why people may or may not engage in online knowledge sharing are complex but fall into a few main categories. First of all, the people involved need to know the opportunity exists. For this marketing may be needed to raise awareness. To engage some people may take part for altruistic or social reasons but to dedicate time users need to see some kind of advantage or benefit. Farmer and agronomists would check if the information is relevant to their needs and useful but also form a value judgment if the information is reliable. To raise awareness and ensure the system meets users' requirements CROPROTECT has been working with a group of pioneer farmers and agronomists right from the start of the project. Researchers need to be incentivized to provide information for online systems connecting to end users. Academic career progression is based on publication in scientific journals especially if these articles are referred to by other academic researchers and there is less reward for engagement with other beneficiaries although this is changing with many sponsors increasingly requiring academics to demonstrate "pathways to impact" for their research. Finally systems need to be accessible and user friendly and have high-quality relevant content presented in an attractive way. Something that is difficult to access or make sense of will not be used so much. Many people spend time on social media because they enjoy the interaction and the content. Agricultural knowledge systems need to be developed so that they are a pleasure and not a pain to use.

Lessons Learnt So Far

Online knowledge sharing systems bridging research and user communities are probably in their infancy and will develop rapidly in the next decade. General lessons and guiding principles that have been learnt so far fall into three categories of relevance, ease of use and marketing:

Relevance

- Provide content which the users find relevant and useful
- Make information meaningful
- Tackle the research – user gap
- Provide local contextualization – tackle spatial/temporal variation
- Provide up to date information

Ease of use

- Have a user friendly interface
- Do not make it more complicated than necessary – keep it simple
- Provide ongoing support
- Make key information easy to access and find

Marketing

- Make sure users are aware that the system exists
- Address legislation farmers have to comply with
- Incentive to use (e.g., compliance with legislation)
- Bring farmers into design

Conclusions

The role of communication in agricultural innovation systems deserves more attention (Leeuwis and Aarts 2011). Knowledge exchange is vital for innovation (Cooke 2001) and innovation is vital for improving agricultural productivity (Alston 2010). While sharing technical information can help we must remember that the final decision about land management practices is with the famer, however, farmer decision making is likely to be improved when the farmer is better informed. Given these requirements and the need to improve agricultural productivity in a sustainable way, there is a policy debate about how to stimulate adequate advisory systems and bridge the gap between people separated spatially and in different organizations (Klerkx and Proctor 2013). The Internet was originally invented for information exchange in a military context but has expanded into many diverse applications and is continuing to evolve (Leiner et al. 2009). Defined as, "a world-wide broadcasting capability, a mechanism for information dissemination, and a medium for collaboration and interaction between individuals and their computers without regard for geographic location" (Leiner et al. 2009), it is easy to see that online tools could help solve 21st century agricultural knowledge exchange challenges.

The capacity exists but the computers will not do it by themselves and people are needed to imagine, develop, and use the systems. There is a need for "knowledge brokers" that can span different areas as has been pointed out earlier even before the emergence of online systems (Gibbons 1994). Ideally hybrid systems coupling the individual interactions of social media with the quality control of scientific papers that filter out the relevant information for each farm is needed. Also, for application of agricultural knowledge there does need to be "regard for geographic location" and, as pointed out by Wood et al. (2014), farmers seek knowledge that can be applied to their individual farm and are concerned with particular rather than general findings. The farmer will make decisions based on her or his knowledge of the local conditions supported by information received from external sources. Farmer confidence is important and trusted sources of online content need to be established. To promote systems which translate and share the latest research developments incentives need to be found to engage researchers in the process, perhaps for them to demonstrate their pathway to impact. Finally, knowledge exchange systems to engage with policymakers could support evidence based policy, more open policy making and joined-up policies.

Acknowledgments

This work was supported by the BBSRC-NERC SARIC (NE/M016676/1) project. Rothamsted Research receives grant-aided support from Biotechnology and Biological Sciences Research Council (BBSRC). The author thanks Dr David Rose, University of Cambridge, for holding a workshop meeting which provided insight into factors that make decision support systems successful.

Conflict of Interest

None declared.

References

Alston, J. M. 2010. The benefits from agricultural research and development, innovation, and productivity growth. OECD Publishing, Paris, France.

Amudavi, D. M., Z. R. Khan, J. M. Wanyama, C. A. O. Midega, J. Pittchar, I. M. Nyangau, et al. 2009. Assessment of technical efficiency of farmer teachers in the uptake and dissemination of push–pull technology in Western Kenya. Crop Prot. 28:987–996.

Andersons Centre. 2014. The effect of the loss of plant protection production products on UK agriculture and horticulture and the wider economy. The Andersons Centre, Melton Mowbray, U.K.. Available at: http://www.cropprotection.org.uk/media/89364/andersons_final_report.pdf.

Baulcombe, D., I. Crute, B. Davies, J. Dunwell, M. Gale, J. Jones, et al. 2009. Reaping the benefits: science and the sustainable intensification of global agriculture. The Royal Society, London.

Beddington, J. 2009. Food, energy, water and the climate: a perfect storm of global events. In Conference presentation given to the Sustainable Development UK Annual Conference, QEII Conference Centre, London, 19 March 2009. Available at: http://webarchive.nationalarchives.gov.uk/20121212135622/http://www.bis.gov.uk/assets/goscience/docs/p/perfect-storm-paper.pdf (accessed 24 August 2015).

Bruce, T. J., and L. E. Smart. 2009. Orange wheat blossom midge, *Sitodiplosis mosellana*, management. Outlooks Pest Manag. 20:89–92.

Chambers, R. G., A. Pacey, and L. A. Thrupp. 1989. Farmer first: farmer innovation and agricultural research. ITDG Publishing, London.

Cooke, P. 2001. Regional innovation systems, clusters, and the knowledge economy. Ind. Corp. Change 10:945–974.

Defra. 2013. Computer usage by farmers in England, 2012. Available at: https://www.gov.uk/government/uploads/system/uploads/attachment_data/file/181701/defra-stats-foodfarm-environ-fps-statsrelease2012-computerusage-130320.pdf.

European Parliament. 2009. Directive 2009/128/EC of the European Parliament and of the Council of 21 October 2009 establishing a framework for Community action to achieve the sustainable use of pesticides. Official Journal of the European Union 52:71–86.

Foley, J. A., N. Ramankutty, K. A. Brauman, E. S. Cassidy, J. S. Gerber, M. Johnston, et al. 2011. Solutions for a cultivated planet. Nature 478:337–342.

Foresight. 2011. The future of food and farming: challenges and choices for global sustainability. Government Office for Science, Foresight, Final Project Report, London.

Garforth, C., T. Rehman, K. McKemey, R. Tranter, R. Cooke, C. Yates, et al. 2004. Improving the design of knowledge transfer strategies by understanding farmer attitudes and behaviour. J. Farm Manag. 12:17–32.

Gibbons, M. 1994. Transfer sciences: management of distributed knowledge production. Empirica 21:259–270.

Hillocks, R. J. 2012. Farming with fewer pesticides: EU pesticide review and resulting challenges for UK agriculture. Crop Prot. 31:85–93.

Kikulwe, E. M., E. Fischer, and M. Qaim. 2014. Mobile money, smallholder farmers, and household welfare in Kenya. PLoS One 9:e109804.

Kindred, D. 2015. Effective exchange between practice and research – using farmer networks to enable researchers to learn. Association of Applied Biologists/Food and Energy Security conference, Knowledge exchange: from research to the food supply chain, 17–19 June 2015, University of Lancaster, Lancaster, U.K.

Klerkx, L., and A. Proctor. 2013. Beyond fragmentation and disconnect: networks for knowledge exchange in the English land management advisory system. Land Use Policy 30:13–24.

Klerkx, L., B. van Mierlo, and C. Leeuwis. 2012. Evolution of systems approaches to agricultural innovation: concepts, analysis and interventions. Pp. 457–483 *in* I. Darnhofer, D. Gibbon and B. Dedieu, eds. Farming systems research into the 21st century: the new dynamic. Springer, Dordrecht.

Kogan, M. 1998. Integrated pest management: historical perspectives and contemporary developments. Annu. Rev. Entomol. 43:243–270.

Leach, M. C., and S. L. Hobbs. 2013. Plantwise knowledge bank: delivering plant health information to developing country users. Learned Publ. 26:180–185.

Leeuwis, C., and N. Aarts. 2011. Rethinking communication in innovation processes: creating space for change in complex systems. J Agric. Educ. Ext. 17:21–36.

Leiner, B. M., V. G. Cerf, D. D. Clark, R. E. Kahn, L. Kleinrock, D. C. Lynch, et al. 2009. A brief history of the internet. ACM SIGCOMM Comput. Comm. Rev. 39:22–31.

Licker, R., M. Johnston, J. A. Foley, C. Barford, C. J. Kucharik, C. Monfreda, et al. 2010. Mind the gap: how do climate and agricultural management explain the 'yield gap' of croplands around the world? Glob. Ecol. Biogeogr. 19:769–782.

Phillipson, J., P. Lowe, A. Proctor, and E. Ruto. 2012. Stakeholder engagement and knowledge exchange in environmental research. J. Environ. Manage. 95:56–65.

Pretty, J. N., and R. Chambers. 1994. Towards a learning paradigm: new professionalism and institutions for a sustainable agriculture. Pp. 182–202 *in* I. Scoones and J. Thompson, eds. Beyond farmer first. Intermediate Technology Publications, London.

Pretty, J., W. J. Sutherland, J. Ashby, J. Auburn, D. Baulcombe, M. Bell, and H. Campbell. 2010. The top 100 questions of importance to the future of global agriculture. International journal of agricultural sustainability 8(4):219–236.

Scoones, I., and J. Thompson. 1994. Beyond farmer first: rural people's knowledge, agricultural research and extension practice. Intermediate Technology Publications, London.

Scoones, I., and J. Thompson. 2009. Farmer first revisited: innovation for agricultural research and development. Pp. 3–30 *in* I. Scoones, J. Thompson and R. Chambers, eds. Farmer first revisited. Practical Action Publishing, Warwickshire, U.K..

Ward, V., S. Smith, A. House, and S. Hamer. 2012. Exploring knowledge exchange: a useful framework for practice and policy. Soc. Sci. Med. 74:297–304.

Wood, B. A., H. T. Blair, D. I. Gray, P. D. Kemp, P. R. Kenyon, S. T. Morris, et al. 2014. Agricultural science in the wild: a social network analysis of farmer knowledge exchange. PLoS One 9:1–10.

16

Integrating islands of knowledge for greater synergy and efficiency in crop research

Andrew Borrell[1] & Matthew Reynolds[2]

[1]Queensland Alliance for Agriculture and Food Innovation (QAAFI), Hermitage Research Facility, University of Queensland, Warwick, Queensland 4370, Australia

[2]International Maize and Wheat Improvement Centre CIMMYT, Int. Apdo. Postal 6-641, 06600 Mexico, Mexico

Keywords
agricultural research funding, benchmarking, crop production, efficiency, focus, food security, Global Plant Council, impact pathways, integration across scales, integrators, Sustainable Development Goals, synergy, teamwork, translational research, transnational collaboration.

Correspondence
Andrew Borrell, Queensland Alliance for Agriculture and Food Innovation (QAAFI), Hermitage Research Facility, University of Queensland, Warwick, QLD 4370, Australia.

E-mail: a.borrell@uq.edu.au

Abstract

A transnational approach to plant science is required to address global challenges such as food security and climate change. Action from the plant, crop and agricultural science communities will be required to meet these challenges. Islands of knowledge (e.g. crop, discipline, scale, environment, organization) need to be better integrated. Mapping is needed to identify gaps and make connections between these islands. Translational research, spanning from the molecular to farm scale, should be the focus. We need more `integrators' to manage complex global projects and integration should be recognized as an important skill in itself. We need to advocate teamwork over individuality - and reward teams. Impact should be taken much more seriously as an outcome and we need to better understand impact pathways. Focusing on key game-changers will more likely lead to impact rather than attempting to do everything.

Introduction

Significant challenges relating to food, nutritional, and water security in the coming decades are no longer doubted (Wiebe et al. 2016). The United Nations Sustainable Development Goals (SDGs) address the eradication of poverty and hunger along with the attainment of clean water, increased health, innovation, responsible production, climate action, and sustainable communities. Action from the plant, crop, and agricultural science communities will be required to achieve these goals. The developed and developing world face common challenges (as stated in the SDGs), common environments (e.g., rain-fed crops such as sorghum grown in northern Australia, sub-Saharan Africa, and central-western India frequently experience drought), grow common crops of interest (including maize, wheat, rice, sorghum, barley, tubers, legumes), and live on a common planet (addressing climate change is a global issue). The SDGs highlight how the developed and developing world will need to work together to address these common challenges.

Meeting these challenges will require a transnational approach. For example, when comparing G20-focused priorities for wheat research (www.wheatinitiative.org) with those of less developed countries served by the Consultative Group for International Agricultural Research (CGIAR, www.cgiar.org), 80% of the traits identified are common (www.wheat.org), providing obvious opportunities for international collaboration. While single crop-focused research still achieves economically significant genetic gains, given the transnational nature of many agricultural problems, it is generally accepted that crop scientists will need to work across borders to maintain global food security. Crop research should benefit from a more internationally oriented approach to achieve better leverage of technology, expertise, and infrastructure (Reynolds et al. 2016).

Important outcomes have already been achieved in less-developed countries through their national agricultural research programs in collaboration with international entities such as the CGIAR, the Bill and Melinda Gates Foundation (www.gatesfoundation.org), and various advanced research institutes. These collaborations can be further developed by fostering an environment where productivity constraints of a transnational nature are routinely investigated from a global perspective, integrating research across crops and agro-ecosystems, with routine sharing of data and other resources. Some potential actions over a range of timescales (short, medium, and long term) will be discussed in this study.

Integrating Islands of Knowledge

Funders of agricultural research frequently favor novel or exciting ideas without considering how outputs may link to previous or parallel projects, creating islands of relatively isolated knowledge. An effort to map islands of knowledge could help identify productive synergies. Examples of islands are disciplines, crops, environments, scales, organizations, or major projects (Fig. 1). It is important to define (or recognize) islands in order to identify opportunities to build bridges among relatively isolated research thrusts using comparative biology (e.g., Valluru et al. 2015), translational research linking upstream work to solve productivity problems, interdisciplinary approaches, and integration across biological scales (e.g., Passioura 2010).

Prioritization of investments by funding agencies on the basis of linkage, rather than funding a new island of knowledge, would be a valuable criterion when evaluating research proposals. At the very least, if a new island is funded – then funders need to make sure it is linked to other islands of knowledge. Funding agencies could consider investing in linkages to achieve synergy, rather than speculating on novelty, single-discipline approaches, or projects with limited scale. Rather than expecting automatic outcomes from relatively specialized teams, impacts will be more likely if all appropriately qualified groups are funded considering comparative advantage.

Focus and Integration

To be most effective, we also need to strike a balance between focus (or reductionism) and integration. Linkages can be made between focused projects by reviewing outputs at regular intervals to identify pertinent linkages. For example, projects should consider the immediate level of integration above and below the focus of the project (Passioura 2010). This would go a long way to opening opportunities for sharing concepts and technologies between different levels of integration from within the same discipline. In this context, a research team studying drought adaptation in cereals at the organ level (e.g., leaf growth) should integrate their findings 'up' a scale to the whole plant level and 'down' a scale to the tissue level, thereby better understanding how leaf dynamics fit into a larger framework (Fig. 2). This approach should also encourage scientists to collaborate across scales (e.g., molecular biologists, biochemists, plant physiologists, crop physiologists, agronomists, and plant breeders) to achieve both increased understanding and greater impact.

Similarly for work in model species that is funded on the basis of its translational potential, confirmation that mechanisms and genes can be extrapolated to crops should be sought at the earliest opportunity, certainly before 'exciting' results – that may not in fact be relevant to productivity constraints – cause researchers to go off on

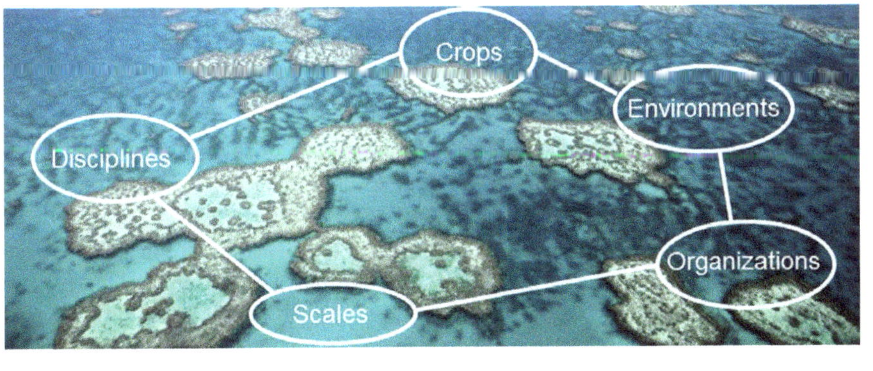

Figure 1. Building bridges between islands of knowledge should enhance the impact of global research addressing food security (background image from www.world.time.com).

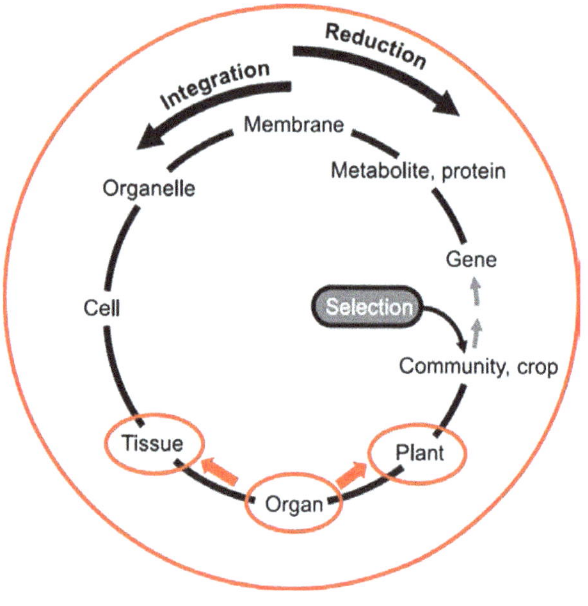

Figure 2. Modified from Figure 1 in Passioura (2010). Levels of organization in crop plants, represented as a loop in which clockwise flow represents reduction, the search for mechanistic understanding at finer and finer scales, whereas anticlockwise flow represents functional integration, the roles of various structures and processes in transmitting genes to the next generation. In the sense that selection of individual genotypes at the crop level (or, in natural systems, the community) determines the genetic makeup of the next generation, the 'loop' is a helix when viewed across generations.

costly tangents. The importance of this point is underscored by the generally poor association found between field and controlled environments (Poorter et al. 2016).

However, to achieve integration, rather than having large unmanageable projects with substantial transaction costs, the scientific community should identify and support bigger-picture thinkers who have the appropriate breadth of experience to identify ways to integrate outputs of projects. We need to invest specifically in the integrators. While in theory, research directors and managers have this role, in practice time for such 'big-picture' observation and thinking is eroded by ever increasing transaction costs associated mainly with administrative compliance and more restrictive funding opportunities. Furthermore, integration should be recognized as an important skill in itself, and all individuals should be encouraged to highlight their work in the broader context, and to promote better communication and understanding of the bigger picture.

Recognizing Teams

We need to promote team effort rather than individual achievement. This will be facilitated by less hierarchical models in favor of team evaluation and multisource assessments. While authorship of scientific publications does highlight a trend from single or few authors per paper 30 years ago, to many authors per paper now, we still have a way to go. Unfortunately, many of the accolades in science go to the individual rather than the team. This does not mean that such accolades are not worthwhile – indeed they do recognize great achievement and they do put science in the spotlight. Nonetheless, we need to recognize the importance of team dynamics as well.

It is important to recognize the value of all disciplines in each context rather than considering one discipline as inherently more valuable or 'scientific'.

Impact is Important

Impact, as well as publication record, should be taken much more seriously as an outcome of research. The magnitude of impact is generally related to the degree of integration of a research study and the extent to which islands of knowledge are linked; however, impact does not always just happen. Careful planning, adequate resourcing and precision in execution will all help to enhance the usefulness of any data collected. From the outset, transnational scientific endeavors which actually target enhanced food security (not just high-quality plant science) should strategically plan a clear pathway to impact. Often, focusing on a few key game changers will result in greater impact than attempting to do everything. The planning process should include benchmarking, identifying strategic goals for crop improvement and management, and developing a detailed project plan. In bilateral partnerships, pairing of researchers (one from each country) from common disciplines (breeding, physiology, modeling, information technology, etc.) can work well in this planning process. Breaking down barriers enhances impact.

Benchmarking – Beginning with the End in Mind

Benchmarking is a fundamental component of planning transnational collaborations. In crop improvement programs, data should be collected and collated to provide a detailed description of the starting point. All progress at the end of the project should be assessed relative to this initial benchmark, enabling impact to be objectively assessed. Greater clarity can be achieved by benchmarking individual components of crop improvement programs (e.g., breeding, physiology, grain quality, information technology, modeling, etc.), and can help identify where disciplines or expertise are missing.

Successful National and International Collaborations as Case Studies

Global wheat partnerships

There are some good examples of agricultural science collaborations at the national and international levels, including successful multilateral and bilateral partnerships, and potential benefits of more global integration of agricultural research have been documented (Reynolds et al. 2012). At the international level, the CGIAR now supports a number of collaborative approaches. Networks for disseminating and evaluating new wheat germplasm, for example, emerged in the Green Revolution (Braun et al. 2010), and continue to this day (Fig. 3). Now with modern GIS tools, greater precision of data collection is possible and statistical and simulation models can be used to effectively interpret such databases (Lobell et al. 2011; Gourdji et al. 2012).

A good example of an effective international partnership is the International Wheat Improvement Network (IWIN), which embraces a global collaboration of wheat scientists testing approximately 1000 new well-adapted, disease-resistant wheat lines each year at over 700 field sites. Breeding is directed toward 12 different megaenvironments, representing a range of temperature, moisture, and disease profiles as described in Braun et al. (2010), based on up-to-date GIS and economic information (Gbegbelegbe et al. 2017). As well as testing widely for yield and adaptive phenology, in Kenya, for example, wheat lines are screened for a new and highly virulent stem-rust fungus, in a systematic effort to avert a devastating pandemic (Singh et al. 2011). Also, in partnership with CGIAR, the International Wheat Yield Partnership (IWYP) utilizes a research platform in a region of high wheat productivity in Mexico, where outputs of basic research from IWYP's international projects can be tested and translated into breeding products before being distributed internationally via the network. The IWIN has resulted in massive phenotypic datasets (Braun et al. 2010; Gourdji et al. 2012) and delivered germplasm that is estimated to be worth several billion dollars in extraproductivity to hundreds of millions of farmers in less-developed countries, annually (Pingali 2012), and by raising yields has saved more than 20 Mha of land from being brought under cultivation (Stevenson et al. 2015).

Translating photosynthesis research from laboratory to field

At the national level, the Australian Research Council (ARC) has recently funded a 7-year Centre of Excellence for Translational Photosynthesis (CoETP; http://photosynthesis.org.au/). This venture is designed to integrate islands of knowledge through multiorganizational (ARC, The Australian National University, The University of Queensland, Western Sydney University, The University of Sydney, CSIRO and International Rice Research Institute), interdisciplinary (plant breeding, crop physiology, genomics, molecular biology, biochemistry, engineering, bioinformatics, and simulation modeling), multiscale (subcell, cell, organ, whole plant, crop), and translational

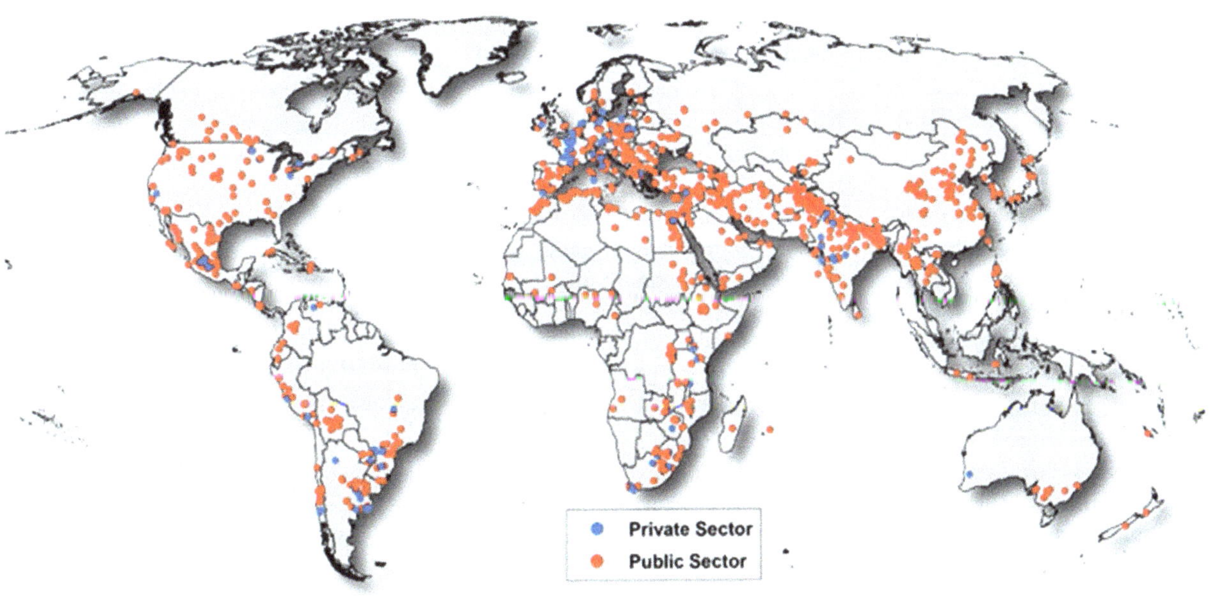

Figure 3. Testing sites of the International Wheat Improvement Network (IWIN).

(spanning molecular to farm levels) approaches. The Centre brings together researchers to tackle one of the grand challenges facing humanity – feeding more people with less resources in the face of climate change. Photosynthesis is the most important biological process to have arisen in the evolution of the Earth, producing the oxygen we breathe, the food we eat, and the climate we live in today (http://photosynthesis.org.au/; Hohmann-Marriott and Blankenship 2011). Improving photosynthesis is now recognized as the new frontier for increasing crop yields (Long and Ort 2010; Zhu et al. 2010; Ainsworth et al. 2012).

The aim of the Centre is to contribute to the development of new crop varieties that are more productive than existing strains, particularly in wheat, rice, and sorghum. This includes varieties that produce high yields of grain in conditions of less water, higher temperature, and higher concentrations of carbon dioxide in the atmosphere. By improving the ability of major food plants to convert sunlight, air, and water into leaf and seed production through photosynthesis, the Centre can contribute to a new yield revolution for food crops. The key word is *translational* – taking scientific discoveries from the laboratory and translating them to useful yield outcomes in the field. It is early days yet – but there is already a palpable sense of excitement about the opportunities and potential outcomes arising from the synergies created in this project.

Increasing the rate of genetic gain for sorghum in sub-Saharan Africa

Successful bilateral partnerships highlight the capacity to combine focus and integration. An example of this is a partnership between the Ethiopian Institute of Agricultural Research (EIAR, Ethiopia) and the University of Queensland (UQ, Australia) funded by the Bill and Melinda Gates Foundation and the Australian Centre for International Agricultural Research (ACIAR). The project addresses food security issues in sub-Saharan Africa by seeking an enduring increase and stability in productivity of sorghum in water-limited production systems through the development of more effective local crop improvement programs. The partnership is outcome driven, focused on strategic step changes, yet aware of the big picture. This project is an example of linking islands of knowledge in terms of disciplines (plant breeding, crop physiology, simulation modeling, bioinformatics), environments (low, mid, and high altitude), scales (organ, plant, crop), and organizations (EIAR, UQ, various Ethiopian universities).

One key insight driving this project is the importance of focusing on a few key game changers, rather than trying to do everything. As this partnership is modeled on the Australian Sorghum Improvement Program, the key game changers were identified by experienced scientists who were able to pick likely winners (Hammer and Jordan 2007). A logical framework was used to dissect the rate of genetic gain, identifying where the greatest returns on investment could be achieved. Some of the game changers proved to be relatively simple, yet profoundly impacted outcomes. For example, the concept of partial replication was introduced to increase the number of genotypes that could be evaluated, smart phones replaced hand-written field-books to revolutionize data capture, and bar codes automated field measurements and seed processing. Linking these technologies has the capacity to significantly enhance the rate of genetic gain in EIAR's sorghum breeding program.

Actions to Increase Synergy and Efficiency in Crop Research

Short term

A key role for the Global Plant Council (GPC) (http://globalplantcouncil.org/) could be to map islands of knowledge to help identify productive synergies. It would be important to develop an inventory of investments in crop research internationally, in order to identify gaps that may be filled with opportunities in comparative biology, translational research, and other interdisciplinary approaches. GPC could do some research on what interdisciplinary models have been most effective in achieving real-life outcomes from agricultural R&D, and then make recommendations to the donor community on more effective deployment of funding strategies. This research could also include economic analysis.

The role of integrators needs to be further investigated. GPC could explore how integrators are a catalyst for creativity by stimulating innovation at the interface between various disciplines, crops, institutions, environments, and scales. GPC could also consider how research organizations and funders might better recognize and invest in integrators.

Short-medium term

Global Plant Council could recommend that funding agencies prioritize investments on the basis of linkage, rather than funding new islands of knowledge. In other words, funding models should consider integration and interdisciplinary approaches over 'magic bullet' type approaches which tend to encourage the opportunist approach over scientific rigor. The GPC blog could list and highlight work/projects/articles that explicitly demonstrate integration of disciplines and synergistic approaches that are outcome and impact oriented. GPC could assist funding agencies in defining and identifying islands using the inventory mentioned above, and propose mechanisms for integration to create synergies and increase efficiencies.

The balance between focus and integration, and the concept of integrating across levels, are worthy of more consideration. GPC could play a role in further understanding integration across levels.

Global Plant Council could have an advocacy role with journals and funding bodies to encourage approaches (aims and scopes, special issues, calls, etc.) that showcase multidisciplinary and outcome-focused work. GPC could also nurture this agenda with groups of young people, high schools, undergraduate groups, etc. It would be beneficial to take advantage of recognized leaders to inspire a new generation to think about the broader context of disciplinary efforts and integration; concept papers or newspaper and magazine articles could also be written about integrative approaches. GPC could also facilitate access to:

- Platforms to share tools and encourage the sharing of resources such as manuals, methods, links to tried and tested manufacturers of technologies, etc.
- Shared intellectual property, for example, through "patent pools" to encourage crosslinkages.
- Long-term and/or multilocation experiments can provide excellent platforms for integration. Rothamsted provides a good example in the Broadbalk experiment which started in 1843.
- Crop-related models that help integration of disciplines and make use of islands of knowledge.
- Other "linking" technologies such as genome resources that link trait biology with breeding data, and phenotyping approaches that are applicable across many environments, crops, etc.
- Crop modeling is an area that depends on interdisciplinary linkage and would achieve much greater impacts as a result of many of the suggestions made herein, including shared data platforms, common research standards, and multilocation and other comparative research approaches.

Medium term

Global Plant Council could investigate models that favor teamwork over individuality. What are the benefits of teamwork? How do we reward teams? How do we sell the merits of team research compared with individualism? There are definite synergies arising from well-integrated team research. For example, some individuals are more likely to publish better science and have greater impact if they are part of a well-integrated team rather than operating in isolation.

Long term

Impact should be taken much more seriously as an outcome. Many organizations and funders focus on publications and citations more than impact. In the face of significant global challenges such as food security, resource scarcity, and climate change, we need scientists to actually apply what has been documented in papers. This is not an either/or scenario. Impact and publications should go hand-in-hand. They are not mutually exclusive. There could be a role for GPC to explore how R&D impacts major global issues around crop production – with a focus on impact. GPC could also do some research on impact pathways, identifying the key steps required to enhance impact and outcomes. GPC can try to promote a 'culture change' in crop research away from speculating on novelty and emphasizing a particular discipline because it is in vogue in favor of integration, problem solving, and favorable outcomes for humanity. Finally, to enable impact to be taken more seriously and implemented cost effectively, GPC could encourage organizations and funders to think more about measurement and metrics at the start of a project, rather than acting retrospectively.

Conclusions

A transnational approach to plant science is required to address global challenges such as food security and climate change. Islands of knowledge (crop, discipline, scale, environment, organization, etc.) need to be better integrated. Mapping is needed to identify gaps and make connections between these islands. Translational research, spanning from the molecular to farm scale, should be the focus. We need more 'integrators' to manage complex global projects and integration should be recognized as an important skill in itself. We need to advocate teamwork over individuality – and reward teams. Impact should be taken much more seriously as an outcome and we need to better understand impact pathways. Focusing on key game changers will more likely lead to impact rather than attempting to do everything. There is a role for the Global Plant Council to facilitate and empower the changes required to bring about greater synergy and efficiency in crop research worldwide.

Acknowledgments

This paper is based on outcomes from a Stress Resilience Symposium held in Brazil in October 2015 organized by the Global Plant Council and Society for Experimental Biology. The authors would like to thank the Society of Experimental Biology for funding support for this symposium.

Conflict of Interest

None declared.

References

Ainsworth, E. A., C. R. Yendrek, S. Sitch, W. J. Collins, and L. D. Emberson. 2012. The effects of tropospheric ozone on net primary productivity and implications for climate change. Annu. Rev. Plant Biol. 63:637–661.

Braun, H. J., G. Atlin, and T. Payne. 2010. Multi-location testing as a tool to identify plant response to global climate change. Pp. 115–138 *in* M. P. Reynolds, ed. Climate change and crop production. CABI Climate Change Series, Oxfordshire, U.K.

Gbegbelegbe, S., D. Cammarano, S. Asseng, R. Robertson, U. Chung, M. Adam, et al. 2017. Baseline simulation for global wheat production with CIMMYT Mega Environment specific cultivars. Field. Crop. Res. 202:122–135.

Gourdji, S. M., K. L. Mathews, M. P. Reynolds, J. Crossa, and D. B. Lobell. 2012. An assessment of wheat breeding gains in hot environments. Proc. R. Soc. B Biol. Sci. 280:1752–1760.

Hammer, G. L., and D. R. Jordan. 2007. An integrated systems approach to crop improvement. Chapter 5. Pp. 45–61 *in* J. H. J. Spiertz, P. C. Struik and H. H. van Laar, eds. Scale and complexity in plant systems research: gene-plant-crop relations. Springer, Netherlands.

Hohmann-Marriott, M. F., and R. E. Blankenship. 2011. Evolution of photosynthesis. Annu. Rev. Plant Biol. 62:515–548 (eds S.S. Merchant, W.R. Briggs, & D. Ort).

Lobell, D. B., M. Bänziger, C. Magorokosho, and B. Vivek. 2011. Nonlinear heat effects on African maize as evidenced by historical yield trials. doi:10.1038/nclimate1043.

Long, S. P., and D. R. Ort. 2010. More than taking the heat: crops and global change. Curr. Opin. Plant Biol. 13:241–248.

Passioura, J. B. 2010. Scaling up: the essence of effective agricultural research. Funct. Plant Biol. 37:585–591.

Pingali, P. L. 2012. Green Revolution: impacts, limits, and the path ahead. Proc. Natl Acad. Sci. USA 109:12302–12308.

Poorter, H., F. Fiorani, R. Pieruschka, T. Wojciechowski, W. H. van der Putten, M. Kleyer, et al. 2016. Pampered inside, pestered outside? Differences and similarities between plants growing in controlled conditions and in the field. New Phytol. 212:838–855.

Reynolds, M. P., J. Hellin, B. Govaerts, P. Kosina, K. Sonder, P. Hobbs, et al. 2012. Global crop improvement networks to bridge technology gaps. J. Exp. Bot. 63:1–12.

Reynolds, M. P., E. Quilligan, P. K. Aggarwal, K. C. Bansal, A. J. Cavalieri, S. C. Chapman, et al. 2016. An integrated approach to maintaining cereal productivity under climate change. Glob. Food Sec. 8:9–18.

Singh, R. P., D. P. Hodson, J. Huerta-Espino, Y. Jin, S. Bhavani, P. Njau, et al. 2011. The emergence of Ug99 races of the stem rust fungus is a threat to world wheat production. Annu. Rev. Phytopathol. 49:465–481.

Stevenson, J. R., N. Villoria, D. Byerlee, T. Kelley, and M. Maredi. 2015. Green Revolution research saved an estimated 18–27 million hectares from being brought into agricultural production. Proc. Natl Acad. Sci. USA 110:8363–8368.

Valluru, R., M. P. Reynolds, and T. Lafarge. 2015. Food security through translational biology between wheat and rice. Food Energy Secur. 4:203–218.

Wiebe, K., D. Mason-D'Croz, T. Sulser, N. Cenacchi, S. Islam, and R. Robertson. 2016. Agriculture and food security in 2050: socioeconomic and climate challenges and options. International Food Policy Research Institute (IFPRI), Washington, DC.

Zhu, X.-G., S. P. Long, and D. R. Ort. 2010. Improving photosynthetic efficiency for greater yield. Annu. Rev. Plant Biol. 61:235–261.

17

Bioenergy from agroforestry can lead to improved food security, climate change, soil quality and rural development

Navin Sharma[1], Babita Bohra[2], Namita Pragya[2], Rodrigo Ciannella[1], Phil Dobie[1] & Sarah Lehmann[3]

[1]World Agroforestry Centre, UN Gigiri, Nairobi, Kenya
[2]World Agroforestry Centre, NASC Complex, New Delhi, India
[3]Intern at GIZ, Untergasse 15, 65510 Idstein, Germany

Keywords
Agroforestry, bioenergy, biofuels, food security, livelihood.

Correspondence
Babita Bohra, World Agroforestry Centre, NASC Complex, New Delhi, India.
E-mail: b.bohra@cgiar.org

Funding Information
No funding information provided.

Abstract

Well-designed bioenergy systems can contribute to several objectives, such as mitigating climate change, increasing energy access, and alleviating rural poverty. With adequate technical assistance and land management, farm yields and income can be increased, food security strengthened, carbon sequestration improved, and pressure for land clearing reduced. There are, nonetheless, risks involved on bioenergy production and several initiatives worldwide have failed to achieve proposed positive outcomes. Overreliance on monoculture plantations, negative land-use change impacts, and use of cereal crops as feedstocks are among the main causes. Agroforestry systems and practices can address most of these risks and thus play an important role in sustainable production of several bioenergy outputs, including efficient solid biomass, biogas, liquid biofuels, and dendro power. This article assesses the potential of such integrated approaches to provide multiple benefits, including the coproduction of food, animal feed, and organic fertilizers, while respecting economic, social, and environmental sustainability indicators. Building on experiences from sub-Saharan Africa, developing Asia, and Latin America, promising perennial species, production models, and value chains are analyzed. Finally, key challenges and potential solutions for larger scale adoption of integrated food-energy approaches are also identified and discussed.

Introduction

Globally, 1.2 billion people lack access to electricity, while 2.7 billion depend on traditional fuels (e.g., firewood, charcoal, and dung) for cooking (IEA, 2011); a figure linked to approximately 4.3 million deaths a year from exposure to household air pollution (WHO). Most of these people can be found in rural areas of developing Asia or sub Saharan Africa. They constitute a large part of the one billion global workforces whose livelihoods depend on agriculture (FAO), as well as the 500 million smallholder farmers who produce 80 % of the food in these regions (Nwanze 2011). Not by chance, many of them are also part of the 980 million people living in extreme poverty in rural areas worldwide (IFAD, 2011).

Improving access to affordable and reliable modern forms of energy services is essential to reduce poverty and promote economic growth in the developing world (Malla 2013). Energy is required not only for basic human needs, such as for lighting, cooking and heating, but also for improving agricultural productivity, allowing the use of advanced machinery from sowing to postharvest processing. Decreasing costs and scaling-up of several technologies for decentralized renewable energy (RE) production is undoubtedly essential to address this issue. Although enhancing energy access, efficiency and the use of renewables might have been missed by the Millennium Development Goals; this target is nowadays largely acknowledged as crucial to the global development agenda. In particular, the Sustainable Development Goals (SDGs) launched in 2015 has adopted a specific goal to "ensure access to affordable, reliable, sustainable and modern energy for all" (UNDP). The Sustainable Energy for All initiative pioneered by the United Nations and the World Bank is also dedicated to achieving similar objectives (The Secretary-General's High-Level Group, 2012).

On the other hand, what is often overlooked by policy makers and governments worldwide is the potential to reduce poverty by encouraging smallholder farmers to become, in addition to consumers, also producers of renewable energy. Traditional agricultural markets (for food, feed, and fiber mainly) account for only 3 % of global GDP (FAO). They are thus too small to sustain so many people whose livelihoods depend on this sector. Rural development requires not only encouraging energy provision to meet local needs, be it to provide cleaner cooking fuel or to intensify agricultural production. Moreover, smallholder farmers need alternatives to enhance their incomes. Energy markets, through sustainable bioenergy production, provide such opportunities. Out of the 7.7 million people estimated to be employed (directly or indirectly) in the renewable energy sector, approximately 40 % are in the bioenergy segment – producing liquid biofuels, modern biomass, or biogas (Ferroukhi et al. 2015). No other renewable energy source has the potential to generate so many jobs for farmers in developing countries. Solar technologies, for instance, have created a significant number of jobs (approximately 3.2 million globally), but these are highly concentrated in a few industrialized countries, in addition to focusing on urban areas.

Well-designed bioenergy production systems can contribute to several objectives, such as mitigating climate change, increasing energy access, and alleviating rural poverty (Casillas and Kammen 2010). With adequate technical assistance and land management, farm yields and income can be increased, food security strengthened, carbon sequestration improved, and pressure for land clearing reduced (Dale et al. 2011). There are, nonetheless, risks involved with bioenergy production and several initiatives have failed to achieve these positive outcomes. The benefits of *Jatropha curcas*, for instance, were clearly oversold in developing countries such as Tanzania (Sosovele 2010). Particular features of this species, such as the toxicity of its seedcake, require further investment in research before larger scale field deployment. Many governments have also encouraged large-scale monoculture of bioenergy feedstocks in poor regions. Ethiopia, for instance, set aside 17.2 million ha of land for biofuel production in 2010 (Sosovele 2010). Land-use changes had negative impacts in some cases. In Indonesia, for instance, conversion of peat swamp forests to oil palm led to significant losses in terms of biodiversity and aboveground biomass carbon, in addition to annual emissions of about 4.6 million Mg of belowground carbon from peat oxidation (Koh et al. 2011). In other instances, the diversion of food crops toward fuel production played a relevant role in food price increases, jeopardizing food security in poor food-importing countries (HLPE, 2013).

Agroforestry systems and practices can address most of these risks and thus play an important role in sustainable bioenergy production. It is well-known that models that integrate woody perennials (trees, shrubs, palms, bamboos, etc.) with crops and/or animals can contribute to climate change adaptation (Lin et al. 2008) and mitigation (Budiadi and Ishii 2010), improve soils (Glover et al. 2012) and natural resource management (Ong and Kho 2015), increase yields (Bayalaa et al. 2012), income (Sood 2006), and thus strengthen food security (Sanchez et al. 1997). Furthermore, such systems allow for food, feed, and several types of bioenergy to be produced respecting economic, social, and environmental sustainability indicators. In an integrated strategy together with improved kilns and stoves, agroforestry can sustainably provide woodfuel, reducing wood harvest pressures in forests in sub-Saharan Africa (Iiyama et al. 2014). It can also provide modern alternatives to cooking fuel, such as biogas produced from waste biomass materials. This approach has been largely adopted in countries such as Vietnam and Nepal (Bogdanski 2012). As Sri Lanka's experience demonstrates, agroforestry can also supply biomass for economical electricity (dendro power) generation (Kulatunga 2012). Regarding liquid biofuels, it is worth noting that global investment in this sector has plummeted in the last 5 years, and most of it is now concentrated on advanced technologies (IREA, 2016), such as second-generation ethanol produced from lignocellulosic biomass or third-generation fuels derived from algae, which are coming to age (UNCTD, 2016). However, as section Programme for the Development of Alternative Biofuel Crops – An ongoing project of the World Agroforestry Centre (ICRAF) discusses with examples in India, Kenya, and Brazil, biofuels can still be sustainably and competitively produced through agroforestry systems that tap into the potential of native (and sometimes less popular) oilseed tree species. These systems have low production cost, not only because the technologies deployed are relatively simple and well-established, or due to the sustainable supply of cheap raw materials, but also because all components of the resulting biomass, including by-products and waste, are used in a bio refinery system (Gupta et al. 2014).

The objective of this article is to evaluate the potential of bioenergy from agroforestry systems to improve livelihoods and food security, to assess progress made in this area and to identify knowledge gaps. It explores the role of trees in providing energy and suggests ways of making bioenergy a mainstream energy source. It cites examples existing of national and regional programs that require farmers to incorporate trees into agricultural systems and describes various activities undertaken in these programs.

The paper also answers specific questions relating to constraints to and motivations for, including trees in farming systems.

Agroforestry, Gender, Health, and Energy Production – Increasing the Quality of Life

The World Agroforestry Centre (ICRAF) defines agroforestry as "a collective name for land-use systems and practices where woody perennials (trees, shrubs, palms, bamboos, etc.) are deliberately integrated with crops and/or animals on the same land management unit in some form of spatial arrangement or temporal sequence. There are three broad types of agroforestry: (i) agrosilviculture which integrates annual crops and trees, (ii) silvopastoral which defines the integration of livestock and trees and (iii) agrosilvopastoral which integrates annual crops, plus livestock, and trees. Furthermore, there are different forms of agroforestry systems, such as alley cropping systems, intercropping, or hedgerow systems.

There are normally both ecological and economic interactions between the woody and non-woody components in agroforestry" (Lundgren and Raintree 1982). For example, agroforestry integrates perennials on farmlands. Trees and crops provide mutual benefits. The roots of crops force the tree roots to colonize a deeper soil horizon, protecting the tree from droughts and affording it access to otherwise unused nutrients. The same roots allow for a much better percolation of rainwater into the deep soil, reducing waterlogging and storing much needed water that becomes available through capillary action to crops during dry spells. The tree roots on the other hand bind the soil, reduce soil erosion, especially on sloping grounds. The trees slow down winds and reduce soil desiccation. They increase soil moisture and reduce nutrient leaching. Their roots act as a pump, bringing nutrients to the surface from deep soil horizons. These become available to crops through leaf, twig and tree-root litter, all of which also contributes to soil organic matter.

Smart agroforestry systems can improve livelihoods, provide food and nutrition security, and clean energy in the form of bioenergy (refer Fig. 1). In this way they contribute to sustainable rural development. Trees provide goods, including timber for fuel, fodder for animals, and food for people (fruits, vegetables, nuts, and oils), which are essential to meeting nutrition requirements, and trees also provide ecosystem services as they help improve the water management and soil fertility, for example, through maintenance of the nitrogen cycle. They support further environmental services, such as carbon storage, water purification, and biodiversity conservation.

Wood-based fuels have been associated with poverty and are believed to pollute the environment. However, when trees are replanted and woodfuels applied in modernized technologies, they can be clean and truly renewable. Trees have the potential to meet the urgent need for energy among poor people. The inclusion of tree-based energy in policy discussions is therefore required.

Agroforestry reduces pressure on forests as trees grown on farms provide fuelwood, helping to meet local energy needs. Forests have been exploited for centuries to provide

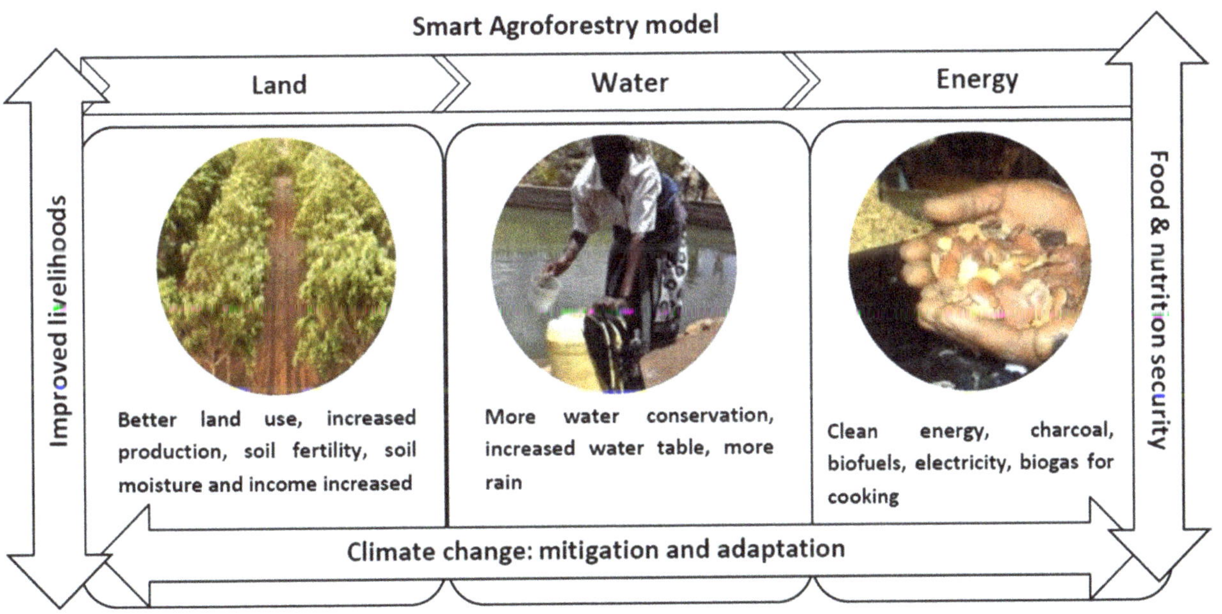

Figure 1. Integration of energy and food security through smart Agroforestry model. Source: Babita Bohra, ICRAF

the basic energy needs of human beings. This remains unabated. In particular, charcoal production for urban energy is one of the major causes of the systematic degradation of dryland forests worldwide (Ouya 2013). To fight forest degradation, it is imperative to foster more efficient production and use of firewood and charcoal, while also promoting alternative sustainable bioenergy options.

In order to address issues related to agriculture, forestry and livelihoods simultaneously, adopting a landscape approach is essential. Landscapes are complex systems with sets of social, biophysical, human, ecological, and economic dimensions that interact with each other. Such interactions happen at multiple levels: the plot, field, farm, and beyond. Understanding and building on interactions and feedback loops is thus important for success (Selvarajah 2013). Furthermore, the landscape approach places forest-dependent communities at the center.

Agroforestry holds potential in climate change mitigation and adaptation. Interestingly, global climate change initiatives tend to target reforestation and forest protection and largely do not mention agroforestry. However, agroforestry provides a win-win solution to a seemingly difficult choice between reforestation and agricultural land-use changes and is thus a powerful way to address mitigation. According to estimates by the International Panel on Climate Change (IPCC), land-use change activities generate 1.6 billion tons of carbon annually (Denman 2007). Modern, well-planned, wisely-implemented, and locally adapted bioenergy systems hold promise in addressing increased carbon emissions.

Agroforestry can further help farmers and communities to adapt to changes in climate through maintaining and enhancing environmental services. It can provide opportunities for diversification that build resilience and generate additional income. There are innumerable examples of traditional land-use practices involving the growing of trees with agricultural crops to augment food production and buffer farmers against the risks of crop failure by providing alternate sources of income. The incorporation of livestock and bioenergy also provide diversification opportunities to farmers. Agroforestry interventions are uniquely placed to achieve multiple objectives of food and environmental security (refer Fig. 2).

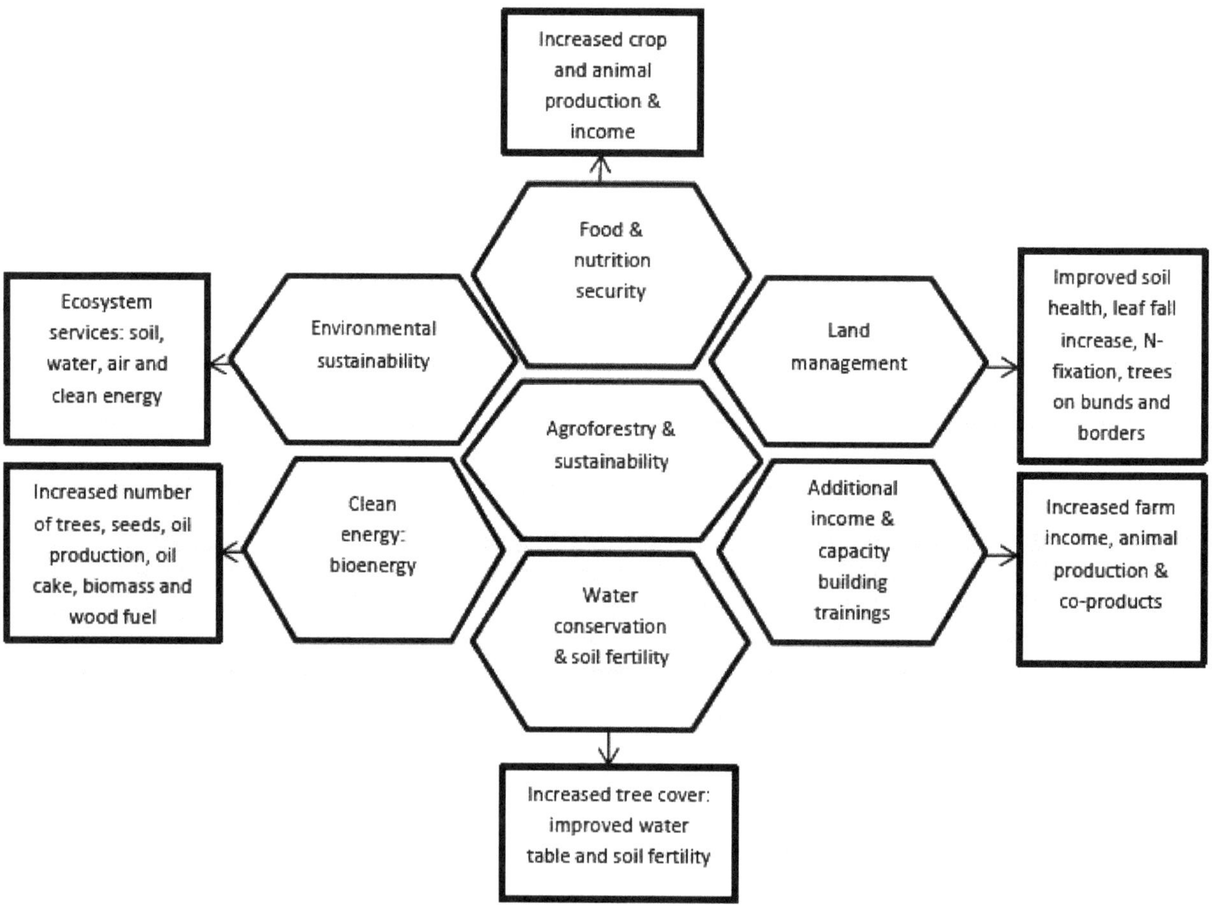

Figure 2. Multiple benefits from Agroforestry system. Source: Babita Bohra, ICRAF

The FAO projects that global food production will need to increase by 60% by 2050 (from 2005/07 levels) to meet the world's demand under a business-as-usual scenario (Millennium Institute, 2013). When the landholding size is small, such as in Africa and the Indian subcontinent, combining traditional agriculture with appropriate trees might be a good solution to optimize farm productivity. Thus, agroforestry interventions might act as potent instrument in achieving the required growth in agriculture.

Agroforestry is a "social forestry" that helps people to meet their economic, environmental, and social needs on their own land (refer Fig. 2). Through providing tree product-based economic opportunities, agroforestry can also help to create rural employment.

Agroforestry as a source of energy

Trees can provide feedstocks for production of various forms of bioenergy, for eg. solid biomass can be used as firewood, as charcoal and for electricity generation. Oilseeds can be used for production of liquid biofuels like biodiesel, while lignocellulosic biomass can be used to produce ethanol. Residues like leaves and oilseed cake can be used for biogas production (refer table 1 for various tree based technologies). Selecting the appropriate tree species to provide the desired benefits and integrating them into traditional agriculture without compromising and, if possible, improving agricultural productivity, remains a challenge. Agroforestry approaches will increasingly find a central place in these bioenergy initiatives.

Forms of bioenergy that can be derived from agroforestry systems

Fuelwood and charcoal

Fuelwood plays an important role in the lives of the poor and rural families, by providing a primary source of energy. Fuelwood has both domestic and industrial uses and is used in rural and urban regions of most economies of the developing world (Dovie et al. 2004). Seventy-nine percent (79%) of the total traditional energy (fuelwood, cow dung, biomass etc.) consumed in developing countries is fuelwood and between 60% and 69% of this is in sub-Saharan Africa (Adebimpe 2013). About 70% of the energy consumed in India is met by fuelwood collected from forests and marginal lands.

Despite such widespread reliance on fuelwood, there are frequently no adequate policies in place that could make fuelwood use more sustainable and efficient. Unsustainable harvesting of fuelwood from forests may lead to negative impacts like habitat fragmentation and loss of environmental services, the encroachment of unprotected natural forests, and finally to deforestation. If fuelwood species are grown on fallow land, such negative impacts can be avoided. Trees can then provide both fuel resources and help to address energy crises within a country (Jain 1994).

Agroforestry can make a significant contribution toward the provision of fuelwood. Although accurate estimates are difficult to obtain, it is assumed that agroforestry contributes to fuelwood production anywhere between 20% in Africa to 70% in Asia.

Charcoal is a vital source of energy for cooking for a large part of the rural population in developing countries (Mwampamba et al. 2013). Charcoal industry provides lucrative business opportunities worth millions of dollars (Mwampamba et al. 2013). Compared to untreated biomass, charcoal has a high energy density and a low water content which makes it easier to store and transport. According to a study done by Mwampamba et al. (2013), charcoal is used by people in wide range of income category and since it reduces indoor air pollution in comparison to firewood, it is more preferred by urban and people in higher income group. To reduce deforestation related to charcoal production, agroforestry systems can be enhanced to support local production of the source material wood. This may also provide the opportunity for employment in local charcoal production.

Liquid biofuel

Liquid biofuels have received increasing attention worldwide as renewable fuels with the potential to help improve energy security, mitigate climate change, and revitalize agricultural economies (Scovronick and Wilkinson 2014). Bioethanol and biodiesel are the two global liquid biofuels that hold promise for replacing gasoline and diesel fuel, respectively (Demirbas 2011).

Demand for ethanol is expected to more than double in the next 10 years. To fill the demand-supply gap, new feedstock and technologies must be developed and achieve commercial scale (Demirbas 2011). Trees and agroforestry systems will have an increasing role to play in the global production of ethanol. Although most of the current ethanol production comes from annual crops such as sugarcane and maize, many trees provide sugar-rich sap or fruit pulps which can be fermented and processed into ethanol. Examples include simarouba (*Simarouba glauca*), which is a promising raw material both for biodiesel (produced from its seed) and ethanol (produced from its fruit pulp) (Joshi and Joshi), as well as sugar palm (*Arenga pinnata*) and nypa palm (*Nypa fruticans*). Trees also provide lignocellulosic biomass that can be used for the production of second-generation ethanol, although these advanced technologies are only slowly reaching commercial scale.

Table 1. Summary of various tree-based energy technologies.

	Technology	Features	Uses & benefits	Successful examples	Remarks
Fuelwood	Direct combustion of fuelwood in fuel-efficient cookstoves Smokeless ovens	The Fuel Value Index (FVI) is used for screening suitable fuelwood species. High moisture content and ash reduce the FVI values and heat of combustion	Cooking Provides livelihood activity for poor rural communities	Fuel-efficient wood burning stoves by TIDE (Technology Informatics Design Endeavour), India Smokeless ovens of Vivekanand trust, Karnataka state, India Project Surya for improved cookstove intervention, India	There is a strong link between fuelwood uses and livelihoods, especially in low- income communities. To avoid forest decline, fuelwood should be harvested responsibly and under proper monitoring. Fast-growing trees should be planted and used trees should be replaced. Tree species like *Ixora parviflora, Schleichera trijuga, Parkinsonia aculeata*, and the shrubs like *Capparis aphylla, Dodonaea viscosa, Combretum ovalifolium, Vitex negundo, Rubus ellipticus, Coriaria nepalensis, Mimosa rubicaulis*, and *Antidesma ghaesembilla* make excellent firewood.
Charcoal	Slow pyrolysis of biomass in kiln	Roughly one-third of dry feed (based on mass) is converted to charcoal and the remaining into syngas and bio-oil/tar. Compared to untreated biomass, charcoal has high energy density and low water content and thus is easier than fuelwood to store and transport (Larsson et al. 2013).	Increases the efficiency of fuelwood Produces less smoke and requires little or no preparation before use Has higher energy content per unit mass and can be reused when left over after cooking. Can be easily transported and stored	Mafia Pilot project in Tanzania, initiated in 1985 and funded by the German development agency, GTZ Malawi Charcoal Project, funded by the World Bank and the Malawi Government was initiated in 1986 Dandora Charcoal Plant, established by the Kenya Planters Co-operative Union	The Mafia Pilot project uses coconut industry waste, the Malawi Charcoal project uses wood waste, and Dandora Charcoal Plant uses coffee husk as feedstock. Examples show that fuelwood supply limitations have now led to the use of other agricultural and forestry residues for charcoal preparation. Efficient kilns are required to improve production efficiency.
Liquid biofuels	*Bioethanol:* Cellulosic ethanol From fermentation of sugar	Ethanol is produced from bio-matter such as sugar cane, corn, or other grains, and is therefore also referred to as bioethanol.	Saves gasoline Reduces GHG emissions Reduces food-fuel competition	Project LIBERTY, US	Project LIBERTY converts baled corn cobs, leaves, husk, and stalk into renewable fuel; therefore, this new technology will help to reduce pressure on food crops like sugarcane, and soybean etc.
	Biodiesel: Transesterification	Biodiesel is usually produced from vegetable oils or animal fats. It can either be used directly as fuel with some engine modification, or in blends with petroleum diesel with few or no modifications in the diesel engine (Demirbas 2011)	Saves petro-diesel Cleaner than petro-diesel and thus leads to GHG emission reductions By-products like oil cake can either be used as fertilizer or as animal feed and the husk can be burned to release energy.	Indian Railways Bio-fuel Park, Hassan, India	Biofuel-park of Hassan is agroforestry model which grows seven species, Pongamia, Madhuka, Azadirachta, Simarouba, Jatropha, Calophyllum, and Aphanamixis. This system does not hinder the normal agricultural process and also provides extra income for farmers from the production of biofuel and its by-products.

(Continued)

Table 1. (Continued)

	Technology	Features	Uses & benefits	Successful examples	Remarks
Bio-electricity	Biomass gasifier plant	Biomass includes dedicated energy crops, agricultural residues, and forest residues as well as other biomass sources such as landfill gas, animal and human waste (Evans et al. 2010). There are two types: *Dedicated biomass-based power plant:* Generally smaller in size due to limited feedstock availability and high transportation costs (Bertrand et al. 2014). Have high investment cost and a low conversion efficiency of 25% (Bertrand et al. 2014). *Cofiring of biomass in coal-based power plant:* Requires little or no investment. The conversion efficiency is on average 36% in OECD countries.	Saves fossil fuel Reduces GHG emissions Promotes sustainable approach in meeting rural energy needs Electricity for irrigation, lighting, and small industries Enhances green cover, carbon sequestration, and can reclaim waste land.	BERI (Biomass Energy for Rural India) project, Tumkur district of Karnataka, India AVANI project, Pithoragarh, Uttarakhand, India Biomass gasifier by Ankur Scientific in Sankheda Taluka of Vadodara district, Gujarat, India SRE (Saran Renewable Energy), Garkha Village, Bihar, India	Has great potential in meeting the electricity needs of rural villages. Continuous supply of raw material is the first prerequisite of all such electrification projects. To ensure the same, BERI uses fast-growing species like Prosopis *juliflora, Lantana camara,* Epil-epil (Subabool), glyciridia, and bamboo. AVANI uses waste pine needles, while Ankur and SRE use waste from the agricultural fields. Further research is required to increase the conversion efficiencies of these projects.
Biogas	Anaerobic digestion/ bio-methanation of biomass in biogas plants Biogas stove	The main components are methane (50–80%) and carbon dioxide. Types of biogas plants: *KVIC (Floating Dome):* Consists of a pit on which a mild steel or plastic dome rests. Suitable for most areas except very cold climates. *Deenbandhu Model:* Has an underground masonry structure, which guards it against cold temperatures. Excavation is difficult in hard bedrock and needs high investment. *Plug Flow model:* Comprised of a cylindrical HDPE film. Suitable for warm climates. Has low cost with good gas production	Saves LPG or firewood as cooking and heating fuel Power generation and energy production for vehicles. Reduces exposure to thick smoke and chances of lung diseases (Bi and Haight 2007). The solid residue or sludge can be used as organic compost.	Biotech plants, Pathanapuram, Kerala, India The Biogas Support Programme (BSP), Nepal Arti biogas plant, India Promotion of Sustainable Agricultural Activities through Demonstration of Bio-gas Plants and Other Allied Activities, by Sarvangeen Vikas Samiti, Gagha Villages, India	Biogas digesters use food and agricultural waste and thus have particular potential, in areas where there are severe health hazards due to the production of large volumes of organic waste and pollution of water supplies. To realize this potential in the Panchayats, there is a need to work on creating awareness and secure the required finances.

Biodiesel is usually produced from animal fat or from several types of vegetable oils. There are many trees around the world with large potential for straight vegetable oil (SVO) and biodiesel production. Oil palm (*Elaeis guineensis*) and jatropha (*J. curcas*) are common examples. Oil palm has proven potential for transformation into a number of products and fuels, including biodiesel (Chew and Bhatia 2008). The original expected potential of jatropha has not been realized; so far, its yield has been poor and gives poor economic returns under smallholder management despite its ability to grow on marginal land (Ilyama et al. 2013). However, a few authors claim that for good and profitable yield from jatropha, its basic agronomic properties need to be properly understood and it needs to be cultivated with proper and careful agricultural practices (Pandey et al. 2011; Behera et al. 2010; Openshaw 2000).

Moreover, trees that could provide liquid biofuels will become more important as techniques for converting the lignin and cellulose of trees directly into ethanol improve. Trees can address many bioenergy forms which are summarized in Table 1. Clearly, our current knowledge suggests that traditional first-generation biofuel crops, especially those grown in monocultures, may not be suitable vehicles for realizing biofuel production at scale. Not only because they have contributed to negative food versus fuel land-use changes and impacts, but also because they have marginal to negative GHG emission benefits. By incorporating biofuel production in agroforestry systems, a positive effect on the overall carbon footprint may be achieved.

Electricity generation from woody biomass

Woody biomass can easily be used in gasification systems that provide fuel to drive machinery and generate electricity. These systems are cost-effective and already widely used in both developed and developing countries. These systems can be effective at both large scale (the electricity generating woody biomass plants) and small scale (community scale installations that are already in use in the developing countries).

The use of biomass to generate electricity not only saves fossil fuel, it also reduces the amount of harmful emissions entering the atmosphere (Ryabov et al. 2006). Dedicated energy species, such as fast-growing trees for wood pellet production, agricultural residues, forest residues, and other biomass sources like animal and human waste, are a few common sources of biomass used for the generation of electricity (Evans et al. 2010).

There has been a steady increase in global biomass-based electricity generation –, on average 13 TWh per year between 2000 and 2008. The market share of biomass-based electricity in total global generation is approximately 2% and this has been maintained over the last 20 years.

Agroforestry systems can be designed for such places to ensure sustainable and efficient production of biomass for local power generation without affecting food production, and in some cases even improving it. Ideal tree species would usually be fast growing and drought tolerant, have coppice capability, be suitable for short gestation periods of 2–3 years, and have high calorific value and less ash content, among other features. Several species meet these requirements, such as *Casuarina junghuhniana*, *Casuarina equisetifolia*, *Eucalyptus*, *Melia dubia*, *Leucaena leucocephala*, *Prosopis juliflora*, and bamboo (Shankar 2012).

Biogas for cooking or electricity generation

Biogas can be produced by anaerobic digestion of human and animal manure, as well as biomass such as agroforestry residues (Food and Agricultural Organisation of United Nations). The main components of biogas are methane and carbon dioxide, with a composition of about 50–80% methane, the balance being mainly carbon dioxide. Biogas can be used for heating, cooking, power generation, and energy production for vehicles (Coimbra-Araújo et al.).

Oilseed cakes left over from SVO and biodiesel production are among the residues which can significantly enhance biogas production. Anaerobic digestion has the potential to drastically reduce GHG emissions when replacing fossil fuels (Weiland 2010).

Under optimal conditions, about 90% of feedstock energy is converted into biogas which can readily be used for cooking and lighting purposes. The sludge produced during the digestion process is nontoxic and odorless in nature. A major portion of its nitrogen and other nutrients remain intact during the process, making it a good source of fertilizer. When compared to cattle manure left in the fields to dry under the sun, the slurry from the digester has higher nitrogen content. Some of this nitrogen also gets converted to urea, which is more readily accessible by plants than many nitrogen compounds.

The dual yield of anaerobic digestion of biomass in the form of energy and fertilizer makes biogas plants an ideal candidate for rural villages (Food and Agricultural Organisation of United Nations). Biogas also offers improved health benefits due to its cleaner combustion compared to fossil fuels and other biomass used in the domestic environment (Food and Agricultural Organisation of United Nations). The chances of exposure to thick smoke and lung diseases are reduced by shift from traditional fuels to biogas (Bi and Haight 2007).

Value chain development of agroforestry products

The major difficulty in demonstrating the value of a biomass-based energy system has been a lack of proper value chains. There have been many efforts to date to provide woodfuel in a sustainable way, but few projects have succeeded in creating value chains that simultaneously address energy needs and forest degradation. Where value chains have been addressed, attention is often only paid to one part of the chain, such as cookstoves or reforestation. In other attempts, livelihood aspects have been overlooked. This way, significant structural changes have not been induced. The entire wood energy value chain, right from production to conversion, transport and marketing to consumption, should be considered.

Factors which affect the formulation of an effective and inclusive wood energy value chain include: lack of reliable baseline data; lack of intersectoral cooperation and coordination; and a lack of adequate consideration in sector policies, mainly due to capacity deficits. Other factors include weak law enforcement capacities, lack of commitment and ownership from respective governments, corruption and oligopolistic structures of the value chain (Sepp et al. 2014).

Support to inclusive value chains should take place at two levels: institutional and technical development. Institutional development will help formalize and organize the value chain and also lead to capacity development of stakeholders, creation of networks of rural and urban wood energy markets, and support to wood energy depots. Technology improvement would help promote further development of kilns, stoves, fuel types etc. (Sepp et al. 2014).

The complex supply chain of biomass for energy generation, unpredicted quality of feedstocks, variability and uncertainty in the supply chain, economic conditions, and market fluctuations affect the amount and the cost of the energy that is produced. While many efforts have been made to make these initiatives economically viable, only a few have succeeded in creating value chains that simultaneously address energy needs and improvements in the livelihoods of smallholders.

The issue of rural energy is generally dependent on national programs which are often technology-centric or end-use-based, and in the majority of cases, without any interlinkages with the rural ecosystem. Such a system is likely to experience challenges in sustainability. In bioenergy production, it is essential that upfront forward business linkages are established, for example, direct market linkages, as this is one of the main drivers for growers to adopt agroforestry.

Practical Examples of Agroforestry Production Models

Benefits of leguminous tree species

Leguminous trees have an important role in agroforestry systems as they restore nutrient cycling and improve soil fertility. The rhizobium bacteria in their roots have the ability to extract nitrogen from air and accumulate it biologically. These nitrogen-fixing trees are often deep rooted and can thus easily access nutrients in subsoil layers. Their extensive root systems also stabilize soil. Leguminous trees provide many other benefits, including food, animal fodder, fuelwood, timber, shade, and can serve as living fence (Elevitch 2008). There are many such species, including *Acacia auriculiformis*, *Cajanus cajan*, *Faidherbia albida*, *Gliricidia sepium*, *L. leucocephala* which have successfully been integrated into various agroforestry systems.

Faidherbia albida

Faidherbia albida, which is indigenous to Africa, has a unique characteristic which makes it suitable for intercropping systems. The tree, also known as winterthorn, has a number of properties which make it ideal for agroforestry systems. As a legume *F. albida* provides a habitat for symbiotic bacteria, able to fixate nitrogen and therefore providing the tree with valuable nutrients, which are passed on to the surrounding soil when the tree sheds its leaves. *F. albida*'s phenology is such that it loses its leaves during the rainy season, hence not shading adjacent crops during their prime growing phase. Its deep reaching taproots make it very drought resistant and protect the landscape against soil erosion. The tree's pods and foliage can be used as animal fodder or even for human nutrition. The flower of *F. albida* is widely known to be a preferred pasture for honey bees. Furthermore, the tree's bark and leaves are used for traditional remedies against a number of diseases (Tijani et al. 2008).

Some 500,000 farmers in Malawi, Tanzania, and Zambia cultivate their crops alongside *F. albida* in agroforestry systems. In some cases, maize yields are reported to have doubled or tripled (World Agroforestry Centre).

In Niger, farmer-managed natural regeneration (FMNR) of degraded landscapes with agroforestry systems based on *F. albida* intercropped with sorghum and millet already covers more than 5 million hectares (World Agroforestry Centre). The success of this system is due to the multiple benefits it provides, including fuelwood and fodder production as well as improved soil fertility which generates higher crop yields.

Gliricidia sepium

A long-term trial conducted by the World Agroforestry Centre in Malawi which intercropped maize with *Gliricidia sepium* [a tree that can provide several bioenergy products (World Agroforestry Centre, 2009)] achieved yields three times higher than when maize was planted alone and left unfertilized (Akinnifesi et al. 2006). Other experiments in India demonstrate that the use of *Millettia pinnata* (Pongamia) (World Agroforestry Centre, 2009) seedcake (a coproduct from SVO or biodiesel production) as organic fertilizer can lead to yield improvements of between 49% and 87% for soybean and maize (Wani and Sreedevi).

In Sri Lanka, a novel agroforestry system involving *Gliricidia* and coconut cultivation is being trialed for electricity generation (Kulatunga 2012). This is an integrated approach involving tree-based food (feed)-energy systems and has been demonstrated at the commercial level. The approach overcomes concerns that growing crops for bioenergy might compete for resources with food production. The system enables energy, fertilizer-fodder-fuelwood trees to be incorporated into agricultural fields to provide feedstock for power generation. The project started with a capacity of 1 MW and only when the system was fully functional, was the capacity enhanced to 10 MW. Next to quality fodder, additional income stems from lopping (wood).The system enabled growers to increase their milk production and generate additional income from wood sales without losing the income from coconuts.

Leucaena leucocephala

Leucaena leucocephala is one of the most widely used species in alley cropping, where it is planted in hedges along contours at intervals of 3–10 m with crops in between. It is a valuable fodder tree in the tropics. The quality of the leaves compares with alfalfa or Lucerne in feed value (Orwa et al. 2009). It is planted, for example, in an agroforestry system with sorghum. Sorghum yields can be high even in poor rainfall conditions.

Leucaena is also being considered as an option for electricity generation due to its fast growth rates and higher biomass yield. *Leucaena* can yield around 2–20 MT per ha per annum. There has been some pilot project to evaluate its potential; however, there are few power plants yet using *Leucaena*.

Cajanus cajan

The GIZ Integrated Food Security Programme (IFSP) implemented in Mulanje district, Malawi, between 1996 and 2004 (FAO) promoted pigeon pea *Cajanus cajan* (pigeon pea) intercropped with maize, the local staple food. This was complemented with the introduction of energy-saving clay stoves. Through the program, smallholders were able to harvest about 5 tons of biomass per ha, helping to satisfy cooking energy needs through their own means, as they used the woody pigeon pea stalks as their main source of fuel. The farmers also have an additional protein-rich food crop and have decreased the need for chemical fertilizer inputs while increasing agricultural yields naturally; *Cajanus cajan* improves soil fertility through nitrogen fixation, mulching of leaves (improving water retention), and decrease soil erosion (prolonged ground cover).

Acacia auriculiformis

About 140 km east of Kinshasa on the Batéké plateau, 8000 hectares of *Acacia auriculiformis* were initially planted in an EU-funded project implemented by the Dutch company HAI. From 1994, the activities were taken over by the German Hans Seidel Foundation, supported by the NGO CADIM. Later CIRAD implemented a research development project about the fuelwood sector around Kinshasa.

Early during the project's life span, the so called Mampu plantation was divided into plots of 25 hectares and allocated to 320 farming families. Since 2009, farmers own official land titles and a farmers' union manages the plots independently. Corn and cassava are intercropped with *Acacia auriculiformis*. *Acacia auriculiformis* fixes nitrogen from the atmosphere and restores it in the soil. This way, higher crop yields can be achieved. Three years after planting, the acacias reach a height of three meters. After around 10 years, a veritable acacia forest has developed. Every 4 meters, a one meter wide strip of soil is left unfarmed, so that acacia seeds can germinate and reconstitute the future forest stand. Until the crop rotation can start (after 10 years) farmers generate income mainly from beekeeping (FAO, 2010).

Acacia auriculiformis is being used for charcoal production. The total charcoal production from the plantation varies from 8000 to 12000 tons per year, corresponding to 5 t/ha of wood and a charcoal yields after carbonization of 20%. In addition, the production system yields about 10 000 tons a year of cassava, 1200 tons of maize, and 6000 tons of honey per year. Some of the farmers can earn up to US$ 4000 a year. The charcoal production satisfies around 2 per cent of the demand of Kinshasa (FAO, 2010).

Benefits of nonleguminous tree species

Eucalyptus species

In Australia, a system to integrate eucalyptus with wheat crop is being implemented. Growing the Mallee variety is profitable and easy to integrate into farming systems in semi-arid regions of southern Australia (Wildy et al. 2003). The trees are integrated into wheat belts to: (i) reduce soil salinity; (ii) give shade and shelter for animals; (iii) to reduce erosion by acting as windbreaks; and (iv) to store carbon. The Commonwealth Scientific and Industrial Research Organization (CSIRO) has evaluated eucalyptus grown in this system for producing liquid biofuels and electricity. Electricity produced by eucalyptus biomass from such systems was found to be economically viable compared to liquid biofuels.

An example of a successful approach is the GIZ Village-based Individual Reforestation Project in Madagascar (Sepp et al. 2014). *Eucalyptus* is used in plantations to reforest degraded land patches. Communes allocate individual land titles to farmers for reforestation purposes, this way the wood produced on the land is the property of the farmers involved. The project addresses the entire value chain, from individual afforestation schemes at village level to harvesting, processing, conversion, distribution, and marketing, all the way to end-consumers and related technology. As a result, it is estimated that the project increases the income of the local population by about 40% compared to the average income in rural areas. It will also provide greater opportunities for women and strengthen their economic position in society as they play a very important role in the wood energy sector as producers, users, and collectors and sometimes also deal in trading of woodfuels. This offers many prospects for promoting gender equality. Moreover, it offsets the unregulated exploitation of more than 72,000 ha of natural forests, which would otherwise have been cleared for charcoal production (Sepp et al. 2014).

Dendrocalamus strictus

Dendrocalamus strictus (Bamboo), in particular, is proving successful as a feedstock for local power generation due to its fast-growing nature; it can grow up to 18 inches in 1 day and can be harvested in 2–4 year cycles. Bamboo also has the ability to regreen degraded lands while releasing 35% more oxygen than the equivalent volume of other trees. A bamboo plantation can absorb 12 tons of carbon dioxide per hectare annually. Bamboo groves also serve as windbreaks and acoustic barriers, and bamboo is said to be superior to regular wood in terms of certain physical and mechanical properties (Siraj 2014).

Bamboo, intercropped with sesame or chickpea as an intercrop was studied by the Indian Agricultural Research Institute in the semi-arid region of central India. Bamboo was planted at 10 × 10 m and 12 × 10 m spacing. The total bamboo culms were higher when bamboo was grown without intercrops. The grain yield of sesame and chickpea were reduced. Yields decreased especially close to the bamboo clumps. However, at more than 3 m distance from the bamboo clump, there was no reduction in crop yields. The income generated through bamboo can compensate the monetary losses of the intercrop. There are multiple benefits related to the planning of bamboo: the soil pH is raised, organic carbon and available phosphorus increases. In drought prone areas, bamboo-based agroforestry can increase the resilience of farmers, conserve and enrich the soil (Ahlawat 2014).

Programme for the Development of Alternative Biofuel Crops – An ongoing project of the World Agroforestry Centre (ICRAF)

The World Agroforestry Centre (ICRAF) is implementing the 4-year *Programme for the Development of Alternative Biofuel Crops* in partnership with centers of excellence in Asia, Africa, and Latin America. Launched in 2013 with seed capital from the International Fund for Agricultural Development (IFAD), this initiative was designed to conduct research in development along several biofuel value chains, while at the same time strengthening food security and improving the livelihoods of smallholders. Pursuing a landscape approach, the program has already started operations in India, Brazil, and Kenya, where it is developing sustainable agroforestry systems with integrated food and energy production. Some of the initiatives under the program are presented in the following boxed text for India, Brazil, and Kenya (World Agroforestry Centre).

India

In India, the Programme is being implemented mainly in the states of Karnataka and Maharashtra. It is supporting the improvement and scaling-up of a promising biofuel initiative from Karnataka (refer Picture 1). Essentially, smallholders in energy-deprived villages are provided with quality planting material and technical assistance for growing native or locally adapted bioenergy trees, such as *Millettia pinnata* (pongamia), *Simarouba glauca* (simarouba), *Madhuca longifolia* (mahua), *Azadirachta indica* (neem) and other species in the borders and bunds of their plots, in addition to communal marginal lands. These agroforestry systems improve food security and alleviate energy poverty, since all their products are consumed locally; the SVO, biodiesel, and biogas are used for running tractors, irrigation pumps and for cooking via adapted stoves, while the biofertilizer derived from the seedcake improves

Picture 1. Integrated food and energy model in Hassan district of Karnataka, India: a pilot site of the ICRAF's program (ICRAF, Biofuels team)

the productivity of traditional crops. This pilot builds on existing partnerships between ICRAF, the University of Agricultural Sciences - Bangalore (UASB), the Karnataka State Biofuel Development Board (KSBDB), and the PDKV Akola University in Maharashtra. In a nutshell, this initiative has been spearheaded by UASB and KSBDB in Karnataka, and has already attracted significant attention nationally and internationally, with other states and countries such as Maharashtra and Nepal starting to develop similar models.

ICRAF has been conducting studies to evaluate this model and its potential for improvement. Preliminary data suggest that the efforts undertaken in Karnataka have a positive impact on the livelihoods of the poorest, and that the model implemented in the state can be scaled-up, subject to some adjustments. It was found, for instance, that smallholders adopting such agroforestry systems can achieve an increase in annual income of up to 36% in the long term under the current scenario. The extra income tends to come during the nonharvesting periods of the year when it is most needed. With further investments, it is estimated that the potential annual income increase in the long term could reach 60%. However, to fully realize this potential, several gaps still need to be addressed, such as through further investments in germplasm collection, nursery techniques, technologies for oil expelling and processing, and value addition at the village level. The model also needs to transition into financially sustainable and bankable proposals, allowing for reduction in public subsidies. IFAD and ICRAF are working with the Government of Karnataka and other partners toward achieving these objectives, while facilitating the scaling-up of the model into Maharashtra and other states.

Brazil

In Brazil, the Programme is addressing research gaps that limit the development and scaling-up of pro-poor *Acrocomia aculeata* (macauba) value chains in Brazil's northeastern region. *Acrocomia aculeata*, a palm species native to the tropical regions of the Americas, is a highly productive tree with oil-bearing fruits (refer Picture 2). It can be integrated into existing pastures without reducing grass yields. The tree often grows on poor soils and has high tolerance to drought. Its fruits are nontoxic; so, *Acrocomia aculeata* coproducts can be used for many different purposes along the value chain, including the production of animal feed, fuel briquettes, activated charcoal, and cosmetics.

ICRAF is partnering with the Brazilian Agricultural Research Corporation (Embrapa) to develop silvopastoral and agroforestry systems that integrate food crops and/or livestock with *Acrocomia aculeate*, benefiting smallholder producers (World Agroforestry Centre). In addition to improving livestock production, these systems will facilitate the integration of smallholder producers from several states into the National Biodiesel Production and Use Programme (PNPB), an initiative that has benefitted more than 100,000 family farmers in 10 years (Ubrabio).

Kenya

In Kenya, the Programme is partnering with Eco Fuels Kenya Ltd. (EFK), a social enterprise that produces liquid biofuel (straight vegetable oil), organic fertilizers, briquettes and poultry feed through a sustainable, no-waste manufacturing process, based entirely on the nut from *Croton megalocarpus* (croton). This is an abundant, indigenous tree found across central and west Kenya, as well as in

Picture 2. *Acrocomia aculeata*-based biofuels model in Brazil: a pilot site of the ICRAF's program (ICRAF, Biofuels team)

other countries in East Africa (refer Picture 3). Until recently, its nut had no commercial value (since it is not edible) and it is still a largely wasted natural renewable resource. *Croton's* potential for multiple uses has long been researched by ICRAF, and its value for energy production has been recognized by the Government of Kenya several years ago (Endelevu Energy & Energy for Sustainable Development, 2008). Nonetheless, it was only more recently that the species started attracting international interest, private capital, and professional management toward developing its value chain.

Founded in 2012, EFK sources the croton nut from rural communities around its factory, which harvest them from local forests and farmlands. Thus, EFK provides an assured market to the farmers. The oil is used mainly to run water pumps and off-grid generators, while the briquettes are targeted at rural and urban dwellers for cooking and heating purposes. Based on the EFK's business model, the Programme is assessing the sustainability of the croton value chain in Kenya and addressing research gaps to support its upscaling in East Africa, while maximizing its positive impact on livelihoods of the rural poor. In particular, it will contribute to increased additional income for approximately 5000 collectors by the end of 2016, in addition to improving the supply of affordable and cleaner energy options in the central region of the country.

Some of the gaps already identified by ICRAF and EFK, which are being tackled by the Programme, include the need to support the selection and growing of superior

Picture 3. *Croton megalocarpus*-based biofuels model in Nairobi, Kenya: a pilot site of the ICRAF's program (ICRAF, Biofuel team)

croton trees through establishment of germplasm, including natural seed collection, as well as to promote capacity development activities, such as training sessions on croton planting and management, targeted at smallholder producers. Although a previous analysis from ICRAF (Ndegwa et al.) indicates that the cultivation of croton can be profitable even when the tree is planted for hedging (as is customary in central Kenya) the project is conducting trials of different agroforestry systems that incorporate the species, so as to identify options and combinations that maximize its value and potential for energy poverty alleviation.

Gender Role in Agroforestry-Based Bioenergy

Women and men are affected in different ways by poverty. In many developing countries, women experience severe levels of deprivation due to several reasons, among which is inadequate access to clean energy options (Karekezi et al. 2002). Women and children are exposed to indoor smoke through the use of woodfuels in traditional cookstoves. Most rural households use biomass energy and women in poor households spend more time searching for firewood than those in households with higher incomes (Energia, 2008). In areas where there is decreased vegetation cover, women and children walk long distances in search of firewood. A study (Thorlakson and Neufeldt 2012) shows that, for example, in the Lower Nyando District of Kenya, households spend on average 9 hours per week on fuelwood collection, with some women reporting that they walked more than 20 km/day to purchase fuelwood in neighboring districts. After agroforestry interventions, the weekly average came down to 6 hours per week. The mature trees on their land made them feel safe and gave access to more stable fuelwood supply. Women with mature trees on their land felt that they now have access to a safer and more stable supply of fuel wood. These women found that once fuelwood was available nearby, they could devote more time to other income-generating activities (Thorlakson and Neufeldt 2012). Moreover, increasing the efficiency of cookstoves may easily reduce the health risks of women and children. Improved cookstoves are available for domestic use that can reduce fuel use and particulate matter emissions when compared to traditional cookstoves (World Bank, 2011).

In another case, during World Congress on Agroforestry in February 2014, held in New Delhi in India, a lady named Prabhavati, from Bijapur taluk of Karnataka, narrated her story on neem seed collection activity. Each woman used to collect about 4–5 tons of neem seeds every season and sell it for only Rs 2–3/kg (Rs 8000–15,000 per year). After the implementation of biofuel program by KSBDB (Karnataka State Biofuel Development Board) and intervention by an NGO called Biofuel Lead, they could access more information on marketing and prices. This helped them to directly sell their seeds to oil mills and also gave them a bargaining power to establish better rates for their seeds. They formed a self-help group of about 150 women, who collect and sell neem seeds and each woman makes, on average, about Rs 30,000/year. This additional income has helped them to improve their standard of living and quality of life.

Key Constraints for Setting up Agroforestry Systems

The adoption of agroforestry still faces major constraints. The domination in agriculture of monocultures and short-term benefits is a challenge and trees in agricultural fields can hamper mechanized operations. A key bottleneck hindering the development of agroforestry is poor access to the resources needed to establish such systems, such as capital, labor, farming inputs, land, extension services or markets. Inadequate legislation, regulations, and policies can further jeopardize agroforestry development. For instance, in many countries and regions, agroforestry has no clear status and falls between agricultural and forestry sectors, leaving its regulation in a gray area.

Lack of suitable policy

Although agroforestry has been practiced for thousands of years in many parts of the world, and despite all its benefits, the sector lacks favorable policies, legal frameworks, and coordination between different government mandates, such as agriculture, forestry, rural development, environment, and trade. Many governments (e.g., India, Indonesia, and Myanmar (Butler 2012)) do not allow free harvest and trade of timber, even for trees grown on private land, imposing several restrictive legal provisions for harvesting and transportation of lumbered trees. This has not been addressed sufficiently in policy formulation, nor has it been adequately integrated into land-use planning or rural development programs.

Thus, the potential of agroforestry to enrich farmers, communities, and by extension, national economies has not been fully exploited. Although motivated by a desire to increase tree cover, the Government of India has come out with a specific policy on Agroforestry (Down To Earth, 2014) which is likely to bolster research and development in the sector. India has been the first country in the world to adopt such a policy, which makes particular reference to the energy sector. It is expected that the policy will encourage other countries to follow suit in their developmental goals.

Unavailability of planting material

Research and development investments in trees have long lagged behind traditional agricultural crops. Several hybrid and suitable varieties of the latter have been developed, in addition to advanced techniques, ensuring that farmers obtain uniform stand and yield. In contrast with some horticultural crops (e.g., mango and citrus), most perennial species have so far relied on new domestication and selection efforts. The result has been that planting materials have a large genetic variation, are unregulated and their productivity is often very low and growers in remote areas lack access to them. Jatropha, for instance, was (Singh 2015) promoted in several countries without proper domestication or varietal development. In some cases, such as in India, large plantations of the crop were encouraged by government entities. As a result, most biofuel projects and policies focused on this crop failed worldwide, including in India. There is a need to have a mechanism in place, similar to that which exists for agricultural crops, which certifies planting material of tree species. Either an institutional mechanism needs to be developed for registration of nurseries or private players need to be encouraged to sell planting material under a "private label" or "self-certification". This could occur through accredited agencies which allow governments to recognize certification.

Research and extension services

One of the major constraints to the promotion and large-scale adoption of agroforestry innovations by farmers and communities is the lack of, or inadequate knowledge, of tree-growing practices. Agroforestry requires a careful study of all the factors that influence its adoption. Moreover, there are thousands of potentially useful trees and crop species that can be mixed in many different ways and local agroecological conditions can change the effects significantly. Considering the benefits trees provide for bioenergy, creative and challenging research is needed. Research results need to be used to confront and overcome the barriers to increased support from extension services. Several international organizations, including the World Agroforestry Centre, are engaged in research and development of trees and agroforestry systems. However, extension services tend to be undertaken by government agencies (in developing countries) which are often under-resourced or ineffective, and which seldom focus on trees. Furthermore, agroforestry is mostly about management techniques and not about products such as seed, fertilizer, and other inputs that can be sold. This way, few are lobbying for its wider application. The consequence is that poor farmers are left without proper advice. There are cases, however, in which these difficulties have been overcome with private sector support, for example, the pulp and paper industry in India has very strong R&D and extension services. There is a need to learn from these successful cases and understand how to replicate them.

Long gestation time

Another reason is that the benefits of integrating trees into landscapes are not fully known. Instead, farmers perceive the extra work and the long-growing periods.

The long lag time between planting and harvesting of tree products hinders the adoption of agroforestry by smallholders, especially those who struggle to meet their basic requirements on a day-to-day basis and prefer to invest in areas where returns are short term. There is a need to select and domesticate fast-growing native species that are coppiciable and also develop technologies for early flowering, such as grafting. Lessons from successful case studies, such as the cocoa project led by the World Agroforestry Centre, can be incorporated into the development of bioenergy feedstock strategies.

Access to market and value chains

Despite the fact that tree-based products, like fruits, medicines, fuelwood, and oilseeds etc., are part of day to day life in rural areas, scant information is available on their demand and potential supply in many developing countries. The market remains a cottage-style industry and middlemen often exploit growers. This leads to low profits to poor producers and production of low-quality products since there is no direct link between product and market. Except for a few examples, marketing systems for agroforestry products remains unorganized. Agroforestry forestry products frequently have not yet developed the same market linkages as for traditional agricultural crops as they are usually not a priority for farmers and communities. To ensure better development and utilization of forest and tree resources, there is a need for strong linkages between the various institutions in the sector. Marketing infrastructure similar to that for agricultural commodities - with assured markets by way of assured prices (e.g., minimum support price) - has to be introduced in order to mainstream agroforestry into agricultural systems.

For example, ITC (Indian tobacco Company) introduced a new concept named as "Agro-forestry Model". Under this model, farmers plant pulpwood trees on 25% of their land and on the remaining 75% of land, they grow food crops. The farmers can harvest the trees every 4 years and ITC provides them the market. In this way they generate income from both the food crops and the trees. This program started in April, 2010 and so far 2427 hectares of land has been covered under this model (Environmental performance, ITC Limited).

Access to credit

There are cumbersome procedures involved that hinder poor households accessing credit from the banking sector. One of the reasons why poor households cannot access credit is that they cannot present guarantees to the bank. Coupled with this, agroforestry investments do not attract the attention of the banking sector. Therefore, financial packages for rural energy programs, along with agroforestry, have to be included in comprehensive planning approaches if they are to have the desired impact. Credits are needed especially for agroforestry as the upfront investments can be higher than for annual crops. Capital is needed for buying seedlings and labor (acquiring and planting trees, caring for the young trees). Initially, trees can take space that was previously devoted to crops.

Land tenure

Secure land rights are necessary if farmers are to invest in agriculture and related activities. In Africa, customary tenure agreements exist in many countries, especially in Sub-Saharan countries. A family gains access to land as a social right; however, the land rights are governed by local and traditional authorities. Lack of permanent land tenure is a big deterrent to the adoption of long-term activities such as agroforestry. Without secure land rights, farmers might fear their land will be taken away by more powerful actors once its quality has been improved. Target groups will be more likely to be interested in agroforestry provided they have long-term access to land and rights to plant and use the products of trees. Land tenure issues also deter financial institutes from providing credit to farmers (see above).

Example of one such technique which takes care of land tenure issues is Farmer managed natural regeneration (FMNR). It is a land restoration technique, which involves the systematic regeneration and management of trees and shrubs from tree stumps, roots, and seeds. For successful implementation of this technique, FMNR ensures that farmers get control over their land and what they grow on it. For this, FMNR projects incorporate assistance to broker agreements between the relevant stakeholders – for example, farming communities, government agents, and other forms of local bodies (Francis et al. 2015).

Policy interventions are also required to change inadequate land tenure systems. USAID is investigating this important issue in Zambia, where a baseline survey of 4000 households from 315 villages is being carried out to see how strongly land tenure contributes to agroforestry interventions (USAID, 2014). It is expected that if land tenure issues are resolved, there would be increased uptake of agroforestry interventions for livelihood improvements.

Despite the above challenges, trees provide abundant opportunities for national economies, especially in the renewable energy sector. Similar to the agriculture sector, agroforestry also deserves more attention from governments, civil society, and the private sector to have its full potential realized.

Conclusions

Access to energy is crucial to realize development objectives. Bioenergy is an option to increase access to energy in rural areas of developing countries. It will however only be successful if the supply of feedstock is cost-effective and creates employment as well as additional incomes for smallholder farmers at the same time. The benefits of bioenergy feedstock production will be negated if agricultural land is compromised for production of nonfood crops or diversion of these crops for energy purposes. With dwindling agricultural resources in some developing countries and increased vulnerability of smallholders due to climate change, this remains a challenge.

Scientifically developed, locally adapted bioenergy systems may address a variety of challenges at the same time, food and energy security for a growing population, mitigating and adapting to climate change. Agroforestry and conservation agriculture should be key elements as they can improve the productivity of smallholder agricultural systems under climate change by strengthening their resilience.

To conclude, we would like to summarize the following findings of this article:

1. Among all types of modern renewable energy, bioenergy can be subject to adequate choices of technology and production models – an affordable option for smallholders in developing countries to locally produce and consume energy. It therefore has great potential for promoting sustainable agricultural growth, reducing energy poverty, driving rural development; creating jobs, and additional income for the poor.
2. Despite the significance of bioenergy in the livelihoods of rural populations, investments in this sector have largely been overlooked by international development players.
3. In many developing countries, bioenergy still has negative impacts such as forest degradation, indoor pollution, and food insecurity which are hindering its development.
4. Bioenergy can be sustainably produced and have positive environmental, social, and economic impacts. Solid scientific evidence and successful case studies in the developing world have proven this.
5. Smart agroforestry systems with integrated food and energy production are among the main solutions. Their potential benefits include:
    a. Production of different bioenergy sources within one comprehensive system, including sustainable charcoal, briquettes, straight vegetable oil, biodiesel, bioethanol, biogas, and electricity from the gasification of fuelwood. This leads to immediate energy poverty alleviation with cleaner alternatives to fossil fuels.
    b. Coproduction of organic fertilizers and biopesticides, as well as improved soil fertility with N_2 fixation from trees and reduced soil erosion, leading to enhanced food crop productivity.
    c. Creation of jobs, generation of additional income, and savings from fossil fuels, as well as integration of smallholders into new value chains.
    d. Strengthened food security due to the simultaneous production of food and through effects of fertilizer trees on productivity.
    e. Climate change mitigation with reduced GHG emissions, carbon sequestration, and storage in addition to adaptation services, with improved resilience of smallholder producers.
    f. No land-use change effects (either direct or indirect).
    g. Decreased risks for the farmer through the use of multiple species rather than monoculture plantations.
    h. Watershed protection and biodiversity conservation.
6. A number of bioenergy projects involving trees that have not taken into account a landscape approach in agroforestry models have not been successful. The main reasons for failure include poorly structured value chains, inability to ensure constant and economical supply of feedstock and lack of adequate market linkages.
7. Several successful agroforestry models have shown potential for sustainable charcoal, electricity and biofuel production using different feedstock such as *Gliricidia*, bamboo, and oil-bearing tree species.
8. Some bottlenecks remain for mainstreaming agroforestry for bioenergy production. These are mainly related to a lack of suitable policies, adequate planting material and extension services, land rights, long gestation periods, and difficulties with access to credit and markets.
9. With political will and improved investments in R&D and scaling-up, it is possible to achieve useful results. A few countries are already demonstrating this, for example, India has become the first nation to launch a national agroforestry policy and is accelerating its

investment in the sector. It is hoped other countries will be encouraged to promote agroforestry systems for bioenergy generation.

Acknowledgment

This study was financed by BMZ/GIZ.

Conflict of Interest

None declared.

References

Adebimpe, A. A. 2013. Population dynamics and infrastructure: meeting the millennium development goals in Ondo State, Nigeria. Afr. Popul. Stud. 27:229–237.

Ahlawat, S. P. 2014. Bamboo based agroforestry for livelihood security and environmental protection in semi-arid region of India, [Online]. Available at http://wca2014.org/abstract/bamboo-based-agroforestry-for-livelihood-security-and-environmental-protection-in-semi-arid-region-of-india/. (Accessed 19 August 2015).

Akinnifesi, F. K., W. Makumba, and F. R. Kwesiga. 2006. Sustainable maize production using gliricidia.maize intercropping in southern Malawi. Expl. Agric. 42:1–17.

Bayalaa, J., G. W. Sileshib, R. Coec, A. Kalinganirea, Z. Tchoundjeud, F. Sinclaire, et al. 2012. Cereal yield response to conservation agriculture practices in drylands of West Africa: a quantitative synthesis. J. Arid Environ. 78:13–25.

Behera, S. K., P. Srivastava, R. Tripathi, J. P. Singh, and N. Singh. 2010. Evaluation of plant performance of *Jatropha curcas* L. under different agro-practices for optimizing biomass – a case study. Biomass Bioenergy 34:30–41.

Bertrand, V., B. Dequiedt, and E. L. Cadre. 2014. Biomass for electricity in the EU-27: Potential demand, CO2 abatements and breakeven prices for co-firing. Energy Pol. 73:631–644.

Bi, L., and M. Haight. 2007. Anaerobic digestion and community development: a case study from Hainan province, China. Environ. Dev. Sustain. 9:501–521.

Bogdanski, A. 2012. Integrated food–energy systems for climate-smart agriculture. Agric. Food Secur. 1:9.

Budiadi, and H. T. Ishii. 2010. Comparison of carbon sequestration between multiple-crop, single-crop and monoculture agroforestry systems of Melaleuca in Java, Indonesia. J. Trop. Forest Sci. 22:378–388.

Butler, R. 2012. LOGGING: Timber certification, trade restrictions, [Online]. Available at: http://rainforests.mongabay.com/1010.htm.

Casillas, C. E., and D. M. Kammen. 2010. The energy-poverty-climate nexus. Science 330:1181–1182.

Chew, T. L., and S. Bhatia. 2008. Catalytic processes towards the production of biofuels in a palm oil and oil palm biomass-based biorefinery. Bioresour. Technol. 99:7911–7922.

Coimbra-Araújo, C. H., L. Mariane, C. B. Júnior, E. P. Frigo, M. S. Frigo, I. R. C. Araujo, et al. 2014. Brazilian case study for biogas energy: production of electric power heat and automotive energy in condominiums of agroenergy. Renew. Sustain. Energy Rev. 40:826–839.

Dale, V. H., K. L. Kline, L. L. Wright, and R. D. Perlack. 2011. Interactions among bioenergy feedstock choices, landscape dynamics, and land use. Ecol. Appl. 21:1039–1054.

Demirbas, A. 2011. Competitive liquid biofuels from biomass. Appl. Energy 88:17–28.

Denman, K. L. 2007. Couplings between changes the climate system and biogeochemistry climate change 2007: The physical science basis. Contribution of working group I to the fourth assessment report of the intergovernmental panel on climate change. University Press, Cambridge.

Dovie, D. B., E. T. Witkowski, and C. M. Shackleton. 2004. The fuelwood crisis in southern Africa – relating fuelwood use to livelihoods in a rural village. GeoJournal 60:123–133.

Down To Earth. 2014. India becomes first country to adopt an agroforestry policy. [Online]. Available at: http://www.downtoearth.org.in/content/india-becomes-first-country-adopt-agroforestry-policy.

Elevitch, C. 2008. Nitrogen fixing trees – the multipurpose pioneers. The Permaculture Research Institute [Online]. Available at: http://permaculturenews.org/2008/09/29/nitrogen-fixing-trees-the-multipurpose-pioneers/.

Endelevu Energy & Energy for Sustainable Development. 2008. A Road map for Biofuels in Kenya - Opportunities and Obstacles. Endelevu Energy & Energy for Sustainable Development, Africa, Kenya.

Energia. 2008. Turning Information into Empowerment: Strengthening Gender and Energy Network in Africa. TIE-ENERGIA Project, Netherland- Energia Secretariat, The Netherlands.

Environmental performance, ITC Limited. [Online]. Available at http://www.itcportal.com/sustainability/sustainability-report-2013/initiative-on-sustainable-agro-forestry.aspx.

Evans, A., V. Strezov, and T. J. Evans. 2010. Sustainability considerations for electricity generation from biomass. Renew. Sustain. Energy Rev. 14:1419–1427.

FAO. 2010. IFES developing countries. FAO, Rome.

FAO. Labour. Food and Agriculture Organization [Online]. Available at: http://www.fao.org/docrep/015/i2490e/i2490e01b.pdf.

FAO. BEFSCI (Bioenergy and Food Security Criteria Indicators). FAO [Online]. Available at: http://www.fao.org/bioenergy/31531-04f3908fc8f82ffc9a978592fc3bf1fa9.pdf.

Ferroukhi, R., A. Khalid, A. L. Pena, and M. Renner. 2015. Renewable energy and jobs annual review 2015. Int. Renew. Energy Agency.

Food and Agricultural Organisation of United Nations. FAO corporate documentary Repository: Bioenergy for development - Technical and environmental dimensions. Food and Agricultural Organisation of United Nations [Online]. Available at: http://www.fao.org/docrep/t1804e/t1804e06.htm. (Accessed 7 October 2014).

Francis, R., P. Weston, and J. Birch. 2015. The social, environmental and economic benefits of Farmer Managed Natural Regeneration (FMNR), World Vision.

Glover, J. D., J. P. Reganold, and C. M. Cox. 2012. Agriculture: plant perennials to save Africa's soils. Nature 489:359–361.

Gupta, V. K., R. Potumarthi, A. O'Donovan, C. P. Kubicek, G. D. Sharma, and M. G. Tuohy. 2014. Chapter 2-bioenergy research: an overview on technological developments and bioresources. Pp. 23–47 *in* V.G. Gupta, Maria Tuohy, Christian P Kubicek, Jack Saddler, Feng Xu, eds. Bioenergy research: advances and applications. Elsevier, Oxford, UK.

HLPE. 2013. Biofuels and food security: a report by the high level panel of experts on food security. Food and Agriculture Organization, Rome.

IEA. 2011. World energy outlook 2011. International Energy Agency, Paris Cedex, France.

IFAD. 2011. Rural poverty report. International Fund for Agricultural Development, Rome.

Iiyama, M., H. Neufeldt, P. Dobie, M. Njenga, G. Ngegwa, and R. Jamnadass. 2014. The potential of agroforestry in the provision of sustainable woodfuel in sub-Saharan Africa. Curr. Opin. Environ. Sustain. 6:138–147.

Ilyama, M., D. Newman, C. Munster, M. Nyabenge, G. W. Sileshi, V. Moraa, et al. 2013. Productivity of *Jatropha curcas* under smallholder farm conditions in Kenya. Agrofor. Syst. 87:729–746.

IREA. 2016. Remap:Roadmap for a renewable energy future. International Renewable Energy Agency, Abu Dhabi.

Jain, R. K. 1994. Fuelwood characteristics of medium tree and shrub species of India. Bioresour. Technol. 47:81–84.

Joshi, S., and S. Joshi. Simarouba glauca. University of Agricultural Sciences, Bagalore, India, [Online]. Available at: http://ageconsearch.umn.edu/bitstream/43624/2/Simarouba%20brochure,%20UAS%20Bangalore,%20India.pdf.

Karekezi, S., K. B. Banda, and W. Kithyoma. 2002. Improving energy services for the Urban Poor in Africa- A Gender Perspective: Energia. Energia News 5:4.

Koh, L. P., J. Miettinen, S. C. Liew, J. Ghazoul, and P. Ehrlich. 2011. Remotely sensed evidence of tropical peatland conversion to oil palm. Proc. Natl Acad. Sci. USA 108:5127–5132.

Kulatunga, G. 2012. Electricity from Gliricidia - an entirely Sri Lankan concept. Daily News. [Online]. Available at: http://archives.dailynews.lk/2012/01/13/fea03.asp.

Larsson, M., M. Gorling, S. Gronkvist, and P. Alvfors. 2013. Bio-methane upgrading of pyrolysis gas from charcoal production. Sustain. Energy Technol. Assess. 3:66–73.

Lin, B. B., I. Perfecto, and J. Vandermeer. 2008. Synergies between agricultural intensification and climate change could create surprising vulnerabilities for crops. Bioscience 58:847–854.

Lundgren, B. O., and J. B. Raintree. 1982. Sustained agroforestry. Pp. 37–49 *in* Nestel, B, eds. Agricultural research for development: potentials and challenges in Asia. International Service for National Agricultural Research, The Hague. Hague, Netherlands

Malla, S. 2013. Household energy consumption patterns and its environmental implications: assessment of energy access and poverty in Nepal. Energy Pol. 61:990–1002.

Millennium Institute. 2013. Global food and nutrition scenarios. Millennium Institute, Washington, DC.

Mwampamba, T. H., A. Ghilardi, K. Sander, and K. J. Chaix. 2013. Dispelling common misconceptions to improve attitudes and policy outlook on charcoal in developing countries. Energy Sustain. Dev. 17:75–85.

Ndegwa, G., V. Moraa, R. Jamnadass, J. Mowo, M. Nyabenge, and M. Liyama. 2011. Potential)for)biofuel) feedstock)in)Kenya. World Agroforetsry Centre, Geoffrey' Ndegwa,'Violet'Moraa,'Ramni'Jamnadass,'Jeremias'Mowo,'.

Nwanze, K. F. 2011. Smallholders can feed the world. International Fund for Agricultural Development (IFAD), Rome, Italy.

Ong, C. K., R. M. Kho. 2015. Tree-crop interactions: agroforestry in a changing climate. CABI, Wallingford, UK..

Openshaw, K.. 2000. A review of *Jatropha curcas*: an oil plant of unfulfilled promise. Biomass Bioenergy 19:1–15.

Orwa, C., A. Mutua, R. Kindt, R. Jamnadass, and S. Anthony. 2009. Agroforestree Database:a tree reference and selection guide version 4.0. [Online]. Available at http://www.worldagroforestry.org/treedb/AFTPDFS/Leucaena_leucocephala.PDF. (Accessed 19 August 2015).

Ouya, D. 2013. Unpacking the evidence on firewood and charcoal in Africa (Agroforestry World Blog). World Agroforestry Centre, [Online]. Available at: http://blog.worldagroforestry.org/index.php/2013/10/03/unpacking-the-evidence-on-firewood-and-charcoal-in-africa/.

Pandey, K. K., N. Pragya, and P. K. Sahoo. 2011. Life cycle assessment of small-scale high-input Jatropha biodiesel production. Appl. Energy 88:4831–4839.

Ryabov, G., D. S. Litun, E. P. Dik, and K. A. Zemskov. 2006. The prospects and problems of utilizing biomass and combustible wastes to produce heat and electricity. Therm. Eng. 53:559–565.

Sanchez, P. A., R. J. Buresh, R. R. Leakey, L. T. Evans, G. D. Anderson, P. Wood, et al. 1997. Trees, soils, and food security [and Discussion]. Philos. Transa. Bio. Sci. 352:949–961.

Scovronick, N., and P. Wilkinson. 2014. Health impacts of liquid biofuel production and use: a review. Glob. Environ. Change 24:155–164.

Selvarajah, R. 2013. The landscape approach: what are ICRAF scientists saying?. World Agroforestry Centre [Online]. Available at: http://worldagroforestry.org/newsroom/highlights/videos-landscape-approach-what-are-icraf-scientists-saying.

Sepp, S., C. Sepp, and M. Mundhenk. 2014. Towards sustainable modern wood energy development, Commissioned by the GIZ Sector Project for the Global Bioenergy Partnership (GBEP).

Shankar, A. V. 2012. Agroforestry - A blog on Agrihortisilviculture: Melia dubia for Biomass power production. [Online]. Available at: http://agrowmania.blogspot.in/2012/05/melia-dubia-for-biomass-power.html.

Singh, S. 2015. Revisting the Jatropha Debate, Forbes India. [Online]. Available at http://forbesindia.com/article/special/revisting-the-jatropha-debate/36187/1.

Siraj, M. A. 2014. Bamboo Power, The Hindu, 11 April 2014. [Online]. Available at: http://www.thehindu.com/features/homes-and-gardens/green-living/bamboo-power/article5900988.ece.

Sood, K. K. 2006. The Influence of household economics and farming aspects on adoption of traditional agroforestry in Western Himalaya. Mt. Res. Dev. 26:124–130.

Sosovele, H. 2010. Policy challenges related to biofuel development in Tanzania/Politische Herausforderungen in Bezug auf Biokraftstoffe in Tansania. Africa Spectr. 45:117–129.

The Secretary-General's High-Level Group. 2012. Sustainable energy for all: A global action agenda [Online]. Available at: http://www.un.org/wcm/webdav/site/sustainableenergyforall/shared/Documents/SEFA-Action%20Agenda-Final.pdf.

Thorlakson, T., and H. Neufeldt. 2012. Reducing subsistence farmers' vulnerability to climate change: evaluating the potential contributions of agroforestry in western Kenya. Agric. Food Secur. 1:15.

Tijani, A. J., M. O. Uguru, and O. A. Salawu. 2008. Anti-pyretic, anti-inflammatory and anti-diarrhoeal properties of *Faidherbia albida* in rats. Afr. J. Biotechnol. 7:696–700.

Ubrabio. Brazilian Program for Biodiesel Production, [Online]. Available at: http://www.ubrabio.com.br/1933/textos/BrazilianProgramForBiodieselProduction_29990/.

UNCTD. 2016. Second generation biofuel markets: state of play, trade and developing country perspectives. United Nations Conference on Trade and Development, Geneva.

UNDP. Sustainable development goals. United Nations Development Programme [Online]. Available at: http://www.undp.org/content/undp/en/home/sdgoverview/post-2015-development-agenda/goal-7.html. (Accessed May 2016).

USAID. 2014. Erc success story: when farmers feel secure-providing the evidence to invest in climate smart agriculture and agroforestry. USAID [Online]. Available at http://usaidlandtenure.net/content/erc-success-story-when-farmers-feel-secure.

Wani, S. P., and T. K. Sreedevi. Pongamia's journey from forest to micro-enterprise for improving livelihoods. International Crops Research Institute for the Semi-Arid Tropics (ICRISAT), Andhra Pradesh, India.

Weiland, P. 2010. Biogas production: current state and perspectives. Appl. Microbiol. Biotechnol. 85:849–860.

WHO. Indoor air pollution. World Health Organization [Online]. Available at: http://www.who.int/indoorair/en/. [Accessed May 2016].

Wildy, D., J. Pate, and J. Bartle. 2003. Silviculture and water use of short rotation Mallee eucalyptus. Rural Industries Research and Development Corporation, Macquarie street, Barton Act, Kingston Act.

World Agroforestry Centre. 2009a. Gliricidia sepium. World Agroforestry Centre [Online]. Available at: http://www.worldagroforestry.org/treedb/AFTPDFS/Gliricidia_sepium.PDF.

World Agroforestry Centre. 2009b. Pongamia pinnata. World Agroforestry Centre [Online]. Available at: http://www.worldagroforestry.org/treedb/AFTPDFS/Pongamia_pinnata.PDF.

World Agroforestry Centre. Faidherbia albida. World Agroforestry Centre [Online]. Available at: http://www.worldagroforestry.org/sites/default/files/F.a_keystone_of_Ev_Ag.pdf.

World Agroforestry Centre. Programme for the development of alternative biofuel crops. World Agroforestry Centre [Online]. Available at: http://worldagroforestry.org/research/alternative-biofuel-crops.

World Agroforestry Centre. Programme for development of alternative biofuel crop: Approach and activities. World Agroforestry Centre [Online]. Available at: http://worldagroforestry.org/research/alternative-biofuel-crops/approach_and_activities.

World Bank. 2011. Household cookstoves, environment, health, and climate change: anew look at an old problem. World Bank, Washington, DC.

Permissions

All chapters in this book were first published in FES, by John Wiley & Sons; hereby published with permission under the Creative Commons Attribution License or equivalent. Every chapter published in this book has been scrutinized by our experts. Their significance has been extensively debated. The topics covered herein carry significant findings which will fuel the growth of the discipline. They may even be implemented as practical applications or may be referred to as a beginning point for another development.

The contributors of this book come from diverse backgrounds, making this book a truly international effort. This book will bring forth new frontiers with its revolutionizing research information and detailed analysis of the nascent developments around the world.

We would like to thank all the contributing authors for lending their expertise to make the book truly unique. They have played a crucial role in the development of this book. Without their invaluable contributions this book wouldn't have been possible. They have made vital efforts to compile up to date information on the varied aspects of this subject to make this book a valuable addition to the collection of many professionals and students.

This book was conceptualized with the vision of imparting up-to-date information and advanced data in this field. To ensure the same, a matchless editorial board was set up. Every individual on the board went through rigorous rounds of assessment to prove their worth. After which they invested a large part of their time researching and compiling the most relevant data for our readers.

The editorial board has been involved in producing this book since its inception. They have spent rigorous hours researching and exploring the diverse topics which have resulted in the successful publishing of this book. They have passed on their knowledge of decades through this book. To expedite this challenging task, the publisher supported the team at every step. A small team of assistant editors was also appointed to further simplify the editing procedure and attain best results for the readers.

Apart from the editorial board, the designing team has also invested a significant amount of their time in understanding the subject and creating the most relevant covers. They scrutinized every image to scout for the most suitable representation of the subject and create an appropriate cover for the book.

The publishing team has been an ardent support to the editorial, designing and production team. Their endless efforts to recruit the best for this project, has resulted in the accomplishment of this book. They are a veteran in the field of academics and their pool of knowledge is as vast as their experience in printing. Their expertise and guidance has proved useful at every step. Their uncompromising quality standards have made this book an exceptional effort. Their encouragement from time to time has been an inspiration for everyone.

The publisher and the editorial board hope that this book will prove to be a valuable piece of knowledge for researchers, students, practitioners and scholars across the globe.

List of Contributors

Umesh Adhikari and Sean A. Woznicki
Department of Biosystems and Agricultural Engineering, Michigan State University, East Lansing, Michigan

A. Pouyan Nejadhashemi
Department of Biosystems and Agricultural Engineering, Michigan State University, East Lansing, Michigan
Department of Plant, Soil, and Microbial Sciences, Michigan State University, East Lansing, Michigan

Eric Hallerman
Department of Fish and Wildlife Conservation, Virginia Polytechnic Institute and State University, Blacksburg, Virginia 24061

Elizabeth Grabau
Department of Plant Pathology, Physiology and Weed Science, Virginia Polytechnic Institute and State University, Blacksburg, Virginia 24061

Katharine R. Howell and Ian C. Dodd
Lancaster Environment Centre, Lancaster University, Lancaster LA1 4YQ, UK

Pitambar Shrestha
Local Initiatives for Biodiversity, Research and Development (LIBIRD), Pokhara, Nepal

Rattan Lal
Carbon Management and Sequestration Center, the Ohio State University, Columbus, Ohio 43210

Sunday E. Obalum, Ifeoma G. Edeh, Ogechi M. Njoku, Ifeyinwa M. Uzoh and Charles A. Igwe
Department of Soil Science, University of Nigeria, Nsukka 410001, Enugu State, Nigeria

Odirichi N. Imoh and Vincent N. Onyia
Department of Crop Science, University of Nigeria, Nsukka 410001, Enugu State, Nigeria

Jose M. Reichert
Soils Department, Federal University of Santa Maria, Av. Roraima, 1000, 97105–900 Santa Maria, Brazil

A. N. M. Rubaiyath Bin Rahman and Jianhua Zhang
School of Life Sciences and State Key Laboratory of Agrobiotechnology, The Chinese University of Hong Kong, Hong Kong, China

José Miguel Reichert and Dalvan José Reinert
Soils Department, Federal University of Santa Maria Av. Roraima 1000, CEP, 97105-900 Santa Maria, RS, Brazil

Miriam Fernanda Rodrigues
Forest Engineering, Federal University of Santa Maria Av. Roraima 1000, CEP, 97105-900 Santa Maria, RS, Brazil

Gabriel Oladele Awe
Department of Crop, Soil and Environmental Sciences, Faculty of Agricultural Sciences, Ekiti State University, Ado Ekiti, Nigeria

Ursino Federico Barreto Riquelme
Universidad Privada del Guairá, Assunción, Paraguay

Douglas Rodrigo Kaiser
Federal University of South Frontier Av. Major Antônio Cardoso 590, 97900-000 Cerro Largo, RS, Brazil

Mónica Elia and Roxana Savin
Department of Crop and Forest Sciences and AGROTECNIO (Center for Research in Agrotechnology), University of Lleida, Av. Rovira Roure 191, 25198 Lleida, Spain

Gustavo A. Slafer
Department of Crop and Forest Sciences and AGROTECNIO (Center for Research in Agrotechnology), University of Lleida, Av. Rovira Roure 191 , 25198 Lleida, Spain
ICREA, Catalonian Institution for Research and Advanced Studies, Barcelona, Spain

Guillermo A. García and Daniel J. Miralles
IFEVA and Department of Plant Production, Faculty of Agronomy , University of Buenos Aires, Av. San Martin 4453 , C1417DSE Buenos Aires, Argentina
CONICET (National Scientifi c and Technical Research Council), Buenos Aires, Argentina

Fernanda G. González
CONICET (National Scientific and Technical Research Council), Buenos Aires, Argentina
EEA Pergamino, INTA (National Institute for Agricultural Technology) Ruta 32 km 4.5, B2700WAA Pergamino, Buenos Aires, Argentina

Ignacio I. Terrile
EEA Pergamino, INTA (National Institute for Agricultural Technology) Ruta 32 km 4.5, B2700WAA Pergamino, Buenos Aires, Argentina

Ariel Ferrante
Queensland Alliance for Agriculture and Food Innovation (QAAFI), The University of Queensland, Toowoomba, Queensland 4350, Australia

Stuart J. Smyth
Department of Agricultural and Resource Economics, University of Saskatchewan, 51 Campus Drive, Saskatoon, Saskatchewan S7N 5A8, Canada

Dionysios Touliatos, Ian C. Dodd and Martin McAinsh
The Lancaster Environment Centre, Lancaster University, Lancaster, UK

Leena Tripathi and Jaindra N. Tripathi
International Institute of Tropical Agriculture, Nairobi, Kenya

Howard Atkinson and Hugh Roderick
Centre for Plant Sciences, University of Leeds, Leeds, UK

Jerome Kubiriba
National Agricultural Research Laboratories, Kampala, Uganda

Ravi Valluru and Matthew P. Reynolds
Global Wheat Program, International Maize and Wheat Improvement Center (CIMMYT), El Batan, Mexico

Tanguy Lafarge
Cirad, UMR AGAP, F-34398, Montpellier, France

René Gommes, Bingfang Wu, Zhongyuan Li and Hongwei Zeng
CAS Key Laboratory of Digital Earth Science, Institute of Remote Sensing and Digital Earth (RADI), Chinese Academy of Sciences, Beijing 100101, China

Matthew Gilliham
Plant Transport and Signalling Lab, ARC Centre of Excellence in Plant Energy Biology, School of Agriculture, Food and Wine, Waite Research Institute, University of Adelaide, Glen Osmond, South Australia 5064, Australia

Scott Chapman
Commonwealth Scientific and Industrial Research Organisation (CSIRO) Agriculture and Food, the University of Queensland, Bioscience Precinct, St. Lucia, Queensland 4067, Australia

Lisa Martin, Sarah Jose and Ruth Bastow
Global Plant Council, Bow House, 1a Bow Lane, London EC4M 9EE, UK

Toby J. A. Bruce
Rothamsted Research, Harpenden, Hertfordshire AL5 2JQ, UK

Andrew Borrell
Queensland Alliance for Agriculture and Food Innovation (QAAFI), Hermitage Research Facility, University of Queensland, Warwick, Queensland 4370, Australia

Matthew Reynolds
International Maize and Wheat Improvement Centre CIMMYT, Int. Apdo. Postal 6-641, 06600 Mexico, Mexico

Navin Sharma, Rodrigo Ciannella and Phil Dobie
World Agroforestry Centre, UN Gigiri, Nairobi, Kenya

Babita Bohra and Namita Pragya
World Agroforestry Centre, NASC Complex, New Delhi, India

Sarah Lehmann
Intern at GIZ, Untergasse 15, 65510 Idstein, Germany

www.ingramcontent.com/pod-product-compliance
Lightning Source LLC
LaVergne TN
LVHW080710210926
848485LV00003B/446
* 9 7 8 1 6 3 2 4 1 4 3 4 2 *

Index

www.ingramcontent.com/pod-product-compliance
Lightning Source LLC
LaVergne TN
LVHW080710210926
848485LV00003B/443

9781639872343